Advances in Biochemistry in Health and Disease

Volume 21

Series Editor

Naranjan S. Dhalla, Institute of Cardiovascular Sciences, St. Boniface Hospital, Winnipeg, MB, Canada

Advances in Biochemistry in Health and Disease focus on the latest developments in biochemical research with implications for health and disease. This book series consists of original edited volumes and monographs, presented by leading experts in the field and provides an up to date and unique source of information for all those interested in the fundamental, biochemical processes of the latest and emerging topics and techniques.

Covering a wide variety of topics, this book series is a valuable source of information from those at the lab bench through to the Health Care workers.

More information about this series at http://www.springer.com/series/7064

Bohuslav Ostadal · Naranjan S. Dhalla
Editors

Sex Differences in Heart Disease

Springer

Editors
Bohuslav Ostadal
Institute of Physiology
Czech Academy of Sciences
Prague, Czech Republic

Naranjan S. Dhalla
St. Boniface Hospital Albrechtsen
Research Centre
Institute of Cardiovascular Sciences
Winnipeg, MB, Canada

Advances in Biochemistry in Health and Disease
ISBN 978-3-030-58676-8 ISBN 978-3-030-58677-5 (eBook)
https://doi.org/10.1007/978-3-030-58677-5

This Springer imprint is published by the registered company Springer Nature Switzerland AG
The registered company address is: Gewerbestrasse 11, 6330 Cham, Switzerland

Dedication

Roberto Bolli, M.D., D.Sc. (Hons)

This book is dedicated to Prof. Dr. Roberto Bolli for his exceptional leadership in cardiovascular science and medicine. He is Professor of Medicine and Director of the Institute of Molecular Cardiology in Louisville, Kentucky. He has been heavily engaged in training of numerous biomedical and clinical fellows from all over the world in the field of cardiovascular medicine for the treatment of heart disease. His dedication for promoting the future generation of cardiovascular clinician/scientists is extraordinary. Indeed, he was able to build a unique Centre of Excellence in Cardiovascular Medicine with respect to both educational and research programmes at his institution.

He is an outstanding investigator in the field of translational cardiology and has published an impressive piece of work in the form of 463 full length papers in high-impact journals; his research has received 35,603 citations with h-factor of 102. His research has focused on the mechanisms responsible for myocardial ischemia/reperfusion injury and on the development of cardioprotective strategies. He established a fundamental role of reactive oxygen species in the pathogenesis of myocardial stunning, and identified the signal transduction pathways as well as the protective genes responsible for the late phase of myocardial preconditioning. He was the first to show that cardiac progenitor c-kit+ cells work via paracrine action, a concept that has changed our understanding of cell therapy.

Dr. Bolli has done an exceptional service in developing Circulation Research in his capacity as Editor-in-Chief, the impact factor of this journal rose from 9.2 to 15.2 during his remarkable editorship (2009–2019). His global leadership in Cardiovascular Research is evident from the fact that he was elected first as Secretary General (1998) and then as President (2004) of a prestigious organization "International Society of Heart

Research". He promoted this organization to its full potentials worldwide in addition to establishing two new awards to recognize the prominent investigators annually. He is currently serving as President of the International Academy of Cardiovascular Sciences. Several years ago, this academy not only recognized his distinguished achievements in cardiovascular medicine by awarding the highest honour "Medal of Merit" but also established the Roberto Bolli Competition for Young Investigators at annual meetings in North America. He has received several honours and awards including Fellowship of the Royal Society (Canada), Gold Medal from the Serbian Physiological Society and Gold Medal from the Institute of Cardiovascular Sciences, Winnipeg Canada.

Preface

It is well known that numerous health problems are affected by sex. Women are more susceptible than men to depression, osteoporosis, asthma, lung cancer due to smoking and autoimmune disease. However, not all medical problems show sex dimorphism, for example, males do not differ from females in terms of their response to infection. And what is the role of sex in cardiovascular diseases? It is first of all necessary to emphasize that the number of clinical and experimental papers dealing with this topic significantly increased during the last 30 years: according to data from Web of Science their amount was negligible still in 1989. This trend is obviously the result of at least two facts: the number of examples of different behaviour of the male and female heart under physiological and pathological conditions is steadily increasing and there were controversial reports on the beneficial or adverse effects of hormonal replacement therapy in women during menopause.

Both clinical and experimental observations support the view that cardiovascular diseases belong to the health problems where the sex differences play an important role. Men are generally at greater risk than are age-matched premenopausal women; this concerns not only ischemic heart disease but also other cardiovascular disorders, such as hypertension, arrhythmias and heart failure. After menopause the incidence of cardiovascular diseases increases in women as well; this observation led to the idea that this increase is associated with the decreasing level of estrogens during menopause. However, the large-scale human subject studies have shown that estrogen replacement in postmenopausal women actually increased the ischemic heart disease suggesting that the mechanisms of sex differences are more complicated.

Cardiovascular diseases are the leading cause of mortality in men and women. Unfortunately, women have traditionally been excluded from clinical trials, and female animals have been used less or sex was not reported in basic research studies. Until recently, consideration of both sexes was not required in clinical and preclinical studies focusing on cardiovascular diseases. This research bias led to the development of cardiovascular therapies that are either less effective or have different side effects in women when compared with men. Sex-specific differences in baseline cardiac function are observed already in healthy adults: baseline differences in cardiac function have been reported both in humans and in animal studies. Sex differences, characteristic of the normal myocardium, create the logical presumption of the different reaction of the cardiac muscle to various pathogenic factors involved in different cardiovascular diseases. Pooling data from males and females may, therefore, obscure important cardiac differences in many parameters under both physiological and pathological conditions. Detailed molecular and cellular mechanisms of these differences are still unknown but one is clear already today: they are so important that they should be considered by the selection of optimum diagnostic and therapeutic procedures in clinical practice.

The increasing interest in sex differences undoubtedly reflects importance of this topic and the urgent need to explain underlying mechanisms, to better understand sex determinants of outcomes and to minimize bias in the management and treatment of cardiovascular diseases in women. This book summarizes some original views on the sex differences in cardiac response to different pathological stimuli; particular attention was paid to the differences in cardiac sensitivity to ischemia/reperfusion injury and to heart failure. With new basic and clinical information on sex differences in cardiac sensitivity to cardiovascular diseases we will be able to offer future patients better evidence-based treatments, better quality of life and lower mortality.

The presentation of the subject matter in the form of 16 manuscripts on sex differences of heart disease, as developed by several investigators for this book, is organized in four parts. Part I, dealing with sex differences in cardiac ischemic injury, includes five chapters on experimental aspects of cardiac ischemia/ reperfusion injury, the role of testosterone and clinical aspects of ischemic heart disease. Part II is devoted to sex differences in heart failure and includes also five chapters. Discussion in this part of the book is centred around the sex differences in heart failure due to volume overload. Part III of this monograph includes four papers on risk factors of cardiovascular diseases, namely, hypertension and obesity, and, finally, three chapters in part IV deal with sex differences of cardiac mitochondria under different pathological conditions. We believe this book will be very useful for cardiovascular scientists, graduate students, postdoctoral fellows and other health professionals.

We are grateful to ing. Mila Markova and ing. Frantisek Papousek from Prague for their help in editing the manuscript. Cordial thanks are also due to Dr. Meran Lloyd-Owen, Ms. Sara Germans-Huisman and Mr. Rajan Muthu for their understanding and patience in developing this monograph. We also wish to express thanks to Dr. Gonzalo Cordova for his efforts in improving the quality of several chapters in this book.

Prague, Czech Republic — Bohuslav Ostadal
Winnipeg, Canada — Naranjan S. Dhalla

Contents

Contributors

Carrie Adase Department of Kinesiology and Nutritional Sciences, California State University, Los Angeles, USA

Nattayaporn Apaijai Cardiac Electrophysiology Research and Training Center, Faculty of Medicine, Chiang Mai University, Chiang Mai, Thailand;
Center of Excellence in Cardiac Electrophysiology Research, Chiang Mai University, Chiang Mai, Thailand

Ana Ayesta Heart Area, Hospital Universitario Central de Asturias, Oviedo, Asturias, Spain

Marta Belmonte Cardiology Unit, Ospedale Maggiore IRCCS Policlinico of Milan, Milan, Italy

Hilary Bews Section of Cardiology, Department of Internal Medicine, Max Rady College of Medicine, Rady Faculty of Health Sciences, University of Manitoba, Winnipeg, Manitoba, Canada

Sukhwinder K. Bhullar Institute of Cardiovascular Sciences, St. Boniface Hospital Albrechtsen Research Centre, Department of Physiology and Pathophysiology, Max Rady College of Medicine, University of Manitoba, Winnipeg, Manitoba, Canada

Consuelo Borrás Freshage Research Group, Department of Physiology, Faculty of Medicine, University of Valencia, Valencia, Spain

Alfredo Cabrera-Orefice Radboud Institute for Molecular Life Sciences, Radboud University Medical Center, Nijmegen, The Netherlands

Nipon Chattipakorn Cardiac Electrophysiology Research and Training Center, Faculty of Medicine, Chiang Mai University, Chiang Mai, Thailand;
Center of Excellence in Cardiac Electrophysiology Research, Chiang Mai University, Chiang Mai, Thailand;
Cardiac Electrophysiology Unit, Department of Physiology, Faculty of Medicine, Chiang Mai University, Chiang Mai, Thailand

Siriporn C. Chattipakorn Cardiac Electrophysiology Research and Training Center, Faculty of Medicine, Chiang Mai University, Chiang Mai, Thailand;
Center of Excellence in Cardiac Electrophysiology Research, Chiang Mai University, Chiang Mai, Thailand;
Cardiac Electrophysiology Unit, Department of Physiology, Faculty of Medicine, Chiang Mai University, Chiang Mai, Thailand

Francisco Correa-Segura Departamento de Biomedicina Cardiovascular, Instituto Nacional de Cardiología, Ciudad de México, Mexico

Naranjan S. Dhalla Institute of Cardiovascular Sciences, St. Boniface Hospital Albrechtsen Research Centre, Department of Physiology and Pathophysiology, Max Rady College of Medicine, University of Manitoba, Winnipeg, Manitoba, Canada

Zdenek Drahota Institute of Physiology, Czech Academy of Sciences, Prague, Czech Republic

Cameron Eekhoudt Institute of Cardiovascular Sciences, Department of Physiology and Pathophysiology, Max Rady College of Medicine, Rady Faculty of Health Sciences, University of Manitoba, Winnipeg, Manitoba, Canada

Vijayan Elimban Institute of Cardiovascular Sciences, St. Boniface Hospital Albrechtsen Research Centre, Max Rady College of Medicine, University of Manitoba, Winnipeg, Manitoba, Canada

Pavel Hamet Department of Medicine, University of Montreal, CRCHUM, Montréal, QC, Canada;
OPTITHERA, Montreal, QC, Canada

Candan Hizel OPTITHERA, Montreal, QC, Canada

Marketa Hlavackova Institute of Physiology, Czech Academy of Sciences, Prague, Czech Republic

Susan E. Howlett Department of Pharmacology, Dalhousie University, Halifax, NS, Canada;
Department of Medicine (Division of Geriatric Medicine), Dalhousie University, Halifax, NS, Canada

Davinder S. Jassal Section of Cardiology, Department of Internal Medicine, Max Rady College of Medicine, Rady Faculty of Health Sciences, University of Manitoba, Winnipeg, Manitoba, Canada;
Institute of Cardiovascular Sciences, Department of Physiology and Pathophysiology, Max Rady College of Medicine, Rady Faculty of Health Sciences, University of Manitoba, Winnipeg, Manitoba, Canada;
Department of Radiology, Max Rady College of Medicine, Rady Faculty of Health Sciences, University of Manitoba, Winnipeg, Manitoba, Canada

Anabela Kim Department of Kinesiology and Nutritional Sciences, California State University, Los Angeles, USA

Elizabeth Lira-Silva Departamento de Farmacología, Instituto Nacional de Cardiología, Ciudad de México, Mexico

Manuel Martínez-Sellés Cardiology Department, Hospital Universitario Gregorio Marañón, CIBERCV, Universidad Europea, Universidad Complutense, Madrid, Spain

Cristina Mas-Bargues Freshage Research Group, Department of Physiology, Faculty of Medicine, University of Valencia, Valencia, Spain

Michael Megaly Abbott Northwestern Hospital, Minneapolis Heart Institute, Minneapolis, MN, USA;
Department of Cardiovascular Medicine, Hennepin Healthcare, Minneapolis, MN, USA

Jawahar L. Mehta Department of Cardiology, University of Arkansas for Medical Sciences, Little Rock, AR, USA

Jan Neckar Institute of Physiology, Czech Academy of Sciences, Prague 4, Czech Republic

Susana Novella Department of Physiology, Faculty of Medicine, University of Valencia, Valencia, Spain

Mohamad Nusier School of Medicine, Jordan University of Science and Technology, Irbid, Jordan

Bohuslav Ostadal Institute of Physiology, Czech Academy of Sciences, Prague, Czech Republic

Petr Ostadal Cardiovascular Center, Department of Cardiology, Na Homolce Hospital, Prague, Czech Republic

Ana B. Paes Department of Physiology, Faculty of Medicine, University of Valencia, Valencia, Spain

Natalia Pavón Departamento de Farmacología, Instituto Nacional de Cardiología, Ciudad de México, Mexico

Franco Romeo Department of Cardiology, University of Rome "Tor Vergata", Rome, Italy

Marwan Saad Warren Alpert Medical School, Cardiovascular Institute, Brown University, Providence, RI, USA

David G. Sapp Department of Pharmacology, Dalhousie University, Halifax, NS, Canada

Marco Schiavone Cardiology Unit, Ospedale Maggiore IRCCS Policlinico of Milan, Milan, Italy

Anureet K. Shah Department of Kinesiology, Nutrition and Food Science, California State University, Los Angeles, CA, USA

Fabiola B. Sozzi Cardiology Unit, Ospedale Maggiore IRCCS Policlinico of Milan, Milan, Italy

Paramjit S. Tappia Asper Clinical Research Institute, St. Boniface Hospital, Winnipeg, Canada

Johanne Tremblay Department of Medicine, University of Montreal, CRCHUM, Montréal, QC, Canada;
OPTITHERA, Montreal, QC, Canada

Sonu Varghese Institute of Cardiovascular Sciences, Department of Physiology and Pathophysiology, Max Rady College of Medicine, Rady Faculty of Health Sciences, University of Manitoba, Winnipeg, Manitoba, Canada

Part I
Sex Differences in Cardiac Ischemia

Chapter 1
Gender-Related Differences in the Pathogenesis and Diagnosis of Ischemic Heart Disease

Marwan Saad, Michael Megaly, Franco Romeo, and Jawahar L. Mehta

Abstract Gender plays an important role in the pathophysiology, clinical presentation, and outcomes of various medical illnesses. Among those, special interest is directed towards ischemic heart disease, being the most common cause of mortality in the United States and the developed world. The prevalence and pathogenesis of traditional cardiovascular risk factors differ between women and men, including hypertension, diabetes, dyslipidemia, as well as smoking and psychosocial risk factors such as depression, emotional stress, and low socioeconomic status. However, the difference in the pathophysiology of atherosclerosis represents the hallmark of gender-related discrepancies in ischemic heart disease. While men are more likely to have obstructive coronary artery disease (CAD), women have a higher prevalence of endothelial dysfunction, more tendency for vasospasm, lower coronary flow reserve, and higher incidence of plaque erosion than rupture compared with men. These differences are, at least in part, attributed to variations in sex hormones. Not only that gender affects the pathophysiology of CAD, but also impacts the difference in its clinical presentation. Women are more likely to present with atypical symptoms, and hence are more likely to present late or to be misdiagnosed. This has been shown to adversely impact the outcomes of CAD in women compared with men. Further, the utilization of various modalities for the diagnosis of CAD in women may be

M. Saad
Warren Alpert Medical School, Cardiovascular Institute, Brown University, Providence, RI, USA
e-mail: marwanssaad@gmail.com

M. Megaly
Abbott Northwestern Hospital, Minneapolis Heart Institute, Minneapolis, MN, USA

Department of Cardiovascular Medicine, Hennepin Healthcare, Minneapolis, MN, USA

F. Romeo
Department of Cardiology, University of Rome "Tor Vergata", Rome, Italy

J. L. Mehta (✉)
Department of Cardiology, University of Arkansas for Medical Sciences, 4301 W Markham Street #532, Little Rock, AR, USA
e-mail: mehtajl@uams.edu

B. Ostadal and N. S. Dhalla (eds.), *Sex Differences in Heart Disease*, Advances in Biochemistry in Health and Disease 21,
https://doi.org/10.1007/978-3-030-58677-5_1

challenging due to multiple factors including limited exercise capacity, lower sensitivity of exercise-induced changes in electrocardiogram, higher prevalence of single-vessel disease, attenuation artifacts from breast tissue, and the higher likelihood of endothelial dysfunction rather than obstructive CAD. Clinicians should pay special attention to such differences aiming to improve outcomes in women with ischemic heart disease.

Keywords Gender · Sex · Women · Ischemic heart disease

Introduction

Among both women and men, ischemic heart disease remains the most common cause of mortality in the United States and the developed world [1]. Furthermore, it represents a heavy burden on the healthcare system in the United States with more than 1.4 million hospitalizations per year [1]. There is emerging evidence that ischemic heart disease presentation and outcome differ between women and men. In this chapter, we highlight the gender-related differences in the pathogenesis and diagnosis of ischemic heart disease.

Gender-Related Differences in the Pathogenesis of Ischemic Heart Disease

(A) Risk factors for coronary artery disease

Hypertension

Among the traditional risk factors for coronary artery disease (CAD), there are important differences in the prevalence and pathogenesis of hypertension between men and women. While the overall prevalence of hypertension appears to be similar among women and men, the prevalence is higher among women over the age of 65 [2], especially black women [3], and could be attributed to hormonal changes [4]. Estrogen plays a beneficial role in systemic blood pressure regulation through vascular vasodilation (either via a direct effect on the endothelial cells or indirectly through promoting nitric oxide release), as well as modulation of the renin-angiotensin system (via increasing the synthesis of angiotensinogen and suppressing the activity of pro-hypertensive angiotensin type 1 receptor and reducing the activity of angiotensin type 2 receptor) [5]. The drop in estrogen levels after menopause may be the basis of higher prevalence of hypertension in elderly women. Higher prevalence of hypertension in women > 65 years may relate to higher rates of CAD in elderly women.

Diabetes, obesity, and insulin resistance

Diabetes is among the most important risk factors for CAD. Diabetic women have comparable outcomes as age-matched men, indicating that diabetes may counteract the advantage of female sex hormones in the pathogenesis of CAD [6]. The peripheral adipose tissue distribution as well as the higher levels of adiponectin in women are associated with improved insulin sensitivity and provides some protection from insulin resistance, the hallmark of metabolic syndrome, compared to men [7–10]. Obesity is an important risk factor for diabetes and CAD. While the burden of obesity affects women and men equally, women above the age of 60 years have higher prevalence of metabolic syndrome [11–13], and among women, black women have the highest prevalence [14]. On the other hand, it is important to recognize that polycystic ovarian syndrome, which is present in about 10% of women, and includes a constellation of obesity, diabetes, and metabolic syndrome-like picture may relate to the higher prevalence of CAD. Treatment of this disease may reverse these risk factors and reduce the risk of cardiovascular disease [15].

Dyslipidemia

While pre-menopausal women (age 20–55 years) have lower cholesterol levels compared to men, the levels increase significantly after the age of 55 years in women. That is usually linked, at least in part, to the increased cardiovascular risk in elderly women [16]. On the cellular level, not only that women have higher levels of high-density lipoprotein (HDL) cholesterol and lower levels of low-density lipoprotein (LDL) cholesterol, very low-density lipoprotein (VLDL) cholesterol, total plasma triglyceride and VLDL triglyceride compared to age-matched men [5], their circulating VLDL particles are smaller in size, while HDL are larger in size, which appears to be inversely associated with cardiovascular disease [17, 18]. Furthermore, hepatic lipase activity in women is much lower compared with men, leading to relatively larger size of HDL and LDL particles. Small HDL and LDL were found to be linked to increased cardiovascular disease risk [19, 20]. Further, low HDL and high triglyceride levels are more predictive of the risk of cardiovascular disease in women compared with men [21–23].

Smoking

Smoking is the single most important modifiable risk factor for CAD and contributed to 480,000 premature deaths in the U.S. from 2005–2009 [2], one-third of which were a result of secondhand smoke. The prevalence of smoking remains higher in men than women, however, the recent decline in tobacco use is less pronounced in women than in men [16]. Tobacco is known to be a major cause of endothelial dysfunction. It promotes inflammatory environment, stimulates platelet aggregation, increases oxidative stress, and induces vasospasm [24, 25]. Interestingly, the impact of smoking as a risk factor for cardiovascular disease was found to be at least 25% greater in women than age-matched men [26]. Smoking cessation is associated with a decline in the risk of cardiovascular disease to the levels of non-smokers in 5–10 years [27, 28].

Psychosocial risk factors

Psychosocial stress plays a role in the pathophysiology of CAD [29]. Among the important psychosocial factors, depression is more prevalent in women and was found to be present in up to 40% of young women presenting with acute myocardial infarction [30]. Not only depression increases the risk of cardiac events in women by at least 50% [31–33], it is also related to poor adherence to medical therapy after cardiac events and worse overall outcomes. In addition, emotional stress, either acute or chronic, is a known risk factor for acute coronary syndromes in women. Multiple studies have shown a relationship between anxiety and marital dissatisfaction with the risk of cardiac events and progression of CAD [34–37]. Importantly, low level of education, poor socioeconomic status, smoking, obesity, metabolic syndrome, and lack of exercise, all of which are highly associated with CAD [38], may be more pronounced in women than in men. The gender differences among different risk factors are summarized in Table 1.1.

(B) **Plaque burden and morphology**

While atherosclerosis is an inflammatory process that involves multiple factors such as dyslipidemia, endothelial dysfunction and oxidative stress [39–43], the resultant atherosclerotic plaques differ in morphology and burden between women and men [13, 44]. Men have been shown to have more obstructive CAD than women

Table 1.1 Gender differences in cardiovascular risk factors

Risk factor	Compared with men, women have
Hypertension	• Higher prevalence of hypertension above age of 65
Diabetes	• More peripheral adipose tissue distribution and higher levels of adiponectin which is associated with improvement in insulin sensitivity
Obesity and metabolic syndrome	• Higher prevalence of obesity and metabolic syndrome above age of 65
	• Polycystic ovarian syndrome associated with obesity and diabetes and metabolic syndrome-like picture
Dyslipidemia	• Lower levels of Cholesterol at age 20–55 years
	• Higher levels of HDL cholesterol
	• Lower levels of LDL cholesterol, VLDL cholesterol, and total plasma triglyceride
	• Smaller VLDL particles
	• Larger HDL particles
	• Less hepatic lipase activity
Smoking	• The recent decline in tobacco use is less pronounced in women than in men
Psychosocial risk factors	• Depression, anxiety, and emotional stress are more prevalent in women than men and are all known risk factors of coronary artery disease and acute coronary syndrome

in multiple studies [45–47]. Atheromatous plaques in men have more volume, are more commonly eccentric [13, 44], more likely to have calcification [48], with more structural and functional abnormalities in epicardial coronary arteries than women [11], while women have less dense fibrous tissue in their plaques [49, 50], and are more likely to have angiographically normal coronaries on angiogram. Women have higher prevalence of endothelial dysfunction, more tendency for vasospasm [2, 4, 51, 52], and lower coronary flow reserve compared with men [11, 44]. Furthermore, compared with men, women exhibit higher incidence of plaque erosion than rupture [53], and increased hyaluronan deposition, as well as elevated levels of serum myeloperoxidase [54, 55].

The differences in plaque burden and morphology among women and men are summarized in Table 1.2.

(C) **Vascular biology and microvascular disease**

Studies relating to vascular biology have led to better understanding of the pathogenesis of atherosclerosis. As mentioned above, while the burden of epicardial CAD may be less in women compared with men, coronary microvasculature plays a major role in the development of atherosclerosis and CAD is women. Women have higher incidence of impaired NO-dependent vasodilation of the coronary microvasculature, resulting in more microvascular spasm and vasospastic angina [56]. Microvascular dysfunction is considered an important part of the pathophysiology of chest pain in the absence of significant coronary obstruction, also known as microvascular

Table 1.2 Gender differences in atherosclerotic plaque burden, plaque morphology, and Pathophysiology of acute coronary syndrome

Criteria	Compared with men, women have
Plaque burden	• Less obstructive CAD and more likely to have normal coronaries on angiogram
	• Higher prevalence of endothelial dysfunction
	• More tendency for vasospasm
	• Lower coronary flow reserve
Plaque morphology	• Less volume
	• Less calcification
	• Less dense fibrous tissue
	• More centrally oriented
	• Increased hyaluronan deposition
	• Elevated levels of serum myeloperoxidase
Pathophysiology of ACS	• Higher incidence of plaque erosion than rupture
	• Higher incidence of SCAD

ACS = acute coronary syndrome
CAD = coronary artery disease
SCAD = spontaneous coronary artery dissection

angina [57, 58]. The authors of the WISE (Women's Ischemia Syndrome Evaluation) study [59], showed a clear role for estrogen in mediating vasodilation of coronary microvasculature through myocyte hyperpolarization, increase in prostacyclin production, and inhibition of myocyte-mediated vasoconstriction induced by calcium and endothelin-1 [59]. Of note, other studies showed that part of estrogen-induced coronary vasodilation is independent of these classical pathways and is mediated through changes in ATP-sensitive potassium or calcium channels [60]. As levels of estrogen decrease, women become more predisposed to microvascular dysfunction. Further etiologies of microvascular dysfunction in women without obstructive CAD include abnormal response of phosphocreatine/ATP to exercise indicating a shift to anaerobic metabolism consistent with myocardial ischemia [2, 52], as well as impairment of coronary flow velocity reserve [61].

Estrogen plays multiple other roles at the cellular level. For example, it accelerates the catabolism of reactive oxygen species through up-regulation of multiple enzymes such as superoxide dismutase and catalase. Thus, it reduces oxidative stress and increases free NO levels [9, 62]. Besides the classic NO/cGMP mediated pathway for vasodilation, estrogen also causes direct smooth muscle relaxation via endothelium derived hyperpolarizing factor (EDHF) [9, 63]. Furthermore, estrogen plays an important role in modulating the shear stress-mediated regulation in arteriolar diameter through G-protein-coupled estrogen receptors (GPER) in the vascular endothelium [18, 64, 65].

The role of estrogen and androgen in vascular biology and pathogenesis of atherosclerosis is shown in Fig. 1.1.

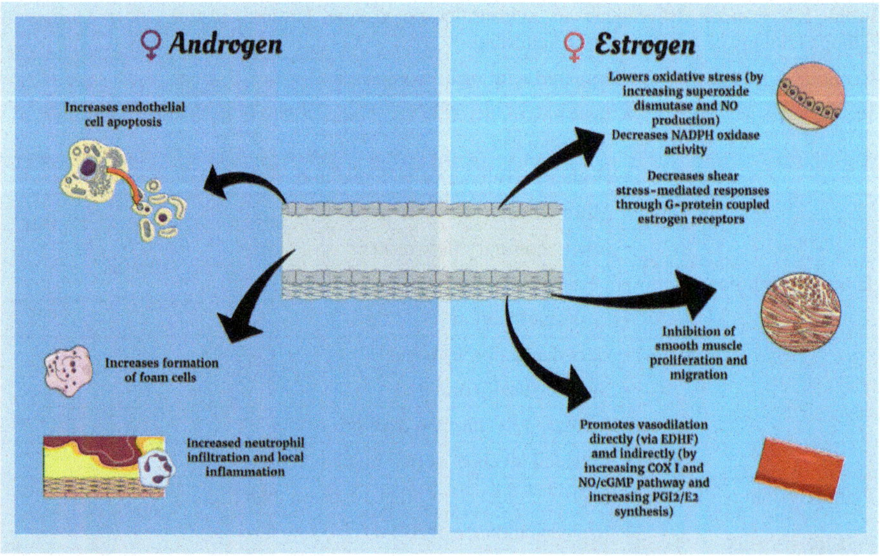

Fig. 1.1 The role of estrogen and androgen in the vascular biology and pathogenesis of atherosclerosis

(D) **Coronary artery dissection**

Although spontaneous coronary dissection (SCAD) was considered to be a rare cause of acute coronary syndrome, evidence is emerging that SCAD is the underlying etiology in 10–25% of patients with myocardial infarction, predominantly in women younger than 50 years of age [66–68] without atherosclerotic CAD. SCAD most commonly affects the left anterior descending artery [68], and is caused either by an intimal tear or bleeding of vasa vasorum leading to intramural hematoma and false lumen between the media and adventitia [69, 70]. While the underlying trigger remains unclear, hormonal changes with subsequent weakening of the arterial walls are among the main theories behind SCAD [71]. Multiple other factors include inflammatory disorders, hemodynamic stress during pregnancy or peripartum period, coronary vasospasm, and connective tissue disorders [72, 73]. In a recent National Inpatient Sample analysis of 7,347 women with SCAD in the period from 2009–2014, in-hospital mortality with acute myocardial infarction in the setting of SCAD was 6.8%, and outcomes appeared to favor conservative management especially in patients with non-ST-segment elevation myocardial infarction (NSTEMI) [74].

(E) **Acute myocardial infarction**

Acute myocardial infarction is a leading cause of death in the United States, with more than 610,000 deaths annually [75]. Development of atherosclerotic plaque is usually the initial step in the pathophysiology of AMI. Disruption of the endothelium overlying these plaques triggers a cascade of pro-inflammatory events that in turn lead to activation of multiple pro-thrombotic factors. Ann et al. described gender-related differences in plaque morphology in patients with STEMI [76]. Yahagi et al. [77] observed that AMI in men was more often associated with thin cap vulnerable plaques than in women. Hence, men are more likely to present with a ruptured plaque than women. Ruptured plaques are characterized by abundant foamy macrophages which are potent stimulator of the coagulation cascade, resulting in more frequently in an occlusive thrombus [78]. Although plaque rupture remains the main etiology of AMI in women, it is more prevalent in men than women [79]. In contrast, plaque erosions are seen more commonly in women than men and are associated with a less dramatic coronary obstruction and micro-embolization [80].

Another gender-related difference is the severity of CAD at the time of AMI. Early studies showed less plaque burden and obstructive CAD in women by angiogram in the setting of acute coronary syndrome, and normal appearing coronaries are not uncommon [47]. Intravascular ultrasound (IVUS) examination of plaques showed less eccentric location, more fibrous tissue, and less prevalence of thin-cap fibroatheromas [80, 81]. Women are more likely to have microvascular dysfunction and impaired coronary flow reserve [82, 83], which may explain acute coronary syndromes in women without significant CAD. SCAD is another major entity resulting in AMI that predominantly occurs in women as mentioned before.

These differences in plaque burden and morphology as well as the pathophysiology of AMI are, at least in part, attributed to variations in sex hormones. On the cellular level, estrogen promotes re-endothelialization, reduces leukocyte adhesion molecules, and inhibits smooth muscle proliferation as well as matrix deposition in response to any vascular injury, hence acts as a major protective agent against atherosclerosis [84–86]. Estrogen also enhances vasodilation of coronary arteries through the production of NO and vasodilator prostaglandins PGE2 and PGI2 through estrogen receptors ERa and ERb [84]. These protective effects of estrogen are believed to be the main reason for the reduced burden of atherosclerosis and the lower incidence of AMI in premenopausal women.

Gender-Related Differences in the Diagnosis of Ischemic Heart Disease

The Symptomatic Conundrum

The symptomatic conundrum is a common term in the current cardiology literature. It describes the fact that women present with different CAD symptoms compared with men. Reasons for higher CAD mortality in women include older age at initial presentation and relatively smaller use of diagnostic tools [87]. Symptoms in women are often "atypical" and may include dyspnea, heartburn, bloating, or generalized fatigue. The symptomatic conundrum often leads to underappreciation of presenting symptoms and feeds the unawareness of association of these symptoms with CAD [88]. In ACS, despite the fact that most women present with typical symptoms including chest and/or arm pain, the likelihood of presenting with atypical symptoms is still high [89, 90].

The symptomatic conundrum not only leads to underappreciation of the symptoms by medical practitioners, but also leads to significant delay of care as the symptoms are also underappreciated by the patient. In a study by Rosenfeld et al. in 2005, 52 US women who presented with AMI were investigated; a large proportion of them attributed their symptoms to alternative causes or minimized their importance [91]. This could be partly explained by the common underestimation of CAD risk in women. In The Berlin Female Risk Evaluation (BEFRI) study, only half of urban women correctly estimated their CAD risk [92].

The symptomatic conundrum can have serious consequences. In stable CAD patients, it might lead to delay of care and lack of appropriate therapy that could help with the patients' symptoms and improve their quality of life. On the other hand, it might lead to late presentation in AMI, especially STEMI, delaying effective reperfusion therapy and leading to more serious long-term sequelae [93].

It is also important to understand that despite the underestimation of CAD risk, women actually have about 20% higher prevalence of angina compared with men. Women also report persistent or worsening symptoms at a rate that is double that of men [94–97]. There is a much larger pool of women with nonobstructive and obstructive CAD who report a heavy burden of symptoms compared with men [98]. Younger women tend to present with absence of chest pain and suffer worse outcomes in comparison with their male counterparts. But, this is attenuated by age, with the difference between genders in the absence of chest pain narrowing or disappearing as age advances [99].

The impact of symptomatic conundrum has led to gender-based differences in the work-up of CAD. Women presenting with CAD symptoms might not be thoroughly investigated as men. Data from Euro Heart Survey in the management and clinical outcomes of stable angina (3779 patients, 42% women) showed that women were less likely to be referred for an exercise test or coronary angiography [100]. Another cross-sectional survey of 1162 patients with angina showed that physicians were more likely to note the risk factors in men and refer them for further work-up [88]. However, an Italian study investigating the use of cardiac procedures in relation to age and sex found that there was an age bias but no gender bias in referral to cardiac catheterization [101]. It is not unreasonable to assume that the underappreciation of CAD because of atypical symptoms in women would result in incorrectly reassuring patients without appropriate diagnostic tests.

Diagnostic Modalities

(A) **Exercise EKG test**

The choice of diagnostic testing and risk stratification in women suspected to have CAD can be challenging. Patients with low pretest probabilities of CAD, both men and women, are less likely to benefit from the addition of stress imaging test [102]. Therefore, for most patients at low-risk, exercise treadmill testing (ETT) without imaging is the initial test of choice for diagnosis and risk stratification. In the low-risk cohort who were able to exercise, the WOMEN trial [What is the Optimal Method of Ischemia Elucidation in Women] showed that the ETT alone was equivalent to ETT plus myocardial perfusion imaging (MPI) [103]. This is supported by guidelines and consensus statements of the ACC (American College of Cardiology) and American Society of Nuclear Cardiology (ASNC) [87, 94, 104, 105].

Exercise stress testing is physiologic and provides a constellation of clinical, electrocardiographic (EKG) and hemodynamic data; therefore, it is generally preferred over pharmacological testing [106]. However, exercise-induced ST depression may be less sensitive in women compared with men [107]. Exercise-induced ST depression in the absence of CAD could be secondary to changes in estrogen levels during the menstrual cycle [108, 109] or due to menopausal hormone therapy which renders EKG changes less accurate in women compared to men [110]. A recent study by Fitzgerald et al. found that the prevalence of false-positive exercise ST depression might be equal in men and women but with less predictable causes in women such as left ventricular hypertrophy and hypertension [111] Although endothelial dysfunction is thought to be a hallmark of atherosclerosis, a study found no relationship between endothelial dysfunction and exercise-induced ST depression in women [112].

Most women referred for stress testing often have limited exercise capacity as they are typically older with significant comorbidities. This is one of the major difficulties as they are often unable to reach a sufficient workload, resulting in submaximal tests of limited sensitivity. In women who are unable to exercise adequately, pharmacological stress testing is reasonable (e.g., adenosine MPI or dobutamine stress echocardiogram).

(B) **Stress imaging**

In symptomatic women who are at intermediate to high risk for CAD and have baseline EKG ST-T abnormalities, stress imaging with echocardiography, radionuclide single-photon emission computed tomography (SPECT), and positron emission tomography (PET) are recommended as the initial diagnostic modality. It is also reasonable to consider stress imaging as the initial modality in women who have poor exercise capacity or an abnormal stress EKG [113–116]. In the WOMEN trial, women with intermediate to high pretest risk, use of myocardial perfusion imaging (MPI) along with ETT was associated with higher accuracy to detect obstructive CAD [103].

Radionuclide single-photon emission computed tomography

Several studies have reported gender-based differences in the diagnostic accuracy of SPECT MPI. The diagnostic accuracy has been reported to be lower in women than men [117–119]. There are multiple reasons for this which include lower exercise capacity, higher prevalence of single-vessel disease, and anterior wall attenuation artifacts from breast tissue. Women also have a smaller heart size which often results in a blurred image affecting the sensitivity of the test [120]. Women also have higher left ventricular ejection fraction (LVEF) [121] leading to higher normal limits of transient ischemic dilation ratio [122]. The use of attenuation correction significantly improves the specificity of SPECT MPI, especially in women with high probability of breast tissue attenuation but does not typically affect the sensitivity of SPECT MPI [123, 124]. Another method to increase accuracy in women is using gender-based normal limits and software interpretation [125]. It is currently recommended by American Society of Nuclear Cardiology (ASNC) to use gender-based normal limits for reporting of LVEF and volumes [104].

The prognostic value SPECT MPI in women has also been debated. However, in a large meta-analysis, women with normal SPECT MPI had prognosis comparable to men with 99% event-free survival over 36 months [126]. A normal study with both normal SPECT MPI and stress EKG has excellent prognosis regarding CAD death or MI [127]. On the other hand, abnormal studies are associated with increases in adverse CAD events and worse prognosis in women [128]. Left ventricular volumes and LVEF on SPECT MPI also provide an added prognostic value in predicting CAD death or MI. Other studies have confirmed the excellent prognostic value of SPECT MPI in women, including elderly women [130] and women of diverse racial and ethnic subsets [131]. It is also important to note that some women have a normal

SPECT with ischemic EKG changes [132]; these patients have worse outcomes with more cardiovascular events despite normal perfusion pattern [133].

Positron emission tomography

The use of PET has been associated with improved diagnostic accuracy compared with SPECT [134, 135]. The use of PET has provided multiple advantages over SPECT including depth-independent attenuation correction, higher spatial and temporal resolution, lower radiation dose with radiopharmaceuticals that have short half-lives. PET also provides the ability to perform absolute quantification of myocardial perfusion [136]. These advantages render PET preferable in women as it negates the higher risk of both false positive and false negative studies given breast tissue attenuation, and amplified partial volume effects in small left ventricles, respectively. The ability to measure absolute myocardial blood flow is specifically relevant in women who have higher incidence of microvascular ischemia that is increasingly recognized to be associated with significant cardiovascular morbidity and mortality [137–139].

Regarding prognosis, a normal PET scan carries an excellent prognosis with 0.4% annual event rate [140] An abnormal PET study is associated with worse prognosis with more perfusion defects being associated with a graded increase in risk [141]. The prognostic value was found to be similar in men and women in the PET Prognosis Multicenter Registry sex-specific sub-analysis [142].

Stress echocardiography

Similar to stress MPI, stress echocardiography significantly improves the specificity and sensitivity for CAD diagnosis. Echocardiography has its limitations in women which include the variability in acoustic windows with breast tissue and being operator-dependent, which affects the ability to capture images at maximal stress. The diagnostic accuracy of stress echocardiography is generally comparable in men and women [143] The echocardiographic evidence of ischemia on pharmacologic stress echocardiography was found to be the only independent predictor of cardiovascular events in 456 women in the study by Cortigiani et al. [144]. Despite its limitations, stress echocardiography is preferred in pregnant and young women due to the absence of ionizing radiation needed for the test.

(C) **CT coronary angiography**

CT coronary angiography (CTCA) has evolved as a reliable gatekeeper of invasive coronary angiogram. A negative CT study reliably excluded obstructive CAD with a high degree of certainty in two independent studies [145, 146]. In patients at low to intermediate risk of CAD, CTCA is highly accurate in excluding the presence of obstructive disease [147], and the accuracy of CTCA is similar in men and women [147–149]. With advancing techniques and the development of CT-derived fractional flow reserve, future application of CTCA might improve outcomes in women suspected to have CAD. The main limitations of CTCA in women include radiation exposure and lower positive predictive value leading to increased downstream testing with invasive testing.

(D) **Coronary angiography in women**

Coronary angiography remains the gold-standard for diagnosis of CAD. In women, a limitation of coronary angiography is the higher prevalence of non-obstructive CAD [150]. Significant obstructive lesion may not be identified in the cardiac catheterization laboratory unless coronary flow reserve is measured [151]. Women generally have smaller left ventricles and small coronary artery sizes. The smaller size along with breast tissue attenuation leads to partial inability to assess smaller mid and distal coronary segments [152, 153]. Moreover, interpretation of the severity of lesions might be different between men and women due to the difference in the area of myocardium supplied. In the FAME trial sub-study, angiographic lesions with similar severity in men and women were less likely to be ischemia-producing in women [154].

The use of fractional flow reserve (FFR) assessment of undetermined lesions have been evolving because of the superiority over angiography-guided PCI [155] However, using similar FFR values to guide therapy in women and men have been debated. In 2014, a study by Lin et al. in 1,090 patients, women who underwent FFR-guided PCI had less favorable long-term outcomes compared with men [156]. This study raised the issue to consider gender-based FFR-guided treatment protocols. More studies are required to further characterize the best women-specific FFR-guided treatment strategy.

The gender-related differences in utilization of various diagnostic modalities for CAD are summarized in Table 1.3.

Table 1.3 Gender differences in clinical presentation and role of diagnostic modalities in coronary artery disease

Criteria	Compared with men, women have the following characteristics
Clinical presentation	• Older age at presentation
	• More likely to present with atypical symptoms (ex. dyspnea, heartburn, bloating, or generalized fatigue) leading to late presentation and significant delay of care
	• Less likely to be referred for an exercise test or coronary angiography
Diagnostic modalities	
1. Exercise stress EKG	• Exercise-induced ST depression may be less sensitive
	• Similar prevalence of false positive exercise ST depression but with less predictable causes in women (e.g. left ventricular hypertrophy, hypertension, etc.)
	• Limited exercise capacity (typically older with significant comorbidities)

(continued)

Table 1.3 (continued)

Criteria	Compared with men, women have the following characteristics
2. SPECT MPI	• Diagnostic accuracy has been reported to be lower in women than men due to (a) lower exercise capacity, (b) higher prevalence of single-vessel CAD, (c) attenuation artifacts from breast tissue, (d) smaller heart size, (e) higher normal limits of transient ischemic dilation ratio
3. PET	• Improved diagnostic accuracy with PET in women as it
	(a) Overcomes the disadvantages of SPECT
	(b) Measures absolute myocardial blood flow (important to detect microvascular ischemia)
	(c) Normal PET carries an excellent prognosis
4. Stress echocardiogram	• Diagnostic accuracy may be affected by poor acoustic windows with breast tissue, however in general is comparable with men
	• Preferred in pregnant and young women (avoid ionizing radiation)
5. CT coronary angiography	• Diagnostic accuracy of CTCA is similar in men and women
	• Limitations in women include radiation exposure
6. Coronary angiogram	• Higher prevalence of non-obstructive CAD
	Small coronary artery size in women leads to partial inability to assess smaller mid and distal coronary segments
	• Possible differences in the outcomes of FFR-guided PCI in women versus men

ACS = acute coronary syndrome
CAD = coronary artery disease
SCAD = spontaneous coronary artery dissection
EKG = Electrocardiogram
SPECT MPI = Single-photon emission computed tomography myocardial perfusion imaging
PET = Positron emission tomography
CTCA = Computed tomography coronary angiography
FFR = functional flow reserve
PCI = percutaneous coronary intervention

Summary

Women have a significant burden of coronary atherosclerosis. This burden increases in post-menopausal years and is ascribed to a decline in estrogen levels; however, the precise etiology is not known. In addition to the atherosclerotic burden, women tend to have a higher prevalence of microvascular angina. Women generally present late and with atypical symptoms of ischemia and hence diagnostic strategies are employed late. There are also variabilities in the use and result of diagnostic strategies in men and women. Reperfusion strategies are used less often in women resulting in adverse outcomes. It is likely that better recognition of these unique gender differences in CAD will lead to improved outcomes in women.

References

1. Kumar A, Cannon CP (2009) Acute coronary syndromes: diagnosis and management, part I. Mayo Clin Proc 84:917–938
2. Benjamin EJ, Blaha MJ, Chiuve SE et al (2017) Heart disease and stroke statistics-2017 update: a report from the American Heart Association. Circulation 135:e146–e603
3. Lloyd-Jones D, Adams RJ, Brown TM et al (2010) Heart disease and stroke statistics—2010 update: a report from the American Heart Association. Circulation 121:e46–e215
4. Yang X-P, Reckelhoff JF (2011) Estrogen, hormonal replacement therapy and cardiovascular disease. Curr Opin Nephrol Hypertens 20:133
5. Mehta JL, McSweeney J (2018) Gender differences in the pathogenesis and management of heart disease. Springer, Berlin
6. Barrett-Connor EL, Cohn BA, Wingard DL, Edelstein SL (1991) Why is diabetes mellitus a stronger risk factor for fatal ischemic heart disease in women than in men?: the Rancho Bernardo Study. JAMA 265:627–631
7. Geer EB, Shen W (2009) Gender differences in insulin resistance, body composition, and energy balance. Gend Med 6:60–75
8. Varlamov O, Bethea CL, Roberts CT Jr (2015) Sex-specific differences in lipid and glucose metabolism. Front Endocrinol 5:241
9. Salas-Salvadó J, Granada M, Bulló M et al (2007) Plasma adiponectin distribution in a Mediterranean population and its association with cardiovascular risk factors and metabolic syndrome. Metabolism 56:1486–1492
10. Yamauchi T, Kamon J, Waki H et al (2001) The fat-derived hormone adiponectin reverses insulin resistance associated with both lipoatrophy and obesity. Nat Med 7:941
11. Moreira GC, Cipullo JP, Ciorlia LAS et al (2014) Prevalence of metabolic syndrome: association with risk factors and cardiovascular complications in an urban population. PLoS ONE 9:e105056
12. Park E, Kim J (2015) Gender-and age-specific prevalence of metabolic syndrome among Korean adults: analysis of the fifth Korean National Health and Nutrition Examination Survey. J Cardiovasc Nurs 30:256–266
13. Vishram JK, Borglykke A, Andreasen AH, et al (2014) Impact of age and gender on the prevalence and prognostic importance of the metabolic syndrome and its components in Europeans. The MORGAM Prospective Cohort Project. PloS one 9:e107294
14. Ogden CL, Carroll MD, Curtin LR et al (2006) Prevalence of overweight and obesity in the United States, 1999–2004. JAMA 295:1549–1555
15. Shaw LJ, Bairey Merz CN, Azziz R et al (2008) Withdrawn: postmenopausal women with a history of irregular menses and elevated androgen measurements at high risk for worsening cardiovascular event-free survival: results from the National Institutes of Health—National heart, lung, and blood institute sponsored women's ischemia syndrome evaluation. J Clin Endocrinol Metab 93:1276–1284
16. Vaccarino V, Badimon L, Corti R et al (2010) Ischaemic heart disease in women: are there sex differences in pathophysiology and risk factors? Position paper from the working group on coronary pathophysiology and microcirculation of the European society of cardiology. Cardiovasc Res 90:9–17
17. Wang X, Magkos F, Mittendorfer B (2011) Sex differences in lipid and lipoprotein metabolism: it's not just about sex hormones. J Clin Endocrinol Metab 96:885–893
18. Magkos F, Mittendorfer B (2009) Gender differences in lipid metabolism and the effect of obesity. Obstet Gynecol Clin North Am 36:245–265
19. Freedman DS, Otvos JD, Jeyarajah EJ et al (2004) Sex and age differences in lipoprotein subclasses measured by nuclear magnetic resonance spectroscopy: the Framingham Study. Clin Chem 50:1189–1200
20. Mora S, Szklo M, Otvos JD et al (2007) LDL particle subclasses, LDL particle size, and carotid atherosclerosis in the Multi-Ethnic Study of Atherosclerosis (MESA). Atherosclerosis 192:211–217

21. Manolio TA, Pearson TA, Wenger NK et al (1992) Cholesterol and heart disease in older persons and women review of an NHLBI workshop. Ann Epidemiol 2:161–176
22. Shai I, Rimm EB, Hankinson SE et al (2004) Multivariate assessment of lipid parameters as predictors of coronary heart disease among postmenopausal women: potential implications for clinical guidelines. Circulation 110:2824–2830
23. Hokanson JE, Austin MA (1996) Plasma triglyceride level is a risk factor for cardiovascular disease independent of high-density lipoprotein cholesterol level: a metaanalysis of population-based prospective studies. J Cardiovasc Risk 3:213–219
24. Handberg EM, Eastwood J-A, Eteiba W et al (2013) Clinical implications of the Women's Ischemia Syndrome Evaluation: inter-relationships between symptoms, psychosocial factors and cardiovascular outcomes. Womens Health 9:479–490
25. Moller-Leimkuhler AM (2010) Higher comorbidity of depression and cardiovascular disease in women: a biopsychosocial perspective. World J Biol Psychiatry 11:922–933
26. Huxley RR, Woodward M (2011) Cigarette smoking as a risk factor for coronary heart disease in women compared with men: a systematic review and meta-analysis of prospective cohort studies. Lancet 378:1297–1305
27. Willett WC, Green A, Stampfer MJ et al (1987) Relative and absolute excess risks of coronary heart disease among women who smoke cigarettes. N Engl J Med 317:1303–1309
28. Kawachi I, Colditz GA, Stampfer MJ et al (1993) Smoking cessation in relation to total mortality rates in women: a prospective cohort study. Ann Intern Med 119:992–1000
29. Yusuf S, Hawken S, Ounpuu S et al (2004) Effect of potentially modifiable risk factors associated with myocardial infarction in 52 countries (the INTERHEART study): case-control study. Lancet 364:937–952
30. Mallik S, Spertus JA, Reid KJ et al (2006) Depressive symptoms after acute myocardial infarction: evidence for highest rates in younger women. Arch Intern Med 166:876–883
31. Wassertheil-Smoller S, Shumaker S, Ockene J, et al (2004) Depression and cardiovascular sequelae in postmenopausal women. The Women's Health Initiative (WHI). Arch Intern Med 164:289–98
32. Rutledge T, Reis SE, Olson MB et al (2006) Depression symptom severity and reported treatment history in the prediction of cardiac risk in women with suspected myocardial ischemia: the NHLBI-sponsored WISE study. Arch Gen Psychiatry 63:874–880
33. Whang W, Kubzansky LD, Kawachi I et al (2009) Depression and risk of sudden cardiac death and coronary heart disease in women: results from the Nurses' Health Study. J Am Coll Cardiol 53:950–958
34. Roest AM, Martens EJ, de Jonge P, Denollet J (2010) Anxiety and risk of incident coronary heart disease: a meta-analysis. J Am Coll Cardiol 56:38–46
35. Orth-Gomer K, Wamala SP, Horsten M et al (2000) Marital stress worsens prognosis in women with coronary heart disease: the stockholm female coronary risk study. JAMA 284:3008–3014
36. Wang HX, Leineweber C, Kirkeeide R, et al (2007) Psychosocial stress and atherosclerosis: family and work stress accelerate progression of coronary disease in women: the stockholm female coronary angiography study. J Int Med 261:245–254
37. Gallo LC, Troxel WM, Kuller LH et al (2003) Marital status, marital quality, and atherosclerotic burden in postmenopausal women. Psychosom Med 65:952–962
38. Moore JX, Chaudhary N, Akinyemiju T (2017) Peer reviewed: metabolic syndrome prevalence by race/ethnicity and sex in the United States, national health and nutrition examination survey, 1988–2012. Prev Chronic Dis 14
39. Grundy SM, Cleeman JI, Daniels SR et al (2005) Diagnosis and management of the metabolic syndrome: an American Heart Association/National heart, lung, and blood institute scientific statement. Circulation 112:2735–2752
40. Grundy SM (2008) Metabolic syndrome pandemic. Arterioscler Thromb Vasc Biol 28:629–636
41. Pucci G, Alcidi R, Tap L et al (2017) Sex-and gender-related prevalence, cardiovascular risk and therapeutic approach in metabolic syndrome: A review of the literature. Pharmacol Res 120:34–42

42. Wenger NK (2012) Women and coronary heart disease: a century after Herrick: understudied, underdiagnosed, and undertreated. Circulation 126:604–611
43. Regitz-Zagrosek V, Oertelt-Prigione S, Prescott E et al (2015) Gender in cardiovascular diseases: impact on clinical manifestations, management, and outcomes. Eur Heart J 37:24–34
44. Cherry COB, Serieux E, Didier M, et al. Prevalence of risk factors for the metabolic syndrome in the middle income Caribbean nation of st. Lucia. *Adv Prev Med*. 2014;2014.
45. Heer T, Schiele R, Schneider S et al (2002) Gender differences in acute myocardial infarction in the era of reperfusion (the MITRA registry). Am J Cardiol 89:511–517
46. Berger JS, Elliott L, Gallup D et al (2009) Sex differences in mortality following acute coronary syndromes. JAMA 302:874–882
47. Rosengren A, Wallentin L, K Gitt A, et al (2004) Sex, age, and clinical presentation of acute coronary syndromes. Eur Heart J 25:663–670
48. Kardys I, Vliegenthart R, Oudkerk M et al (2007) The female advantage in cardiovascular disease: do vascular beds contribute equally? Am J Epidimiol 166:403–412
49. Cameron AJ, Shaw JE, Zimmet PZ (2004) The metabolic syndrome: prevalence in worldwide populations. Endocrinol Metab Clin North Am 33:351–375
50. Aguilar M, Bhuket T, Torres S et al (2015) Prevalence of the metabolic syndrome in the United States, 2003–2012. JAMA 313:1973–1974
51. Dallongeville J, Cottel D, Ferrières J et al (2005) Household income is associated with the risk of metabolic syndrome in a sex-specific manner. Diabetes Care 28:409–415
52. Ong KL, Tso AW, Lam KS, Cheung BM (2008) Gender difference in blood pressure control and cardiovascular risk factors in Americans with diagnosed hypertension. Hypertension 51:1142–1148
53. Regitz-Zagrosek V, Lehmkuhl E, Weickert MO (2006) Gender differences in the metabolic syndrome and their role for cardiovascular disease. Clin Res Cardiol 95:136–147
54. Wong RJ (2015) Trends in prevalence of the metabolic syndrome—reply. JAMA 314:950–951
55. Lovre D, Mauvais-Jarvis F (2015) Trends in prevalence of the metabolic syndrome. JAMA 314:950–950
56. Lidfeldt J, Nyberg P, Nerbrand C, et al (2003) Socio-demographic and psychosocial factors are associated with features of the metabolic syndrome. The Women's Health in the Lund Area (WHILA) study. Diabetes Obes Metab 5:106–112
57. Nugent L, Mehta PK, Merz CNB (2011) Gender and microvascular angina. J Thromb Thrombolysis 31:37–46
58. Lanza GA, Crea F (2010) Primary coronary microvascular dysfunction: clinical presentation, pathophysiology, and management. Circulation 121:2317–2325
59. James PA, Oparil S, Carter BL et al (2014) 2014 evidence-based guideline for the management of high blood pressure in adults: report from the panel members appointed to the eighth Joint National Committee (JNC 8). JAMA 311:507–520
60. O'Donnell E, Floras JS, Harvey PJ (2014) Estrogen status and the renin angiotensin aldosterone system. Am J Physiol Regul Integr Comp Physiol 307:R498–R500
61. Vitale C, Fini M, Speziale G, Chierchia S (2010) Gender differences in the cardiovascular effects of sex hormones. Fundam Clin Pharmacol 24:675–685
62. Yamauchi T, Kamon J, Waki H, et al (2001) The fat-derived hormone adiponectin reverses insulin resistance associated with both lipoatrophy and obesity. Nat Med 7:941–946
63. Ding EL, Song Y, Malik VS, Liu S (2006) Sex differences of endogenous sex hormones and risk of type 2 diabetes: a systematic review and meta-analysis. JAMA 295:1288–1299
64. Freedman DS, Otvos JD, Jeyarajah EJ et al (1998) Relation of lipoprotein subclasses as measured by proton nuclear magnetic resonance spectroscopy to coronary artery disease. Arterioscler Thromb Vasc Biol 18:1046–1053
65. Johnson JL, Slentz CA, Duscha BD et al (2004) Gender and racial differences in lipoprotein subclass distributions: the STRRIDE study. Atherosclerosis 176:371–377
66. Vanzetto G, Berger-Coz E, Barone-Rochette G et al (2009) Prevalence, therapeutic management and medium-term prognosis of spontaneous coronary artery dissection: results from a database of 11,605 patients. Eur J Cardiothorac Surg 35:250–254

67. Roura G, Ariza-Sole A, Rodriguez-Caballero IF et al (2016) Noninvasive follow-up of patients with spontaneous coronary artery dissection with CT angiography. JACC Cardiovasc Imaging 9:896–897
68. Saw J, Aymong E, Mancini GB et al (2014) Nonatherosclerotic coronary artery disease in young women. Can J Cardial 30:814–819
69. Henkin S, Negrotto SM, Tweet MS et al (2016) Spontaneous coronary artery dissection and its association with heritable connective tissue disorders. Heart 102:876–881
70. Alfonso F, Bastante T (2014) Spontaneous coronary artery dissection: novel diagnostic insights from large series of patients. Circ Cardiovasc Interv 7:638–641
71. Tweet MS, Gulati R, Hayes SN (2015) What Clinicians Should Know Alphabout Spontaneous Coronary Artery Dissection. Mayo Clin Proc 90:1125–1130
72. Saw J, Ricci D, Starovoytov A et al (2013) Spontaneous coronary artery dissection: prevalence of predisposing conditions including fibromuscular dysplasia in a tertiary center cohort. JACC Cardiovasc Interv 6:44–52
73. Vijayaraghavan R, Verma S, Gupta N, Saw J (2014) Pregnancy-related spontaneous coronary artery dissection. Circulation 130:1915–1920
74. Mahmoud AN, Taduru SS, Mentias A et al (2018) Trends of incidence, clinical presentation, and in-hospital mortality among women with acute myocardial infarction with or without spontaneous coronary artery dissection: a population-based analysis. JACC Cardiovasc Interv 11:80–90
75. Mozaffarian D, Benjamin EJ, Go AS et al (2016) Heart Disease and Stroke Statistics-2016 Update: A Report From the American Heart Association. Circulation 133:e38-360
76. Ann SH, De Jin C, Singh GB et al (2016) Gender differences in plaque characteristics of culprit lesions in patients with ST elevation myocardial infarction. Heart Vessels 31:1767–1775
77. Yahagi K, Davis HR, Arbustini E, Virmani R (2015) Sex differences in coronary artery disease: pathological observations. Atherosclerosis 239:260–267
78. Dreyer RP, Wang Y, Strait KM et al (2015) Gender differences in the trajectory of recovery in health status among young patients with acute myocardial infarction: results from the variation in recovery: role of gender on outcomes of young AMI patients (VIRGO) study. Circulation 131:1971–1980
79. Lansky AJ, Ng VG, Maehara A et al (2012) Gender and the extent of coronary atherosclerosis, plaque composition, and clinical outcomes in acute coronary syndromes. JACC Cardiovasc Imaging 5:S62-72
80. Schwartz RS, Burke A, Farb A et al (2009) Microemboli and microvascular obstruction in acute coronary thrombosis and sudden coronary death: relation to epicardial plaque histopathology. J Am Coll Cardiol 54:2167–2173
81. White SJ, Newby AC, Johnson TW (2016) Endothelial erosion of plaques as a substrate for coronary thrombosis. Thromb Haemost 115:509–519
82. Gulati M, Shaw LJ, Bairey Merz CN (2012) Myocardial ischemia in women: lessons from the NHLBI WISE study. Clin Cardiol 35:141–148
83. Chandrasekhar J, Mehran R (2016) Sex-Based Differences in Acute Coronary Syndromes: Insights From Invasive and Noninvasive Coronary Technologies. JACC Cardiovasc Imaging 9:451–464
84. Mendelsohn ME, Karas RH (2005) Molecular and cellular basis of cardiovascular gender differences. Science 308:1583–1587
85. Edwards DP (2005) Regulation of signal transduction pathways by estrogen and progesterone. Annu Rev Physiol 67:335–376
86. Rubanyi GM, Kauser K, Johns A (2002) Role of estrogen receptors in the vascular system. Vascul Pharmacol 38:81–88
87. Mieres JH, Shaw LJ, Arai A et al (2005) Role of noninvasive testing in the clinical evaluation of women with suspected coronary artery disease: consensus statement from the Cardiac Imaging Committee, Council on Clinical Cardiology, and the Cardiovascular Imaging and Intervention Committee, Council on Cardiovascular Radiology and Intervention. American Heart Association. Circulation 111:682–696

88. Crilly MA, Bundred PE, Leckey LC, Johnstone FC (2008) Gender bias in the clinical management of women with angina: another look at the Yentl syndrome. J Womens Health 17:331–342
89. Khan NA, Daskalopoulou SS, Karp I et al (2013) Sex differences in acute coronary syndrome symptom presentation in young patients. JAMA Intern Med 173:1863–1871
90. Meischke H, Larsen MP, Eisenberg MS (1998) Gender differences in reported symptoms for acute myocardial infarction: impact on prehospital delay time interval. Am J Emerg Med 16:363–366
91. Rosenfeld AG, Lindauer A, Darney BG (2005) Understanding treatment-seeking delay in women with acute myocardial infarction: descriptions of decision-making patterns. Am J Crit Care 14:285–293
92. Oertelt-Prigione S, Seeland U, Kendel F et al (2015) Cardiovascular risk factor distribution and subjective risk estimation in urban women–The BEFRI Study: a randomized cross-sectional study. BMC Med 13:52
93. Nguyen HL, Saczynski JS, Gore JM, Goldberg RJ (2010) Age and sex differences in duration of prehospital delay in patients with acute myocardial infarction: a systematic review. Circ Cardiovasc Qual Outcomes 3:82–92
94. Shaw LJ, Mieres JH, Hendel RH, Boden WE et al (2011) Comparative effectiveness of exercise electrocardiography with or without myocardial perfusion single photon emission computed tomography in women with suspected coronary artery disease: results from the What Is the Optimal Method for Ischemia Evaluation in Women (WOMEN) trial. Circulation 124:1239–1249
95. Johnson BD, Shaw LJ, Pepine CJ, Reis SE et al (2006) Persistent chest pain predicts cardiovascular events in women without obstructive coronary artery disease: results from the NIH-NHLBI-sponsored Women's Ischaemia Syndrome Evaluation (WISE) study. Eur Heart J 27:1408–1415
96. Figueras J, Domingo E, Ferreira I et al (2012) Persistent angina pectoris, cardiac mortality and myocardial infarction during a 12 year follow-up in 273 variant angina patients without significant fixed coronary stenosis. Am J Cardiol 110:1249–1255
97. Brorsson B, Bernstein SJ, Brook RH, Werkö L (2002) Quality of life of patients with chronic stable angina before and four years after coronary revascularisation compared with a normal population. Heart 87:140–145
98. Jespersen L, Hvelplund A, Abildstrøm SZ et al (2011) Stable angina pectoris with no obstructive coronary artery disease is associated with increased risks of major adverse cardiovascular events. Euro Heart J 33:734–744
99. Canto JG, Rogers WJ, Goldberg RJ et al (2012) Association of age and sex with myocardial infarction symptom presentation and in-hospital mortality. JAMA 307:813–822
100. Daly C, Clemens F, Lopez Sendon JL et al (2006) CLINICAL PERSPECTIVE. Circulation 113:490–498
101. Boccia A, Damiani G, D'Errico MM, et al. Age-and sex-related utilisation of cardiac procedures and interventions: a multicentric study in Italy. Int J Cardiol
102. Hendel RC, Berman DS, Di Carli MF et al (2009) ACCF/ASNC/ACR/AHA/ASE/SCCT/SCMR/SNM 2009 appropriate use criteria for cardiac radionuclide imaging: a report of the American College of Cardiology Foundation Appropriate Use Criteria Task Force, the American Society of Nuclear Cardiology, the American College of Radiology, the American Heart Association, the American Society of Echocardiography, the Society of Cardiovascular Computed Tomography, the Society for Cardiovascular Magnetic Resonance, and the Society of Nuclear Medicine Endorsed by the American College of Emergency Physicians. J Am Coll Cardiol 53:2201–2229
103. Kwok Y, Kim C, Grady D et al (1999) Meta-analysis of exercise testing to detect coronary artery disease in women. Am J Cardiol 83:660–666
104. Taqueti VR, Dorbala S, Wolinsky D et al (2017) Myocardial perfusion imaging in women for the evaluation of stable ischemic heart disease—state-of-the-evidence and clinical recommendations. J Nucl Cardiol 24:1402–1426

105. Gibbons RJ, Balady GJ, Bricker JT et al (2002) ACC/AHA 2002 guideline update for exercise testing: summary article: a report of the American College of Cardiology/American Heart Association Task Force on Practice Guidelines (Committee to Update the 1997 Exercise Testing Guidelines). J Am Coll Cardiol 40:1531–1540
106. Kohli P, Gulati M (2010) Exercise stress testing in women: going back to the basics. Circulation 122:2570–2580
107. Hlatky MA, Pryor DB, Harrell FE et al (1984) Factors affecting sensitivity and specificity of exercise electrocardiography: multivariable analysis. Am J Med 77:64–71
108. Morise AP, Beto R (1997) The specificity of exercise electrocardiography in women grouped by estrogen status. Int J Cardiol 60:55–65
109. Grzybowski A, Puchalski W, Zieba B et al (2008) How to improve noninvasive coronary artery disease diagnostics in premenopausal women?: The influence of menstrual cycle on ST depression, left ventricle contractility, and chest pain observed during exercise echocardiography in women with angina and normal coronary angiogram. Am Heart J 156:964
110. Barolsky SM, Gilbert CA, Faruqui A et al (1979) Differences in electrocardiographic response to exercise of women and men: a non-Bayesian factor. Circulation 60:1021–1027
111. Fitzgerald BT, Scalia WM, Scalia GM (2019) Female False Positive Exercise Stress ECG Testing-Fact Versus Fiction. Heart Lung Circ 28:735–741
112. Sharma S, Mehta PK, Arsanjani R et al (2018) False-positive stress testing: Does endothelial vascular dysfunction contribute to ST-segment depression in women? a Pilot Study. Clin Cardiol 41:1044–1048
113. Iskandar A, Limone B, Parker MW et al (2013) Gender differences in the diagnostic accuracy of SPECT myocardial perfusion imaging: a bivariate meta-analysis. J Nucl Cardiol 20:53–63
114. Murthy VL, Naya M, Taqueti VR et al (2014) Effects of sex on coronary microvascular dysfunction and cardiac outcomes. Circulation 129:2518–2527
115. Greenwood JP, Motwani M, Maredia N et al (2014) Comparison of cardiovascular magnetic resonance and single-photon emission computed tomography in women with suspected coronary artery disease from the Clinical Evaluation of Magnetic Resonance Imaging in Coronary Heart Disease (CE-MARC) Trial Circulation 129:1129–1138
116. Jug B, Gupta M, Papazian J et al (2012) Diagnostic performance of 64-slice multidetector coronary computed tomographic angiography in women. J Nucl Cardiol 19:1154–1161
117. Mullani NA, Caras D, Ahn C et al (2000) Fewer women than men have positive SPECT and PET cardiac findings among patients with no history of heart disease. J Nucl Med 41:263–268
118. Iskandrian AE, Heo J, Nallamothu N (1997) Detection of coronary artery disease in women with use of stress single-photon emission computed tomography myocardial perfusion imaging. J Nucl Cardiol 4:329–335
119. Hansen CL, Crabbe D, Rubin S (1996) Lower diagnostic accuracy of thallium-201 SPECT myocardial perfusion imaging in women: an effect of smaller chamber size. J Am Coll Cardiol 28:1214–1219
120. Taillefer R, DePuey EG, Udelson JE et al (1997) Comparative diagnostic accuracy of Tl-201 and Tc-99m sestamibi SPECT imaging (perfusion and ECG-gated SPECT) in detecting coronary artery disease in women. J Am Coll Cardiol 29:69–77
121. Sharir T, Kang X, Shaw LJ et al (2006) Prognostic value of poststress left ventricular volume and ejection fraction by gated myocardial perfusion SPECT in women and men: gender-related differences in normal limits and outcomes. J Nucl Cardiol 13:495–506
122. Rivero A, Santana C, Folks RD et al (2006) Attenuation correction reveals gender-related differences in the normal values of transient ischemic dilation index in rest-exercise stress sestamibi myocardial perfusion imaging. J Nucl Cardiol 13:338–344
123. Sharma P, Patel CD, Karunanithi S et al (2012) Comparative Accuracy of CT Attenuation-Corrected and Non–Attenuation-Corrected SPECT Myocardial Perfusion Imaging. Clin Nucl Med 37:332–338
124. Masood Y, Liu Y-H, DePuey G et al (2005) Clinical validation of SPECT attenuation correction using x-ray computed tomography–derived attenuation maps: multicenter clinical trial with angiographic correlation. J Nucl Cardiol 13:676–686

125. Van Train KF, Maddahi J, Berman DS et al (1990) Quantitative analysis of tomographic stress thallium-201 myocardial scintigrams: a multicenter trial. J Nucl Med 31:1168–1179
126. Metz LD, Beattie M, Hom R et al (2007) The prognostic value of normal exercise myocardial perfusion imaging and exercise echocardiography: a meta-analysis. J Am Coll Cardiol. 49:227–237
127. Shaw LJ, Hage FG, Berman DS et al (2012) Prognosis in the era of comparative effectiveness research: Where is nuclear cardiology now and where should it be? J Nucl Cardiol 19:1026–1043
128. Hachamovitch R, Berman DS, Kiat H et al (1996) Effective risk stratification using exercise myocardial perfusion SPECT in women: gender-related differences in prognostic nuclear testing. J Am Coll Cardiol 28:34–44
129. Sharir T, Germano G, Kavanagh PB et al (1999) Incremental prognostic value of post-stress left ventricular ejection fraction and volume by gated myocardial perfusion single photon emission computed tomography. Circulation 100:1035–1042
130. Valeti US, Miller TD, Hodge DO, Gibbons RJ (2005) Exercise single-photon emission computed tomography provides effective risk stratification of elderly men and elderly women. Circulation 111:1771–1776
131. Cerci MSJ, Cerci JJ, Cerci RJ, Neto CCP et al (2011) Myocardial perfusion imaging is a strong predictor of death in women. JACC Cardiovasc Imaging 4:880–888
132. Abbott BG, Afshar M, Berger AK, Frans J (2003) Prognostic significance of ischemic electrocardiographic changes during adenosine infusion in patients with normal myocardial perfusion imaging. J Nucl Cardiol 10:9–16
133. Klodas E, Miller TD, Christian TF et al (2003) Prognostic significance of ischemic electrocardiographic changes during vasodilator stress testing in patients with normal SPECT images. J Nucl Cardiol 10:4–8
134. Parker MW, Iskandar A, Limone B et al (2012) Diagnostic accuracy of cardiac positron emission tomography versus single photon emission computed tomography for coronary artery disease: a bivariate meta-analysis. Circ Cardiovasc Imaging 5:700–707
135. Mc Ardle BA, Dowsley TF, Wells GA, Beanlands RS (2012) Does rubidium-82 PET have superior accuracy to SPECT perfusion imaging for the diagnosis of obstructive coronary disease?: A systematic review and meta-analysis. J Am Coll Cardiol 60:1828–1837
136. Dorbala S, Di Carli MF (2014) Cardiac PET perfusion: prognosis, risk stratification, and clinical management. Semin Nucl Med 44:344–357
137. Merz CNB, Shaw LJ, Reis SE et al (2006) Insights from the NHLBI-Sponsored Women's Ischemia Syndrome Evaluation (WISE) Study: Part II: gender differences in presentation, diagnosis, and outcome with regard to gender-based pathophysiology of atherosclerosis and macrovascular and microvascular coronary disease. J Am Coll Cardiol 47:S21–S29
138. Shaw LJ, Merz CNB, Pepine CJ et al (2006) Insights from the NHLBI-Sponsored Women's Ischemia Syndrome Evaluation (WISE) Study: Part I: gender differences in traditional and novel risk factors, symptom evaluation, and gender-optimized diagnostic strategies. J Am Coll Cardiol 47:S4–S20
139. Taqueti VR, Shaw LJ, Cook NR et al (2017) Excess cardiovascular risk in women relative to men referred for coronary angiography is associated with severely impaired coronary flow reserve, not obstructive disease. Circulation 135:566–577
140. Yoshinaga K, Chow BJ, Williams K et al (2006) What is the prognostic value of myocardial perfusion imaging using rubidium-82 positron emission tomography? J Am Coll Cardiol 48:1029–1039
141. Dorbala S, Di Carli MF, Beanlands RS et al (2013) Prognostic value of stress myocardial perfusion positron emission tomography: results from a multicenter observational registry. J Am Coll Cardiol 61:176–184
142. Kay J, Dorbala S, Goyal A et al (2013) Influence of sex on risk stratification with stress myocardial perfusion Rb-82 positron emission tomography: results from the PET (Positron Emission Tomography) Prognosis Multicenter Registry. J Am Coll Cardiol 62:1866–1876

143. Roger VL, Pellikka PA, Bell MR et al (1997) Sex and test verification bias: impact on the diagnostic value of exercise echocardiography. Circulation 95:405–410
144. Cortigiani L, Dodi C, Paolini EA et al (1998) Prognostic value of pharmacological stress echocardiography in women with chest pain and unknown coronary artery disease. J Am Coll Cardiol 32:1975–1981
145. Chow BJ, Abraham A, Wells GA et al (2009) Diagnostic accuracy and impact of computed tomographic coronary angiography on utilization of invasive coronary angiography. Circ Cardiovasc Imaging 2:16–23
146. Meijboom WB, Meijs MF, Schuijf JD et al (2008) Diagnostic accuracy of 64-slice computed tomography coronary angiography: a prospective, multicenter, multivendor study. J Am Coll Cardiol 52:2135–2144
147. Dharampal AS, Papadopoulou SL, Rossi A et al (2012) Computed tomography coronary angiography accuracy in women and men at low to intermediate risk of coronary artery disease. Eur Radiol 22:2415–2423
148. Mieres JH, Gulati M, Bairey Merz N et al (2014) Role of noninvasive testing in the clinical evaluation of women with suspected ischemic heart disease: a consensus statement from the American Heart Association. Circulation 130:350–379
149. Pundziute G, Schuijf JD, Jukema JW et al (2008) Gender influence on the diagnostic accuracy of 64-slice multislice computed tomography coronary angiography for detection of obstructive coronary artery disease. Heart 94:48–52
150. Jespersen L, Hvelplund A, Abildstrom SZ et al (2012) Stable angina pectoris with no obstructive coronary artery disease is associated with increased risks of major adverse cardiovascular events. Eur Heart J 33:734–744
151. Kern MJ, Lerman A, Bech J-W et al (2006) Physiological assessment of coronary artery disease in the cardiac catheterization laboratory: a scientific statement from the American Heart Association committee on diagnostic and interventional cardiac catheterization. Council Clin Cardiol Circ 114:1321–1341
152. Solimene MC (2010) Coronary heart disease in women: a challenge for the 21st century. Clinics 65:99–106
153. Nevsky G, Jacobs JE, Lim RP et al (2011) Sex-specific normalized reference values of heart and great vessel dimensions in cardiac CT angiography. Am J Roentgenol 196:788–794
154. Kim H-S, Tonino PA, De Bruyne B et al (2012) The impact of sex differences on fractional flow reserve–guided percutaneous coronary intervention: a FAME (Fractional Flow Reserve Versus Angiography for Multivessel Evaluation) substudy. JACC Cardiovasc Interv 5:1037–1042
155. Tonino PA, De Bruyne B, Pijls NH et al (2009) Fractional flow reserve versus angiography for guiding percutaneous coronary intervention. N Engl J Med. 360:213–224
156. Li J, Rihal CS, Matsuo Y et al (2013) Sex-related differences in fractional flow reserve–guided treatment. Circ Cardiovasc Interv. 6:662–670

Chapter 2
Sex Differences in Cardiac Ischemia/Reperfusion Injury

Bohuslav Ostadal, Petr Ostadal, and Jan Neckar

Abstract It is now well known that differences in the structure and function of the heart exist between male and female hearts. Several lines of experimental and clinical investigations have reported that there are sex differences in the tolerance to myocardial ischemia, whereby adult male hearts are more susceptible to ischemia/reperfusion (I/R) injury as compared to pre-menopausal female hearts. Experimental studies have also shown that adult female hearts have increased resistance and male hearts are more susceptible to I/R in animals exposed to perinatal hypoxia. Although there is now a large body of evidence which indicates that estrogen is involved in the sex differences with respect to cardiac tolerance to ischemia, the exact mechanisms involved in the cardiac response to ischemia or hypoxia are not fully understood. Accordingly, this chapter is intended to describe some of the known molecular and cellular mechanisms that contribute to sex differences in the susceptibility to I/R injury. With such a new basic information and advancements in the understanding of the mechanisms responsible for sex differences in cardiac sensitivity to ischemic injury, it is hoped that some specific therapeutic strategies will be developed for post-menopausal females for better quality of life, and lower mortality.

Keywords Sex differences · Ischemia/reperfusion injury · Role of estrogen · Cardiac tolerance

B. Ostadal (✉) · J. Neckar
Institute of Physiology, Czech Academy of Sciences, Videnska 1083, 14220 Prague 4, Czech Republic
e-mail: ostadal@biomed.cas.cz

P. Ostadal
Cardiovascular Center, Department of Cardiology, Na Homolce Hospital, Prague, Czech Republic
e-mail: ostadal@biomed.cas.cz

B. Ostadal and N. S. Dhalla (eds.), *Sex Differences in Heart Disease*,
Advances in Biochemistry in Health and Disease 21,
https://doi.org/10.1007/978-3-030-58677-5_2

Introduction

The existence of sex differences in ischemic heart disease (IHD) is now well established. In comparison to males, IHD has been shown to occur much later in females who also have a lower incidence of focal stenotic lesions. Furthermore, the prevalence of IHD increases in post-menopausal women, suggesting an association with the diminished level of estrogen [1–3]. Even though IHD is the major cause of mortality in both women and men, it has largely been viewed as a "male disease" and accordingly, the majority of experimental and clinical studies have been conducted in males. Such an approach is primarily based on the fact that pathophysiology of IHD in males, unlike females, is not influenced by periodic fluctuations in sex hormones.

The information that women are discriminated in diagnostics and treatment of cardiovascular diseases was actually first indicated in the early 1990s. In addition, the number of publications in the topic of "sex and the heart" has only increased significantly in the last thirty years [4] (Fig. 2.1), most likely as a consequence of the observed differences in the responses of healthy and diseased male and female heart as well as of the inconsistencies in the effects of hormonal replacement therapy (HRT) in post-menopausal women. In this regard, the use of estrogen in post-menopausal women has been reported to increase the incidence of IHD [5, 6]. Although several pieces of information on sex-related differences in cardiac structure and function have become available in the literature, the underlying pathophysiological mechanisms have yet to be fully defined. In this article, we provide a brief review of some of the mechanisms that contribute to sex differences of cardiac ischemia/reperfusion (I/R) injury with particular attention to the experimental approach [4, 7–9]. It is pointed out that the term "sex" has been used when reporting biological aspects, while the term "gender" is reserved for the psychosocial and cultural factors.

Sex Differences of the Normal Heart

Sex-related differences in cardiac structure and function have been observed under normal physiological conditions [10]. Although there are no differences in the weight of the cardiac muscle during early ontogenetic development, there occurs an increase of myocardial weight in males at puberty as well as in adulthood; this change makes male hearts 15–30% heavier than female hearts [11]. The initial number of cardiomyocytes is comparable in both sexes; however, during ontogenetic development the number of myocytes in female hearts remains stable whereas the number myocytes in the male heart decreases significantly [12]. It is pointed out that the cardiomyocyte loss in male hearts is accompanied by a reactive hypertrophic response.

The average heart rate for women has been reported to be approximately 3–5 beats/min more than for the males [13–15]. Moreover, the female heart has longer action potential duration, longer QT interval, and a shorter sinus node recovery time as compared to the male heart [16]. In men under the age of 60 years, the average systolic and diastolic pressure is higher by 6–7 mm Hg and 3–5 mm Hg, respectively, as compared to age-matched women. In post-menopausal women, the systolic blood pressure increases, to the extent that the incidence of hypertension is more prevalent in women after menopause than in men [17].

There are several inconsistencies with the data on sex differences in the cardiac contractility, which have been summarized recently by Machuki et al. [18]. For example, Schwertz et al. [19, 20] and Machuki et al. [18] have observed that female cardiomyocytes have a larger contraction and greater Ca^{2+} transient amplitude as compared to male cardiomyocytes, whereas Farrel et al. [21] have failed to confirm these findings. These contrasting observations were suggested by Machuki et al. [18] to be, at least partly, due to the use of whole ventricular myocytes versus left ventricular apical cardiomyocytes, particularly since differences in apical versus basal Ca^2 current have been reported in rabbit hearts [22]. An important element of cardiomyocyte contraction is the cAMP-PKA-L-type Ca^{2+} channel pathway. Machuki et al. [18] have reported that intracellular cAMP, Ca^{2+} channel density and Ca^{2+} transient were larger in female than in male cardiomyocytes. These investigators have also suggested that estrogen can regulate the expression of genes for the cAMP-L-type calcium channel pathway and contribute to sex differences in cardiac contraction.

Over the last few years, the number of studies describing myocardial sex differences at the molecular level has increased; their enumeration exceeds the possibilities of this review. For the purposes of our chapter, sex differences in the mechanisms of cardiac excitation–contraction coupling [21], cardiomyocyte calcium handling, reactive oxygen species (ROS) generation and β-adrenoceptor density [23–26] are described. It may be noted that Ca^{2+} homeostasis is regulated as a function of the estrous cycle [27] and myofilament Ca^2 sensitivity is increased in hearts of ovariectomized female rats. Interestingly, Ca^{2+} homeostasis is also regulated by testosterone, which activates phospholipase C and subsequent production of inositol-3-phosphate, which in turn mediates the release of Ca^{2+} from the sarcoplasmic reticulum and increases intracellular Ca^{2+} [28]. Higher expression of sarcolemmal and mitochondrial ATP-sensitive potassium (K_{ATP}) channels has been reported in the female myocardium; their inhibition during ischemia increases the extent of tissue injury [29]. Estrogen regulates also the expression of phospholamban and ryanodine receptors. In this regard, the higher levels of ryanodine receptors in female cardiomyocytes have been linked to higher Ca^{2+} release from the sarcoplasmic reticulum [30]. Interestingly, no sex differences have been observed in SERCA (Ca^{2+}-pump ATPase) expression levels [18].

Sex differences with respect to cardiac structure, function, and cellular mechanisms during aging have been summarized by Keller and Howlett [31]. Dworatzek et al. [32] have described age–dependent changes in cardiac extracellular matrix remodeling. In this regard, collagenous protein types I, III, and VI were significantly lower in the female heart. Similarly, Arellano et al. [33] revealed a specific down-regulation of Sirt1 and Sirt3 in aged female human hearts, which was accompanied by a decline in the mitochondrial anti-oxidative defense systems. Sex related differences in cardiac mitochondrial function and energy metabolism as well as the possible role of mitochondria in sex differences in tolerance to cardiac I/R injury have been summarized in [34, 35].

Sex Dependent Tolerance to Experimental Cardiac Ischemia/Reperfusion Injury

Sex differences in the normal cardiac structure and function suggest that differences also exist in various cardiac pathologies including I/R injury. Experimental evidence demonstrating that female rat hearts are more tolerant to hypoxia was first published by our group in 1984 [36] (Fig. 2.2). Sex differences in cardiac resistance to I/R injury were later observed in different experimental animals and confirmed higher hypoxic tolerance of the female myocardium [29, 37–40]. Young adult female hearts have smaller infarct size, display greater recovery of contractile function, and undergo less arrhythmias during reperfusion as compared to age-matched males [41, 42]. Similarly, female hearts of transgenic mice overexpressing Na^+/Ca^{2+} exchanger and β_2 adrenergic receptors [43, 44] have less injury following I/R, and enhanced contractility. Moreover, sex differences in the sensitivity to I/R injury exist also in spontaneously hypertensive rats. In this regard, post-ischemic recovery was significantly better in female hearts, despite the comparable elevation in blood pressure [45]. An increase in infarct size has also been observed in obese males as compared to obese females [46]. Furthermore, sex differences in cardiac remodeling after myocardial infarction have been reported [47–50]. These investigators found that male mice had delayed myocardial healing, and a greater incidence of cardiac rupture, indicative of an attenuated structural remodeling in the female heart; this effect was attributed to the premature extracellular matrix degradation induced by metalloproteinases (MMP) activation in male heart.

We have reported earlier that the adaptation to chronic hypoxia increases cardiac tolerance in both sexes, but the increase in tolerance to oxygen deprivation is specifically maintained in female hearts [36]. Turcato et al. [51] have shown that the cardioprotective effect of ischemic preconditioning decreases in females during the aging process [52, 53]. It seems to us, that this observation belongs to the general biological phenomenon: the degree of cardiac protection has obviously its threshold. We have namely observed similar negative effects in highly tolerant neonatal hearts: indeed, adaptation to chronic hypoxia as well as ischemic preconditioning were without any effect [54, 55]. Lieder et al. [56] have observed that sex is not a determinant of a cardioprotection effect afforded by ischemic preconditioning as well as remote ischemic preconditioning in isolated rat hearts. In fact, in the recent CONDI-2/ERIC PPCI Trial no cardioprotective effect of remote ischemic conditioning in patients with acute myocardial infarction and no sex differences were observed [57].

Role of Hormones in Sex-Dependent Variation in Cardiac Sensitivity to Ischemia

Sex hormones are known to exert cardiac, renal, and vascular effects, which in turn, may also determine the severity of ischemic injury. While testosterone activates the renin-angiotensin system (RAS) and induces vasoconstriction as well as cardiac hypertrophy, estrogen modulates RAS activity, but in contrast to testosterone, produces vasodilatation associated with less aggressive remodeling of the heart in different cardiac pathologies, including cardiac ischemia [58]. Accordingly, experimental investigations have primarily focused on the effect of estrogens; however, it should be noted that cardioprotection provided by estrogen in experimental animals is in contrast to the lack of protective effect observed after HRT in women.

There is clear evidence that ovariectomy in female rats increases the infarct size whereas administration of estrogen is protective in males [59]. Most of the cardioprotective effects of estrogen are considered to be mediated through the nuclear hormone receptors as well as through estrogen receptor (ER)-α or ER-β [60]. Experimental data suggest that these receptor types play an important role in the protection against cardiac I/R injury [61–63]. In fact, these receptors are expressed in cardiac myocytes and fibroblasts as well as in vascular smooth muscle and endothelial cells [64, 65]. ER are also located on the cell plasma membrane and the mitochondria [66]. Recently, a third, membrane associated ER, identified as G-protein coupled estrogen receptor (GPER, [67]) has been found to inhibit the opening of the mitochondrial permeability transition pore (PTP [68]), which is known to be involved in the development of ischemic injury.

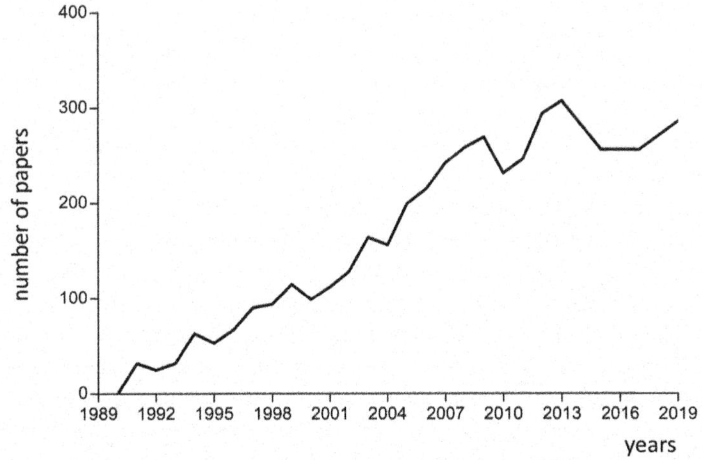

Fig. 2.1 Number of clinical and experimental papers dealing with "Sex and Heart" from 1989 to 2019. *Source* Web of Science. Adapted from [4], with permission

The non-genomic (non-nuclear) processes occur rapidly and independently of protein synthesis [69, 70]. In this regard, while endothelial NO synthase (eNOS) expression is higher in the female heart as compared to that in males, blockade of eNOS activity with L-NAME abolishes the sex differences in tolerance to cardiac I/R injury [44, 53]. It should be mentioned that the higher content of eNOS is associated with S-nitrosylation of L-type Ca^{2+} channels, which, in turn, markedly decreases cardiac I/R injury by attenuating the occurrence of intracellular calcium overload [71]. In addition, estrogen activates phosphatidylinositol 3-kinase (PI3K) activity, which is considered to play a role in cardioprotection in females [72]. Taken together, it can be suggested that the protective effect of estrogen could be attributed to changes in the expression of specific proteins or altered post translational protein modifications.

The mechanisms associated with the augmented tolerance to oxygen deprivation in the female heart are complex and not fully understood. Experimental studies have shown higher expression of the sarcolemmal [29] and mitochondrial [73] K_{ATP} channels as well as increased protein kinase B (Akt) and PKCε levels in the female heart [74]. Indeed, in human studies, younger women have been found to have higher Akt levels as compared to aged men and post-menopausal women [75]. The protective effects of estrogen may also be mediated by the inhibition of the the effects of tumor necrosis factor-α (TNF-α) as well as remodifying the TNF-α expression [76].

Although there is a paucity of information regarding the possible cardiac actions of testosterone, both adverse and beneficial effects of testosterone on the heart have been reported [77]. In this regard, testosterone has been shown to activate androgen receptors, which are known to be expressed in cardiomyocytes. It should be mentioned that like estrogens, androgens are present in both sexes, but in different amounts and proportions. Endogenous androgens, such as dehydroepiandrosterone, androstenedione and testosterone, can be metabolized to estradiol via 17β-hydroxysteroid dehydrogenase and aromatase activities. Interestingly, these enzymes are expressed in cardiomyocytes and treatment with estrogen precursors has been found to increase ER-α, ER-β, and iNOS expressions in a sex-dependent manner [78].

There are inconsistencies in the literature regarding the impact of testosterone on cardiovascular function. It has been suggested that testosterone may increase the susceptibility for IHD in men [79]. In addition, high doses of androgenic steroid have been linked to progression of atheroma and thereby elevating the potential for myocardial infarction [80]. However, there is no compelling evidence to indicate that physiological concentrations of testosterone cause any ischemic lesions. On the other hand, some clinical studies have shown that testosterone can exert beneficial effects on the cardiovascular system. Epidemiological observations have revealed an association between testosterone and reduced cardiovascular events and mortality [81]. In fact, low levels of testosterone have been linked to higher cardiovascular mortality [82]. These relationships may be the consequence of an indirect effect of testosterone through its conversion to dihydrotestosterone or 17β-estradiol. In this regard, testosterone administration to ovariectomized female rats has been reported to attenuate myocardial injury, reduce inflammatory infiltrates and MMP-3 and 13 formations [83]. These researchers suggested that the beneficial effects were not as a consequence of any direct action of testosterone, since concurrent inhibition of aromatase and 5α-reductase enzyme activities did not induce significant changes in the extent of the I/R injury. Thus, it is conceivable that the beneficial effects of testosterone observed in males may be attributed to its conversion to estradiol and its metabolites.

Recently, Ghimire et al. [84] have suggested that low testosterone protects against I/R injury subsequent to cardioplegic arrest in older mice. By employing a mouse model of myocardial infarction, Cavasin et al. [85] have reported that while estrogen prevented chronic remodeling and depressed cardiac function, testosterone negatively impacted healing of the heart muscle (as evidenced by higher cardiac rupture), and thus contributed to cardiac dysfunction as well as to adverse cardiac remodeling. On the other hand, testosterone has been shown to exert cardioprotective effects through an upregulation of the α_1-adrenoceptor; this beneficial action was significantly reduced by androgen receptor inhibitors [86]. Taken together, these inconsistencies, point to the need for further investigations, which should take into consideration the experimental model, form and dosage of steroid hormone, as well as timing for the evaluation of effects. Furthermore, it should be pointed out that steroid hormone receptor function is influenced by a number of co-regulatory proteins to alter transcription [87]; this aspect should also be considered in the interpretation and determination of the significance of the data.

Sex-Dependent Effect of Early Hypoxia on Cardiac Tolerance to Ischemia in Adulthood

There is now epidemiological evidence demonstrating an association between early hypoxia and increased risk of cardiovascular disease in adulthood [88, 89]. It is known that pregnancy at high altitude, pregnancy with anemia, placental insufficiency, as well as heart, lung, and kidney disease can result in hypoxic stress in the developing fetus. Hypoxemia is considered to be one of the most frequent insults during early stages of postnatal development as a consequence of congenital cyanotic heart defects. In this regard, it would be interesting to investigate the possibility of sex differences in tolerance to cardiac I/R injury in adult animals exposed to perinatal hypoxia. To study these late effects of early disturbances, the theory of critical developmental periods, introduced already in 1921 by Stockard [90], should be revived. According to this idea the tissues are most sensitive to injury during the period of intense growth. The perinatal period (late prenatal and early postnatal) is regarded as a typical critical period during ontogenetic development [91].

In a rat model, we have observed that perinatal exposure to chronic intermittent hypobaric hypoxia increases the tolerance to acute cardiac I/R injury in adult females, as evidenced by fewer ischemic arrhythmias. In contrast, the number of arrhythmias increased in males [92]. These results were later confirmed by Xue and Zhang [93]; it was observed that early hypoxia, unlike in female rats, increased cardiac enzyme

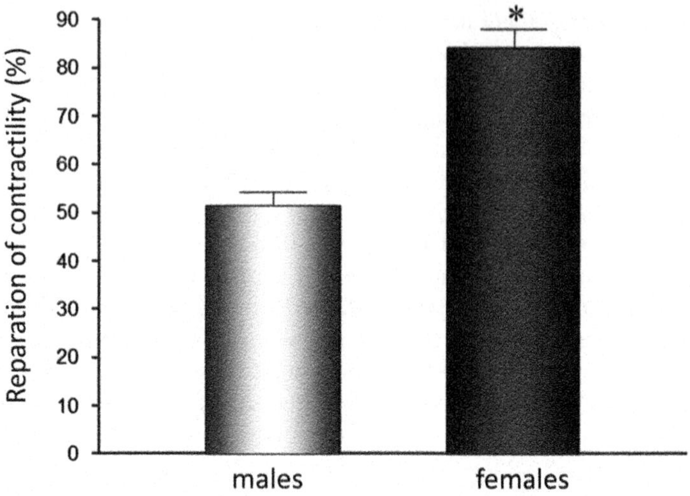

Fig. 2.2 Sex differences in the cardiac tolerance to acute oxygen deprivation in rats (expressed as % of the reparation of contractility of the isolated right ventricle after acute anoxia). * $P < 0.01$, data from [36], with permission

release and infarct size in adult male rats only. Recent study by Thompson et al. [94] revealed increase in mean arterial blood pressure and cardiac and mitochondrial dysfunction specifically in male offspring following exposure of guinea pigs to prenatal hypoxia at 50 days of gestation.

Several mechanisms for the sex-dependent alterations in tolerance to cardiac ischemia in adults following perinatal hypoxia have been proposed. For example, abnormal fetal programming of PKCε gene expression results in a down-regulation of PKCε activity in adult male offspring only [93]. Chronic hypoxia during gestation has been shown to down-regulate PKCε expression in the developing heart, most likely due to epigenetic modifications [95]; in addition, a higher degree of methylation of specificity protein (SP) 1 binding sites and PKCε transcription were also observed. Hypoxia-induced methylation was significantly greater in the heart of male fetuses. It was suggested that this sex-specific effect may be due, in part, to the higher expression level of both ERα and ERβ in the heart of female fetuses [95]. The observation that both ER interacted with the SP1 binding sites at the PKCε promoter in the fetal heart suggests a putative mechanism for the increased protection of SP1 binding sites and PKCε transcription in the female hearts in response to hypoxic stress. Recently, Lv et al. [96] have reported a decrease in the expression of glucocorticoid receptor (GR) mRNA and protein specifically in the heart of adult male offspring, following antenatal hypoxia. On this evidence, it was suggested that antenatal hypoxia reprograms GR gene expression in the heart in a sex-dependent manner, which may constitute a novel mechanism for the explanation of sex differences in the late effects of hypoxia. Thompson et al. [94] proposed that perinatal hypoxia is a programming stimulus that may increase the susceptibility to cardiac and mitochondrial dysfunction later in life, which in turn increases the risk of heart disease in adulthood in sex-dependent manner.

From the aforementioned, it is evident that sex-specific late effects of perinatal interventions on the sensitivity of the adult heart to I/R injury have important clinical implications. Indeed, cardiac ischemic tolerance in adulthood may be influenced by perinatal conditions (e.g. hypoxemia due to operated congenital cyanotic heart disease) in sex-dependent manner. While it may be difficult to translate experimental findings directly to clinical situation in humans, it is plausible that perinatal hypoxia may result in programming of a specific genes in the offspring in a sex-dependent manner, which may influence the susceptibility of the heart to I/R injury. This warrants further investigation.

Conclusions

It is well-established that structural and functional differences exist between male and female hearts. Clinical and experimental data suggest that adult male hearts have a higher risk of I/R injury than hearts of females at least until menopause; however, the mechanisms involved are not consistent and not completely understood. There is a wealth of information that clearly implicates estrogen to have an important contributory role in the sex differences in cardiac tolerance to ischemia, but there is a lack of information regarding the identity of the molecular mechanisms responsible for this clinically highly relevant biological phenomenon. We are convinced that new experimental lines of evidence on sex differences in cardiac sensitivity to ischemic injury will provide the foundations for a women-specific therapeutic strategy for improved quality of life, and lower mortality.

Acknowledgments This study was supported by grant of the Czech Science Foundation 18-03207S and by an institutional grant MH CZ-DRO (Nemocnice Na Homolce-NNH, 00023884), IG150501.

References

1. Regitz-Zagrosek V, Kararigas G (2017) Mechanistic pathways of sex differences in cardiovascular disease. Physiol Rev 97:1–37
2. Leinwand LA (2003) Sex is a potent modifier of the cardiovascular systém. J Clin Invest 112:302–307
3. Regitz-Zagrosek V (2006) Therapeutic implications of the gender-specific aspects of cardiovascular disease. Nat Rev Drug Discov 5:425–439
4. Ostadal B, Netuka I, Maly J et al (2009) Gender differences in cardiac ischemic injury and protection-experimental aspects. Exp Biol Med 234:1011–1019
5. Hulley S, Grady D, Bush T, et al (1998) Randomized trial of estrogen plus progestin for secondary prevention of coronary heart disease in postmenopausal women. Heart and Estrogen/progestin Replacement Study (HERS) Research Group. JAMA 280: 605–613

6. Rossouw JE, Anderson GL, Prentice RL et al (2002) Risk and benefits of estrogen plus progestin in healthy postmenopausal women: principal results From the Women´s Health Initiative randomized controlled trial. JAMA 288:321–333
7. Ostadal P, Ostadal B (2012) Women and the management of acute coronary syndrome. Can J Physiol Pharmacol 90:1151–1159
8. Ostadal B, Ostadal P (2014) Sex-based differences in cardiac ischaemic injury and protection: therapeutic implications. BJP 171:541–554
9. Kolar F, Ostadal B (2013) ExActa: Sex differences in cardiovascular function. Acta Physiol 207:584–587
10. Legato MJ, Leghe JK (2010) Gender and the heart: sex-specific differences in the normal myocardial anatomy and physiology. In: Legato MJ (ed) Principles of Gender Specific Medicine. Elsevier, New York, pp 151–161
11. De Simone G, Devereux RB, Daniels SR, Meyer RA (1995) Gender differences in left ventricular growth. Hypertension 26:979–983
12. Olivetti G, Giordano G, Corradi D et al (1995) Gender differences and aging: effects on the human heart. J Am Coll Cardiol 26:1068–1079
13. Bazet H (1920) An analysis of the time-relations of electrocardiograms. Heart 7:370
14. Czubryt MP, Espira L, Lamoureux L, Abrenica B (2006) The role of sex in cardiac function and disease. Can J Physiol Pharmacol 84:93–109
15. Burke JH, Goldberger JJ, Ehlert FA et al (1996) Gender differences in heart rate before and after autonomic blockade: evidence against an intrinsic gender effect. Am J Med 100:537–543
16. Jochmann N, Stangl K, Garbe E et al (2005) Female-specific aspects in the pharmacotherapy of chronic cardiovascular diseases. Eur Heart J 26:1585–1595
17. Albert CM, McGovern BA, Newell JB, Ruskin JN, Sex differences in cardiac arrest
18. Machuki JO, Zhang HY, Geng J, et al (2019) Estrogen regulation of cardiac cAMP-L-type Ca^{2+}channel pathway modulates sex differences in basal contraction and responses to β_2 AR-mediated stress in left ventricular apical myocytes. Cell Commun Signal 17:1–17
19. Schwertz DW, Beck JM, Kowalski JM, Ross JD (2004) Sex differences in the response of rat ventricle to calcium. Biol Res Nurs 5:286–298
20. Schwertz DW, Vizgirda V, Solaro RJ et al (1999) Sexual dimorphism in rat left atrial function and response to adrenergic stimulation. Mol Cell Biochem 200:43–53
21. Farell SR, Ross JL, Howlet SE (2010) Sex differences in mechanisms of cardiac excitation-contraction coupling in rat ventricular myocytes. Am J Physiol Heart Circ Physiol 299:H36–H45
22. Sims C, Reisenweber S, Viswanathan PC et al (2008) Sex, age, and regional differences in L-type calcium current are important determinants of arrhythmia phenotype in rabbit heart with drug/induced lon QT type 2. Circ Res 102:86–100
23. Chu SH, Sutherland K, Beck J et al (2005) Sex differences in expression of calcium-handling proteins and beta-adrenergic receptors in rat heart ventrikle. Life Sci 76:2735–2749
24. Colom B, Oliver J, Roca P, Garcia-Palmer FJ (2007) Caloric restriction and gender modulate cardiac muscle mitochondrial H_2O_2 production and oxidative damage. Cardiovasc Res 74:456–465
25. Curl CL, Delbridge LM, Wendt JR (2008) Sex differences in cardiac muscle responsiveness to Ca $^{2+}$ and L-type Ca $^{2+}$ channel modulation. Eur J Pharmacol 586:288–292
26. Bell JR, Mellor KM, Wollermann AC, Delbridge LM (2011) Cardiac ischaemic stress: cardiomyocyte Ca $^{2+}$, sex and sex steroids. Clin Exp Pharmacol Physiol 38:717–723
27. MacDonald JK, Pyle WG, Reitz CJ, Howlett SE (2014) Cardiac contraction, calcium transients, and myofilament calcium sensitivity fluctuate with the estrous cycle in young adult female mice. Am J Physiol Heart Circ Physiol 306:H938–H953
28. Vicencio JM, Ibarra C, Estrada M et al (2006) Testosterone induces an intracellular calcium increase by a non-genomic mechanism in cultured rat cardiac myocytes. Endocrinology 147:1386–1395
29. Johnson MS, Moore RL, Brown DA (2006) Sex differences in myocardial infarct size are abolished by sarcolemmal K_{ATP} channel blocade in rat. Am J Physiol Heart Circ Physiol 290:H2644–H2647

30. Bhupathy P, Babu GJ, Ito M, Periasamy M (2009) Threonine-5 at the N-terminus can modulate sarcolipin function in cardiac myocytes. J Mol Cell Cardiol 47:723–729
31. Keller KM, Howlett SE (2016) Sex differences in the biology and pathology of the aging heart. Can J Cardiol 32:1065–1073
32. Dworatzek E, Baczko I, Kararigas G (2016) Effects of aging on cardiac extracellular matrix in men and women. Proteomics Clin Appl 10:84–91
33. De Arellano PS, Kuhl AA et al (2019) Sex differences in the aging human heart: decreased sirtuins, proinflammatory shift and reduced anti-oxidative defense. Aging 11:1918–1933
34. Milerova M, Drahota Z, Chytilova A et al (2016) Sex difference in the sensitivity of cardiac mitochondrial permeability transition pore to calcium load. Mol Cell Biochem 412:147–154
35. Ostadal B, Drahota Z, Houstek J et al (2019) Developmental and sex differences in cardiac tolerance to ischemia-reperfusion injury: the role of mitochondria. Can J Physiol Pharmacol 97:808–814
36. Ostadal B, Prochazka J, Pelouch V et al (1984) Comparison of cardiopulmonary responses of male and female rats to intermittent high altitude hypoxia. Physiol Bohemoslov 33:129–138
37. Murphy E, Steenbergen C (2007) Gender-based differences in mechanisms of protection in myocardial ischemia-reperfusion injury. Cardiovasc Res 75:478–486
38. Booth EA, Lucchesi BR (2008) Estrogen-mediated protection in myocardial ischemia-reperfusion injury. Cardiovasc Toxicol 8:101–113
39. Deschamp AM, Murphy E, Sun J (2010) Estrogen receptor activation and cardioprotection in ischemia-reperfusion injury. Trends Cardiovasc Med 20:73–78
40. Ross JL, Howlet SE (2012) Age and ovariectomy abolish beneficial effects of female sex on rat ventricular myocytes exposed to stimulated ischemia and reperfusion. PLoS ONE 7:1–11
41. Bell JR, Porrello ER, Hugginss CE et al (2008) The intrinsic resistence of female hearts to an ischemic insult is abrogated in primary cardiac hypertrophy. Am J Physiol Heart Circ Physiol 294:H1514–H1522
42. Lujan HL, Di Carlo SE, Sex differences to myocardial ischemia and beta-adrenergic receptor blocade in conscious rats. Am J Physiol Heart Circ Physiol 294: H1523-H1529
43. Cross HR, Lu L, Steenbergen C et al (1998) Overexpression of the cardiac Na^{+}/Ca^{2+} exchanger increases susceptibility to ischemia/reperfusion injury in male, but not female, transgenic mice. Circ Res 83:1215–1223
44. Cross HR, Murphy E, Steenbergen C (2002) Ca^{2+} loading and adrenergic stimulation reveal male/female differences in susceptibility to ischemia-reperfusion injury. Am J Physiol Heart Circ Physiol 283:H481–H498
45. Besik J, Szarszoi O, Kunes J et al (2007) Tolerance to acute ischemia in adult male and female spontaneously hypertensive rats. Physiol Res 56:267–274
46. Clark C, Smith W, Lochner A, Du Toit EF (2011) The effect of gender and obesity on myocardial tolerance to ischemia. Physiol Res 60:291–301
47. Cavasin MA, Tao Z, Menon S, Yang XP (2004) Gender differences in cardiac function during early remodeling after acute myocardial infarction in mice. Life Sci 75:2181–2192
48. Litwin SE (1995) The rat model of postinfarction heart failure. Heart Fail 11:182–195
49. Jain M, Liao R, Podesser BK et al (2002) Influence of gender on the response to hemodynamic overload after myocardial infarction. Am J Physiol Heart Circ Physiol 283:H2544–H2550
50. Shioura KM, Geenen DL, Goldspink PH (2008) Sex-related changes in cardiac function following myocardial infarction in mice. Am J Physiol Regul Integr Comp Physiol 295:R528–R534
51. Turcato S, Turnbull L, Wang GY et al (2006) Ischemic preconditioning depends on age and gender. Basic Res Cardiol 101:235–243
52. Humphreys RA, Kane KA, Parratt JR (1999) The influence of maturation and gender on the anti-arrhythmic effect of ischaemic preconditioning in rats. Basic Res Cardiol 94:1–8
53. Wang M, Crisostomo P, Wairiuko GM, Meldrum DR (2006) Estrogen receptor-alpha, mediates acute myocardial protection in females. Am J Physiol Heart Circ Physiol 290:H2204
54. Ostadalova I, Ostadal B, Kolar F et al (1998) Tolerance to ischaemia and ischaemic preconditioning in neonatal rat heart. J Mol Cell Cardiol 30:857–865

55. Ostadal B, Ostadalova I, Dhalla NS (1999) Development of cardiac sensitivity to oxygen deficiency: comparative and ontogenetic aspects. Phys Rev 79:635–659
56. Lieder HR, Irmert A, Kamler M et al (2019) Sex is no determinant of cardioprotection by ischemic preconditioning in rats, but ischemic/reperfused tissue mass is for remote ischemic preconditioning. Physiol Rep 7:e14146
57. Hausenloy DJ, Kharbanda RK, Moller UK et al (2019) Effect of remote ischaemic conditioning on clinical outcomes in patients with acute myocardial infarction (CONDI-2/ERIC-PPCI): a single-blind randomised controlled trial. Lancet 394:1415–1424
58. Boese AC, Kim SC, Yin KJ et al, Sex differences in vascular physiology and pathophysiology: estrogen and androgen signaling in health and disease. Am J Physiol Heart Circ Physiol 313:H524-H545.
59. Lagranha CJ, Deschamps A, Aponte A et al (2010) Sex differences in the phosphorylation of mitochondrial proteins result in reduced production of reactive oxygen species and cardioprotection in females. Circ Res 106:1681–1691
60. Murphy E (2011) Estrogen signaling and cardiovascular disease. Circ Res 109:687–696
61. Zhai P, Eurell TE, Cooke PS et al (2000) Myocardial ischemia-reperfusion injury in estrogen receptor-alpha knockout and wild type mice. Am J Physiol Heart Circ Physiol 278:H1640–H1647
62. Gabel SA, Walker VR, London RE et al (2005) Estrogen receptor beta medites gender differences in ischemia/reperfusion injury. J Mol Cell Cardiol 38:289–297
63. Nikolic I, Liu D, Bell JA et al (2007) Treatment with an estrogen receptor-beta-selective agoinst is cardioprotective. J Mol Cell Cardiol 42:769–780
64. Arias-Loza PA, Jazbutyte V, Pelzer T (2008) Genetic and pharmacologic strategies to determine the function of estrogen receptor alpha and estrogen receptor beta in cardiovascular system. Gend Med 5(Suppl A):S34–S45
65. Hutson DD, Gurrala R, Ogola BO et al (2019) Estrogen receptor profiles across tissues from male and female Rattus norvegicus. Biol Sex Differ 10:1–13
66. Chen Q, Yager JD, Russo J (2005) Regulation of mitochondrial respirátory chain structure and function by estrogens/estrogen receptors and potential physiological/pathophysiological implications. Biochim Biophys Acta 1746:1–7
67. Deschamps AM, Murphy E (2009) Activation of a novel estrogen receptor, GPER, is cardioprotective in male and female rats. Am J Physiol Heart Circ Physiol 297:H1806–H1813
68. Bopassa JC, Eghbali M, Toro L, Stefani E (2010) A novel estrogen receptor GPER inhibits mitochondria permeability transition pore opening and protects the heart against ischemia-reperfusion injury. Am J Physiol Heart Circ Physiol 298:H13–H23
69. Di Lisa F (2006) A female way to protect the heart: say NO to calcium. Circ Res 98:298–300
70. Knowlton AA, Lee AR (2012) Estrogen and the cardiovascular system. Pharmacol Ther 135:54–70
71. Sun J, Picht E, Ginsburg KS et al (2006) Hypercontractile female hearts exhibit increased S-nitosylation of the L-type Ca^{2+} channel alpha 1 subunit and reduced ischemia-reperfusion injury. Circ Res 98:403–411
72. Tong H, Imahashi K, Steenberger C, Murphy E (2002) Phosphorylation of glycogen synthase kinase-3beta during preconditioning through a phosphatidylinositol-3–kinase–dependent pathway is cardioprotective. Circ Res 90:377–379
73. Lee TM, Su SF, Tsai CC et al (2000) Cardioprotective effects of 17 beta-estradiol produced by activation of mitochondrial ATP-sensitive K^+ channels in canine hearts. J Mol Cell Cardiol 32:1147–1158
74. Bae S, Zhang L (2005) Gender differences in cardioprotection against ischemia/reperfusion injury in adult rat hearts: focus on Akt and protein kinase C signaling. J Pharmacol Exp Ther 315:1125–1135
75. Camper-Kirby D, Welch S, Walker A et al (2001) Myocardial Akt activation and gender: increased nuclear aktivity in females versus males. Circ Res 88:1020–1027
76. Xu Y, Williams SJ, O´Brien D, Davidge ST (2006) Hypoxia or nutrient restriction during pregnancy in rats leads to progressive cardiac remodeling and impairs postischemic recovery in adult male offspring. FASEB J 20:1251–1253

77. Yu Y, Wei SG, Weiss RM, Felfer RB (2018) Sex differences in the central and peripheral manifestation of ischemia-induced heart failure in rats. Am J Physiol Heart Circ Physiol 316:H70-79
78. Grohe C, Kahlert S, Lobbert K, Vetter H (1998) Expression of oestrogen receptor alpha and beta in rat heart: role of local oestrogen synthesis. J Endocrinol 156:R1–R7
79. Van Der Wall EE (2011) Testosterone bad for men, good for women? Neth Heart J 19:1–2
80. Parker MW, Thompson PD (2010) Anabolic-androgen steroids: worse for the heart than we knew? Circ Heart Fail 3:470–471
81. Jones TH, Kelly DM (2018) Randomized controlled trials-mechanistic studies of testosterone and the cardiovascular systém. Asian J Androl 20:120–130
82. Ponikowska B, Jankowska EA, Maj J et al (2010) Gondal and andrenal androgen deficiencies as independent predictors of increased cardiovascular mortality in men with type II diabetes mellitus and stable coronary artery disease. Int J Cardiol 143:343–348
83. Maldonado O, Ramos A, Guapillo M et al (2019) Effects of chronic inhibition of testosterone metabolism on cardiac remodeling after ischemia/reperfusion-induced myocardial damage in gonatectomized rats. Biol Open 8:1–7
84. Ghimire A, Bisset ES, Howlett SE (2019) Ischemia and reperfusion injury following cardioplegic arrest is attenuated by age and testosterone deficiency in male but not female mice. Biol Sex Differ 10:42–55
85. Cavasin MA, Tao ZY, Yu AL, Yang XP (2006) Testosterone enhances early cardiac remodeling after myocardial infarction, causing rupture and degrading cardiac function. Am J Physiol Heart Circ Physiol 290:H2043–H2050
86. Tsang S, Wu S, Liu J, Wang TM (2008) Testosterone protects rat hearts against ischaemic insults by enhancing the effects of alpha(1)-adrenoceptor stimulation Br J Pharmacol 153: 693–709
87. Mendelsohn ME (2005) Molecular and cellular basis of cardiovascular gender differences. Science 335:1583–1587
88. Li G, Xiao Y, Estrella JL et al (2003) Effect of fetal hypoxia on heart susceptibility to ischemia and reperfusion injury in the adult heart. J Soc Gynecol Invest 10:265–274
89. Li G, Bae S, Zhang L (2004) Effect of prenatal hypoxia on heat stress-mediated cardioprotection in adult rat heart. Am J Physiol 286:1712–1719
90. Krecek J (1970) The weanling period as a critical period of development. In: Krecek J (ed) The Postnatal Development of Phenotype. Academia, Prague, pp 33–44
91. Ostadalova I, Babicky A (2012) periodization of the early postnatal development in the rat with particulat attention to the weaning period. Physiol Res 61(Suppl 1):1–7
92. Netuka I, Szarszoi O, Maly J et al (2006) Effect of perinatal hypoxia on cardiac tolerance to acute ischaemia in adult male and female rats. Clin Exp Pharmacol Physiol 33:714–719
93. Xue Q, Zhang L (2009) Prenatal hypoxia causes a sex-dependent increase in heart susceptibility to ischaemia and reperfusion injury in adult male offspring: role of protein kinase C_{ε}. J Pharmacol Exp Therap 330:624–632
94. Thompson LP, Song H, Polster BM (2019) Fetal programming and sexual dimorphism of mitochondrial protein expression and activity of hearts of prenatally hypoxic guinea pig offspring. Hindawi Oxid Med Cell Long Article ID 7210249
95. Patterson AJ, Chen M, Xue Q et al (2010) Chronic prenatal hypoxia induces epigenetic programming of PKC_{ε} gene expression in rat hearts. Circulation Res 107:365–373
96. Lv J, Ma Q, Dasgupta Ch et al (2019) Antenatal hypoxia and programming of glucocorticoid receptor expression in the adult heart. Front Physiol 10:art. 323

Chapter 3
The Roles of Testosterone in Cardiac Ischemia/Reperfusion Injury

Nattayaporn Apaijai, Siriporn C. Chattipakorn, and Nipon Chattipakorn

Abstract Cardiac ischemia/reperfusion (I/R) injury is a serious cardiac complication following acute myocardial infarction, which include myocardial infarct size expansion, left ventricular (LV) dysfunction, and fatal cardiac arrhythmias. Sex hormone deprivation due to aging or other pathological conditions is an independent risk factor for cardiovascular disease. Several scientific research studies have been conducted to investigate the role of sex hormones in myocardial injury during cardiac I/R. Testosterone is the primary sex hormone in men; it regulates male sexual characteristics, and controls muscle and bone mass. In addition, testosterone plays an important role in regulating LV function through multiple mechanisms, including those controlling cellular calcium homeostasis, regulating cardiac mitochondrial function and enhancing antioxidants. Findings from studies regarding the roles of testosterone on the heart during cardiac I/R are controversial; some have reported that decreased testosterone level could impair LV function, whilst others reported the benefits of testosterone deprivation during cardiac I/R. In this chapter, we include evidence regarding the effects of testosterone deprivation and exogenous testosterone administration on myocardial injury in terms of myocardial infarct size, LV function, arrhythmias and molecular alterations. Reports from in vitro, ex vivo, in vivo studies, and clinical reports are summarized and discussed. The contents of this chapter will explain the roles of testosterone during cardiac I/R in preventing cardiac complications. Insights from these reports may help to devise strategies to improve treatment in patients with acute myocardial infarction.

N. Apaijai · S. C. Chattipakorn · N. Chattipakorn (✉)
Cardiac Electrophysiology Research and Training Center, Faculty of Medicine, Chiang Mai University, Chiang Mai 50200, Thailand
e-mail: nipon.chat@cmu.ac.th; nchattip@gmail.com

Center of Excellence in Cardiac Electrophysiology Research, Chiang Mai University, Chiang Mai 50200, Thailand

S. C. Chattipakorn · N. Chattipakorn
Cardiac Electrophysiology Unit, Department of Physiology, Faculty of Medicine, Chiang Mai University, Chiang Mai 50200, Thailand

B. Ostadal and N. S. Dhalla (eds.), *Sex Differences in Heart Disease*,
Advances in Biochemistry in Health and Disease 21,
https://doi.org/10.1007/978-3-030-58677-5_3

Keywords Testosterone · Cardiac ischemia/reperfusion injury · Myocardial infarction · Left ventricular function · Mitochondria

Introduction

Acute myocardial infarction (AMI) remains a global leading cause of death [1]. It is caused by narrowing of the coronary arteries. It has been demonstrated that risk factors associated with AMI are atherosclerosis, hypertension, high cholesterol, diabetes, obesity, smoking and hormonal imbalance [2]. Reperfusion therapy is an effective treatment for AMI as it allows the return of blood to the injured myocardium to resupply essential oxygen and nutrients. However, this intervention has been shown to cause potentially serious adverse effects on the heart, including myocardial infarct size expansion, left ventricular (LV) dysfunction and fatal cardiac arrhythmias, collectively known as cardiac ischemia/reperfusion (I/R) injury [3].

It has been demonstrated that sex hormone dysregulation is associated with the severity of coronary artery disease [4]. Testosterone is a male sex steroid hormone which is secreted mainly by the testicles and, to a lesser extent, by the adrenal glands. Testosterone regulates the development of male reproductive organs and promotes secondary sexual characteristics [4]. In addition, it controls hair growth, muscle growth, collagen synthesis, muscle strength and endurance, memory function, mood, red blood cell production and body mass density [5]. Testosterone deprivation, as seen in male aging, has been associated with the pathophysiology of various organ dysfunctions [4]. In the case of the heart, testosterone deprivation increases the risk of cardiovascular disease by increasing total plasma cholesterol levels and reducing plasma high density lipoprotein (HDL) levels [6]. A report from a clinical study demonstrated that ST–Elevation Myocardial Infarction (STEMI) male patients with hypotestosteronemia had a higher incidence of microvascular obstruction as identified by coronary angiography, compared to STEMI patients with normal plasma testosterone [7]. These findings suggest that testosterone may potentially be involved in the pathogenesis of AMI in male subjects.

In order to understand the mechanisms employed by testosterone in the heart, an animal model of testosterone deprivation has been established. In male rodents, a bilateral orchiectomy (ORX) has been shown to eliminate the endogenous production of testosterone [8, 9]; exogenous testosterone being subsequently given to those rodents as therapy [10, 11]. Exogenous testosterone replacement therapy has been widely used to treat patients with hypogonadism and other low–testosterone related diseases [12]. It has been shown that exogenous testosterone offers various benefits for the human body, for instance it augmented positive mood, increased sexual motivation and performance, increased muscle strength and a lean body mass, reduced body fat mass and improved bone mineral density [5]. However, exogenous testosterone has also been shown to cause adverse effects such as acne, mild fluid retention, increased urination symptoms and worsening of sleep apnea [13]. In accordance with these findings, the Endocrine Society do not recommend treatment with exogenous testosterone in men with heart failure (NYHA class III or IV) [14].

In this chapter, we have summarized and discussed the cumulative data from in vitro, ex vivo, and in vivo studies, and from clinical reports regarding the effects of testosterone deprivation and exogenous testosterone administration on the heart during cardiac I/R injury. This information will provide mechanistic insight for future clinical investigations.

The Effects of Testosterone Deprivation on the Heart with Cardiac I/R Injury

Previous reports have demonstrated that cardiac I/R injury led to myocardial infarct size expansion, LV dysfunction and arrhythmia [10, 15–17]. Cardiac I/R injury has been induced in both ex vivo and in vivo settings [18]. For induction in the ex vivo setting, the heart was removed and connected to a Langendorff system [18]. The myocardial ischemia was induced by either stopping the perfusate for several minutes to induce global ischemia or ligating the left coronary artery to induce regional ischemia [18]. After successful myocardial ischemia, the perfusate was opened to initiate reperfusion to the injured myocardium [18]. In the in vivo setting, a thoracotomy was performed and the left coronary artery or left anterior descending (LAD) coronary artery was ligated to induce myocardial ischemia and reperfusion was induced by a loosening of the knot [8, 10, 15]. The alterations in the electrocardiogram were used to confirm a successful cardiac I/R in vivo [8, 10, 15].

The bilateral ORX is a surgical procedure effective in causing a decrease in plasma testosterone level in the body prior to the induction of cardiac I/R [10]. Findings regarding the effects of testosterone deprivation on the heart with I/R injury are inconsistent; some studies have reported that testosterone deprivation led to an aggravation of myocardial damage after cardiac I/R [10, 19–23], whereas several other studies reported contrary findings [24–32]. A comprehensive summary of these reports is shown in Table 3.1.

With regards to the deteriorating effects caused as a result of testosterone deprivation on the heart with I/R, results from ex vivo and in vivo studies have shown that testosterone deprivation for at least 2 weeks increased myocardial infarct size after cardiac I/R [10, 21–23]. Testosterone deprivation for 3 weeks did not alter LV function [19], however, a worsening of LV dysfunction was initially observed after 7 weeks of ORX as indicated by increased LV end−diastolic pressure (LVEDP), reduced LV developed pressure (LVDP) and ± dP/dt in ex vivo studies [20, 21, 23]. In an in vivo study, after 12 weeks of ORX, increased LVEDP and −dP/dt and decreased stroke volume (SV) were observed [10]. Cardiac arrhythmia following cardiac I/R was investigated in two studies. These both found that the arrhythmia

Table 3.1 The effects of testosterone deprivation on the heart with cardiac I/R injury, in comparison with normal testosterone level

Animal model/ Duration of hormone depletion	I/R protocol	Findings				Interpretation	Refs.
		Infarct size	LV function	Arrhythmias	Molecular findings		
ORX rats 3 weeks	Ex vivo Global ischemia 20 min/45 min	N/A	– ↔ Coronary flow, Aortic flow, CO, Cardiac work, LVDP, ± dP/dt, HR, SBP, DBP	N/A	–↔ LDH, MnSOD, GPX –↓ Catalase	Testosterone deprivation did not alter LV function, but it reduced antioxidants in rats with cardiac I/R	[19]
ORX rats 7 weeks	Ex vivo Global ischemia 20 min/40 min	N/A	– ↑ LVEDP – ↓ LVDP	N/A	–↑ Intracellular Ca^{2+}	Testosterone deprivation aggravated LV dysfunction by increasing intracellular calcium levels in rats with cardiac I/R	[20]

(continued)

Table 3.1 (continued)

Animal model/ Duration of hormone depletion	I/R protocol	Findings				Interpretation	Refs.
		Infarct size	LV function	Arrhythmias	Molecular findings		
ORX rats 8 weeks	Ex vivo LAD ligation 30 min/120 min	↑	– ↔ Coronary flow, HR – ↓ LVDP, ± dP/dt	N/A	–↑ LDH –↔ HSP70, HSF1	Testosterone deprivation increased myocardial infarct size and aggravated LV dysfunction in rats with cardiac I/R, but it was not involved in HSP pathway	[21]

(continued)

Table 3.1 (continued)

Animal model/ Duration of hormone depletion	I/R protocol	Findings				Interpretation	Refs.
		Infarct size	LV function	Arrhythmias	Molecular findings		
ORX rats 9 weeks	Ex vivo Left coronary ligation 30 min/120 min	↑	– ↑ LVEDP – ↓ LVDP, ± dP/dt	–↑ PVB, VT	–↑ LDH	Testosterone deprivation increased infarct size, arrhythmias, and aggravated LV dysfunction in rats with cardiac I/R	[23]
ORX rats 15 days	In vivo Left coronary artery ligation 60 min/4 h	↑	N/A	N/A	N/A	Testosterone deprivation increased infarct size in rats with cardiac I/R	[22]

(continued)

Table 3.1 (continued)

Animal model/ Duration of hormone depletion	I/R protocol	Findings				Interpretation	Refs.
		Infarct size	LV function	Arrhythmias	Molecular findings		
ORX rats 12 weeks	In vivo LAD ligation 30 min/120 min	↑	– ↑ LVEDP, – dP/dt – ↔ HR, LVESP, Pmax, min + dP/dt – ↓ SV	–↑ Arrhythmia score –↓ Time to 1st VT/VF onset	–↑ Bax/Bcl2 ratio –↑ Mito ROS, Mito depolarization, Mito swelling – ↔ Procaspase 3 –↓ p–$Cx43^{ser368}$/Cx43	Testosterone deprivation increased infarct size, arrhythmias and aggravated LV dysfunction by reducing gap junction proteins, increasing apoptosis, and cardiac mitochondrial dysfunction in rats with cardiac I/R	[10]

(continued)

Table 3.1 (continued)

Animal model/ Duration of hormone depletion	I/R protocol	Findings				Interpretation	Refs.
		Infarct size	LV function	Arrhythmias	Molecular findings		
ORX rats 4 weeks	Ex vivo Global ischemia 20 min/30 min	N/A	– ↔ LVDP	N/A	– ↔ TUNEL$^+$ cells	Testosterone deprivation did not alter LV function and apoptosis in rats with cardiac I/R	[27]
ORX mice 20–32 weeks	Ex vivo Global ischemia 90 min/30 min	↓	– ↑ LVDP, ± dP/dt, RPP – ↔ Coronary flow – ↓ Magnitude of contracture	N/A	N/A	Testosterone deprivation reduced infarct size and LV dysfunction in mice with cardiac I/R	[24]

(continued)

Table 3.1 (continued)

Animal model/ Duration of hormone depletion	I/R protocol	Findings				Interpretation	Refs.
		Infarct size	LV function	Arrhythmias	Molecular findings		
ORX rats 3 weeks	Ex vivo Left coronary artery ligation 30 min/150 min	↓	N/A	N/A	– ↔ p–AMPK/AMPK, p–mTOR/mTOR, RAGE –↓ Beclin1, Atg5,12, LC3II/LC3I, Cleaved caspase9,3	Testosterone deprivation reduced infarct size by reducing autophagy and apoptosis in rats with cardiac I/R	[28]
ORX rats 4 weeks	In vivo LAD ligation 25 min/40 min	N/A	N/A	N/A	–↓ myocardial TNFα, IL1β, ICAM1 –↓ Cardiac troponin I	Testosterone deprivation reduced cardiac inflammation and injury in rats with cardiac I/R	[25]

(continued)

Table 3.1 (continued)

Animal model/ Duration of hormone depletion	I/R protocol	Findings				Interpretation	Refs.
		Infarct size	LV function	Arrhythmias	Molecular findings		
ORX rats 7 weeks	Ex vivo Global ischemia 25 min/40 min	N/A	−↑ LVDP ± dP/dt	N/A	−↑ p−Akt/Akt −↑ p−Bad, Bax, Bim, Bcl2 MnSOD −↓ FasL	Testosterone deprivation attenuated LV dysfunction by activating survival proteins and antioxidants in rats with cardiac I/R	[26, 29, 32]
ORX rats 7 weeks	Ex vivo Global ischemia 25 min/40 min	N/A	−↑ LVDP ± dP/dt −↓ LVEDP	N/A	−↓ p−p38/p38, Caspase 1, Caspase 3	Testosterone deprivation attenuated LV dysfunction by reducing cellular stress, inflammation, and apoptosis in rats with cardiac I/R	[31]

(continued)

Table 3.1 (continued)

Animal model/ Duration of hormone depletion	I/R protocol	Findings				Interpretation	Refs.
		Infarct size	LV function	Arrhythmias	Molecular findings		
ORX rats 12 weeks	In vivo LAD ligation 5 min/5 min	N/A	–↓ Cardiac rate BP	N/A	–↑ Mito respiration –↔ Mito Ca^{2+} transport, Mito membrane potential –↓ Muscle fiber separation	Testosterone deprivation attenuated LV dysfunction by reducing cardiac mitochondrial dysfunction in rats with cardiac I/R	[30]
ORX rats with HFD induced obese–insulin resistance 12 weeks	In vivo LAD ligation 30 min/120 min	↔	–↑ LVEDP –dP/dt –↔ HR Pmax, min –↓ LVESP SV	–↔ Arrhythmia score Time to 1st VT/VF onset	–↔ p–$Cx43^{ser368}$/Cx43 –↔ Mito ROS, Mito depolarization, Mito swelling	Although testosterone deprivation did not increase infarct size and arrhythmias, it worsened LV dysfunction in obese–insulin resistant rats with cardiac I/R, in comparison with their sham operation controls	[10]

(continued)

Table 3.1 (continued)

Animal model/ Duration of hormone depletion	I/R protocol	Findings				Interpretation	Refs.
		Infarct size	LV function	Arrhythmias	Molecular findings		
ORX rats with obesogenic diet induced obese–insulin resistance 20 weeks	In vivo LAD ligation 30 min/120 min	↔	N/A	N/A	N/A	Testosterone deprivation did not increase infarct size in obese–insulin resistant rats with cardiac I/R, compared with their sham operation controls	[37]

I/R: ischemia/reperfusion; ORX: orchiectomy; LV: left ventricular; LAD: left anterior descending coronary artery; CO: cardiac output; LVDP: left ventricular developed pressure; LVEDP: left ventricular end diastolic pressure; LVESP: left ventricular end systolic pressure; RPP: rate pressure product; HR: heart rate; SBP: systolic blood pressure; DBP: diastolic blood pressure; LDH: lactate dehydrogenase; MnSOD: manganese superoxide dismutase; GPX: glutathione peroxidase; Ca^{2+}: calcium; HSP: heat shock protein; HSF: heat shock factor; PVB: premature ventricular beat; VT: ventricular tachycardia; VF: ventricular fibrillation; Mito: mitochondria; ROS: reactive oxygen species; Cx: connexin; TNF: tumor necrosis factor; IL: interleukin; N/A: not assessed score [10], the number of premature ventricular beats (PVB) and ventricular tachycardia (VT) increased after 9 weeks of ORX [23]. The mechanisms underlying the deteriorating effects of testosterone deprivation on cardiac I/R were determined, and the data showed that testosterone–deprived rats had an increase in lactate dehydrogenase (LDH) [21, 23], apoptosis and cardiac mitochondrial dysfunction [10]. In addition, gap junction protein expression was decreased in ORX rats when compared to their sham operation controls [10], leading to a disturbance of electrical activity in the heart. A comprehensive summary of these reports is shown in Table 3.1.

Mitochondria are vital organelles which control energy metabolism, cell death and cell survival in physiological and pathological conditions [33]. In addition, mitochondria are largely responsible for reactive oxygen species (ROS) production, thereby resulting in a predisposition to oxidative stress [34]. An excessive level of oxidative stress disrupts mitochondrial membrane integrity which will lead to cardiac mitochondrial membrane depolarization and opening of the mitochondrial permeability transition pores (mPTP), followed by mitochondrial swelling [33]. Cytochrome *c* is released from the mitochondrial intermembrane space to form an apoptosome, which ultimately induces apoptosis [33]. Since cardiac mitochondrial dysfunction was found in cases of both testosterone deprivation and cardiac I/R [15, 17, 35], the severity of cardiac mitochondrial dysfunction was potentially increased in testosterone deprived rats with cardiac I/R [35].

There is only one ex vivo study which reported that testosterone deprivation for 4 weeks did not alter LV function and apoptosis in rats [27]. However, myocardial infarct size and arrhythmia status were not determined in that study [27]. The findings from this ex vivo report were not consistent with the in vivo studies, indicating that the duration of testosterone deprivation as well as the study models could influence the effects of testosterone on the heart.

Although testosterone deprivation has been reported to cause adverse effects on the heart with I/R, inconsistent reports exist. Two studies have reported that myocardial infarct size was decreased in testosterone deprived rodents [24, 28]. Myocardial infarct size was decreased 3 weeks after ORX (in a rat model) [28], and 20 weeks after ORX (in a mouse model) [24]. To the contrary, other studies reported that testosterone deprivation increased myocardial infarct size in rats between 2 and12 weeks after ORX [10, 21–23]. The discrepancy in myocardial infarct size might be due to differences in the duration of testosterone deprivation and cardiac I/R protocol. There is an inconsistency in the data in cases of short–term testosterone deprivation. Two weeks of testosterone deprivation increased myocardial infarct size in an in vivo setting after 60 min of ischemia and 4 h of reperfusion. On the other hand, 3 weeks of testosterone deprivation decreased myocardial infarct size in an ex vivo setting after 30 min of ischemia and 150 min of reperfusion. Thus, the difference in cardiac I/R setting such as ischemia and reperfusion duration might involve a discrepancy of myocardial infarct size in testosterone deprived rats with cardiac I/R. A comprehensive summary of these reports is shown in Table 3.1.

As regards LV function, in ex vivo studies testosterone deprivation for 7 weeks has been shown to have both deteriorating and beneficial effects [20, 25, 26, 29, 31, 32]. Testosterone deprived rats who were subjected to 20 min of ischemia resulted in a worsening of LV dysfunction [20]. However, testosterone deprived rats subjected to 25 min of ischemia showed an improvement in LV function during cardiac I/R, all rats undergoing similar reperfusion duration (40 min) [26, 29, 31, 32]. These findings suggest that the ischemic duration could be a key factor in the mediation of LV function in testosterone deprived rats with cardiac I/R. It has been proposed that testosterone deprivation attenuated LV dysfunction during cardiac I/R by increasing the level of survival protein kinase, enhancing antioxidants, reducing apoptosis [26, 29, 32] and inflammation [31]. This finding was supported by another in vivo study in which a short duration of ischemia (5 min) and reperfusion (5 min) was shown to reduce LV dysfunction in testosterone deprived rats by promoting cardiac mitochondrial respiration [30]. A comprehensive summary of these reports is shown in Table 3.1.

Metabolic syndrome is associated with low testosterone levels [36]. It has been shown that the percentage of body fat has a negative correlation with circulating testosterone level [36]. In obese–insulin resistant rats, testosterone deprivation did not increase myocardial infarct size, or arrhythmias, and it did not alter intracellular molecular signaling, compared with their sham operated rats [10, 37]. However, a worsening of LV dysfunction was reported in one study in which it was shown that testosterone deprivation increased LVEDP and left ventricular end systolic pressure (LVESP) in obese–insulin resistant rats, when compared with their sham operated controls [10]. These data indicated that, in a rodent model of metabolic syndrome, although testosterone deprivation did not increase myocardial infarct size, it aggravated LV dysfunction. A comprehensive summary of these reports is shown in Table 3.1.

The Effects of Exogenous Testosterone Administration on the Heart with Cardiac I/R

Testosterone acts via two major pathways: the genomic and the non–genomic pathways [5]. In the genomic pathway, after testosterone enters the cytosol it can be converted to dihydrotestosterone (DHT) by an enzyme 5α–reductase [5]. DHT then binds with an androgen receptor (AR), which forms a homodimer and undergoes a conformational change [5]. This AR complex then binds with other proteins to facilitate their nuclear translocation [5]. After the AR complex enters the nucleus, it binds to the androgen response elements (ARE), which are found on several promoter sites on the DNA [5]. ARA 70, a coactivator, is subsequently recruited for the transcription of AR-regulated genes, which can exert their action on target organs via their protein products [5]. In the non-genomic pathway, this process requires neither AR nuclear translocation nor AR-DNA-binding. The activated AR in the cytoplasm interacts with other signaling pathways such as PI3K/Akt, Ras/Raf and Src [5]. These signaling molecules activate the mitogen-activated protein kinase (MAPK)/extracellular signal–regulated kinase (ERK) pathway [5]. The activation of the MAPK/ERK pathway is associated with cell proliferation gene expression [5]. In this chapter, we have summarized the effects of exogenous testosterone administration on the heart during cardiac I/R in animals with normal plasma testosterone levels.

In an in vitro study, the isolated cardiomyocytes were subjected to 30 min of ischemia (by incubating the cells with an ischemic solution) and 30 min of reperfusion (by incubating the cells with normal culture media) [38]. Testosterone was added to the cells during ischemia until the end of reperfusion. Cellular electrophysiological studies were performed, and the results showed that testosterone reduced action potential duration (APD90) and the number of premature ventricular contractions (PVC) [38]. However, the resting membrane potential, the action potential amplitude, and Vmax were not affected by the application of testosterone [38]. These findings suggested that testosterone could reduce arrhythmias at the cellular level. A comprehensive summary of these reports is shown in Table 3.2.

In the case of ex vivo studies, pretreatment with a very low dose of testosterone (0.001 nM) in isolated rat hearts was not shown to confer any level of protection on the heart regarding cardiac I/R injury [39]. Pretreatment with testosterone at a dose of 3 μg/kg improved LV function, but it did not reduce LDH levels after cardiac I/R [40]. In addition, pretreatment with a high dose of testosterone (5 mg/kg) increased myocardial infarct size by decreasing antioxidant levels, but it did not alter the level of apoptosis [41]. These data suggested that, in rats with a normal circulating testosterone level, the effects of exogenous testosterone were dependent on the dosage of exogenous testosterone. A summary of these reports is shown in Table 3.2.

The Effects of Exogenous Testosterone Administration on the Hearts of Testosterone Deprived Rats with Cardiac I/R

In one ex vivo study, testosterone deprived animals were given exogenous testosterone prior to cardiac I/R in a range of dosages and for various durations of treatment. The study reported that pretreatment with a single dose of testosterone injection (intraperitoneal injection; IP) at 10 mg/kg 6 h prior to cardiac I/R did not reduce LV dysfunction in rats and apoptosis was increased [27]. Another ex vivo study showed that pretreatment with a single dose of testosterone injection (intramuscular injection; IM) at 500 mg/kg, 14 days prior to cardiac I/R significantly attenuated LV dysfunction by reducing intracellular calcium levels, possibly through an increase in sarcoplasmic reticulum calcium ATPase levels [20]. This showed that a single high–dose of exogenous testosterone administration could protect the heart against cardiac I/R injury in testosterone deprived rats. A comprehensive summary of these reports is shown in Table 3.3.

Table 3.2 The effects of exogenous testosterone administration on a heart with cardiac I/R, in comparison with vehicle group

Animal models	I/R protocol	Testosterone regimen (Dose/Route/Duration)	Results				Interpretation	Refs.
			Infarct size	LV function	Arrhythmias	Molecular findings		
Isolated cardiomyocytes from rabbits	In vitro Simulated ischemic buffer 30 min/30 min	−10, 100 nmol/L −During ischemia until end of reperfusion	N/A	N/A	−↔ RMP, APA, Vmax −↓ APD90, PVC	N/A	Testosterone administration reduced cellular electrophysiological changes in cardiomyocytes with I/R	[38]
Male Wistar rats	Ex vivo Global ischemia 30 min/30 min	−0.001 nM −Mixed with perfusate −Pretreatment	↔	−↔ Perfusion pressure, Coronary Resistance	N/A	N/A	Single testosterone administration prior to cardiac I/R did not alter LV function in rats with cardiac I/R	[39]
Male Wistar rats	Ex vivo Global ischemia 30 min/60 min	−5 mg/kg −IM −30 days −Pretreatment	↑	−↑ LVEDP −↔ LVSP −↓ LVDP, ± dP/dt	N/A	−↔ Bax, Bcl2, Bad SOD1,2,3 GPX1,3 −↓ Catalase	Chronic pretreatment with testosterone aggravated LV dysfunction by decreasing antioxidant enzymes in rats with cardiac I/R	[41]
Male Wistar rats	Ex vivo Global ischemia 40 min/40 min	−3 μg/kg −SC −10 days −Pretreatment	N/A	−↑ Coronary flow −↔ SBP,DBP	N/A	−↔ LDH	Chronic pretreatment with testosterone increased coronary flow, but it did not alter cardiac injury marker in rats with cardiac I/R	[40]

I/R: ischemia/reperfusion; ORX: orchiectomy; IM: intramuscular injection; SC: subcutaneous injection; LV: left ventricular; LVEDP: left ventricular end−diastolic pressure; LVSP: left ventricular systolic pressure; LVDP: left ventricular developed pressure; SBP: systolic blood pressure; DBP: diastolic blood pressure; SOD: superoxide dismutase; GPX: glutathione peroxidase; LDH: lactate dehydrogenase; N/A: not assessed

Table 3.3 The effects of exogenous testosterone administration on the heart from testosterone deprived animals with cardiac I/R, in comparison with vehicle group

Animal models	I/R protocol	Testosterone regimen (Dose/Route/Duration)	Results				Interpretation	Refs
			Infarct size	LV function	Arrhythmias	Molecular findings		
ORX rats	Ex vivo Global ischemia 20 min/30 min	−10 mg/kg −IP −6 h prior to I/R −Pretreatment	N/A	−↔ LVDP	N/A	−↑ $TUNEL^+$ cells	Pretreatment with a single dose of testosterone did not alter LV function, but it increased apoptosis in testosterone deprived rats with cardiac I/R	[27]
ORX rats	Ex vivo Global ischemia 25 min/40 min	−500 mg/kg −Single dose, IM −14 days prior to I/R −Pretreatment	N/A	−↑ LVDP −↓ LVEDP	N/A	−↓ Intracellular Ca^{2+}	Pretreatment with a single dose of testosterone improved LV function by decreasing intracellular calcium levels in testosterone deprived rats with cardiac I/R	[20]
ORX rats	Ex vivo Global ischemia 25 min/40 min	−100 mg −Subcutaneous implantation −3 weeks −Pretreatment	N/A	−↔ ± dP/dt, LVDP	N/A	−↔ p−Akt/Akt, p−Bad, Bcl2, Bim, FasL, MnSOD −↓ Bax, p−STAT3/STAT3	Chronic pretreatment with testosterone did not alter LV function, but it reduced apoptosis in testosterone deprived rats with cardiac I/R	[26, 32]
ORX rats	Ex vivo Global ischemia 20 min/45 min	−100 mg/day −Subcutaneous implantation −3 weeks −Pretreatment	N/A	−↑ Aortic flow, CO, Cardiac work, LVDP, + dP/dt −↔ Coronary flow, −dP/dt, HR, SBP, DBP	N/A	−↑ Catalase −↔ MnSOD, GPX −↓ LDH	Chronic pretreatment with testosterone improved LV function by increasing antioxidant in testosterone deprived rats with cardiac I/R	[19]

(continued)

Table 3.3 (continued)

Animal models	I/R protocol	Testosterone regimen (Dose/Route/ Duration)	Results				Interpretation	Refs
			Infarct size	LV function	Arrhythmias	Molecular findings		
ORX rats	Ex vivo LAD ligation 30 min/120 min	–2 mg/kg –SC –8 weeks –Pretreatment	↓	–↑ LVDP, ± dP/dt –↓ LVEDP	–↓ PVB, VT	–↓ LDH	Chronic pretreatment with testosterone reduced infarct size, improved LV function, and arrhythmias by decreasing cardiac damage in testosterone deprived rats with cardiac I/R	[23]
ORX rats	Ex vivo LAD ligation 30 min/120 min	–2 mg/kg –SC –8 weeks –Pretreatment	↓	–↑ LVDP, ± dP/dt –↔ HR, Coronary flow –↓ LVEDP	N/A	–↔ HSP70, HSF1 –↓ LDH	Chronic pretreatment with testosterone reduced infarct size and improved LV function by decreasing cardiac damage in testosterone deprived rats with cardiac I/R	[21]
ORX mice	In vivo LAD ligation 1 h/30 days	–346 ng/kg –SC –Every 72 h for 30 days –Pretreatment	↓	N/A	N/A	–↓ Cellular infiltration, MMP3	Chronic pretreatment with testosterone reduced infarct size by reducing inflammation in testosterone deprived rats with cardiac I/R	[42]
ORX rats	In vivo LAD ligation 30 min/120 min	–2 mg/kg –SC –4 weeks –Pretreatment	↓	–↑ SV –↓ LVEDP, –dP/dt	–↑ Time to 1stVT/VF onset –↓ Arrhythmia score	–↑ p–Cx43/Cx43, Procaspase 3 –↓ Bax/Bcl2, Mito ROS, Mito membrane depolarization, Mito swelling	Chronic pretreatment with testosterone reduced infarct size and arrhythmias, and improved LV function by increasing gap junction proteins, reducing apoptosis and cardiac mitochondrial dysfunction in testosterone deprived rats with cardiac I/R	[10, 35]
ORX rats	In vivo LAD ligation 60 min/4 h	–0.02, 0.2, 2.02 μg/kg –IV –During ischemia	↓	N/A	N/A	N/A	Single dose of testosterone administration during ischemia reduced infarct size in testosterone deprived rats with cardiac I/R	[22]

(continued)

Table 3.3 (continued)

Animal models	I/R protocol	Testosterone regimen (Dose/Route/Duration)	Results				Interpretation	Refs
			Infarct size	LV function	Arrhythmias	Molecular findings		
ORX Obese−insulin resistant rats	In vivo LAD ligation 30 min/120 min	−2 mg/kg −SC −4 weeks −Pretreatment	↓	−↑ SV −↓ LVEDP, −dP/dt	−↑ Time to 1st VT/VF onset −↓ Arrhythmia score	−↑ p−Cx43/Cx43, Procaspase 3 −↓ Bax/Bcl2, Mito ROS, Mito membrane depolarization, Mito swelling	Chronic pretreatment with testosterone reduced infarct size, arrhythmias, and improved LV function by increasing gap junction protein, reducing apoptosis and cardiac mitochondrial dysfunction in testosterone deprived obese−insulin resistant rats with cardiac I/R	[10]
ORX Obese−insulin resistant rats	In vivo LAD ligation 45 min/120 min	−2 mg/kg −Subcutaneous implantation −4 weeks −Pretreatment	↓	N/A	N/A	−↓ CKMB	Chronic pretreatment with testosterone reduced infarct size via reducing cardiac damage in testosterone deprived obese−insulin resistant rats with cardiac I/R	[37]
Naturally aged rats	Ex vivo Global ischemia 30 min/60 min	−1 mg/kg −SC −30 days −Pretreatment	↑	−↑ LVDP, ± dP/dt CI −↔ LVSP LVEDP	N/A	−↔ GSK3β, Caspase3, Bcl2, GRP78, CHOP, eIF2α −↓ p−Akt/Akt	Although chronic pretreatment with testosterone improved LV function, it increased infarct size in aged rats. These effects were mediated by a reduction of survival kinase, however, it did not alter apoptosis and ER stress	[43]

I/R: ischemia/reperfusion; ORX: orchiectomy; IP: intraperitoneal injection; IM: intramuscular injection; SC: subcutaneous injection; LV: left ventricular; LVEDP: left ventricular end−diastolic pressure; LVESP: left ventricular end systolic pressure; LVSP: left ventricular systolic pressure; LVDP: left ventricular developed pressure; SV: stroke volume; CO: cardiac output; CI: cardiac index; HR: heart rate; SBP: systolic blood pressure; DBP: diastolic blood pressure; PVB: premature ventricular beat; VT: ventricular tachycardia; VF: ventricular fibrillation; ROS: reactive oxygen species; Mito: mitochondria; HSP: heat shock protein; HSF: heat shock factor; LDH: lactate dehydrogenase; MMP: matrix metalloproteinase; Cx; connexin; MnSOD: manganese superoxide dismutase; GPX: glutathione peroxidase; LDH: lactate dehydrogenase; CKMB: creatine kinase MB isoenzyme; GSK3β: Glycogen synthase kinase 3 beta; GRP78: Glucose regulatory protein 78; CHOP: C/EBP homologous protein; eIF2α: Eukaryotic initiation factor 2 alpha; N/A: not assessed

As regards chronic treatment, testosterone was subcutaneously implanted in testosterone deprived rats (100 mg, 3 weeks) prior to cardiac I/R [19, 26, 32]. The results showed that, in the setting of ex vivo cardiac I/R (25 min ischemia and 40 min reperfusion), testosterone attenuated apoptosis but not LV dysfunction [26, 32]. Another study showed that, in an ex vivo setting of cardiac I/R (25 min ischemia and 45 min reperfusion), testosterone attenuated LV dysfunction and increased antioxidant enzyme level [19]. These findings suggested that, with similar doses of testosterone, the duration of reperfusion may affect LV function in testosterone deprived rats with cardiac I/R. A comprehensive summary of these reports is shown in Table 3.3.

The method of regional ischemia performed in some ex vivo studies was ligation of the LAD [21, 23]. Exogenous testosterone (2 mg/kg) was given to the testosterone deprived rats daily for 8 weeks prior to cardiac I/R, and the results demonstrated that testosterone could reduce myocardial infarct size and LV dysfunction [23]. In addition, the numbers of PVC and VT were decreased in testosterone deprived rats with cardiac I/R [23]. Another study showed that testosterone could reduce LDH but it did not affect the heat shock proteins HSP70 and HSF1 [21]. HSP70 and HFS1 are stress inducible proteins which are excessively expressed during cardiac I/R [21]. An activation of the HSP70–HSF1 complex could initiate apoptosis and an inflammatory response [21]. However, the beneficial effects of testosterone were not mediated by this protein. A comprehensive summary of these reports is shown in Table 3.3.

In in vivo studies, the effects of acute and chronic exogenous testosterone have been investigated in testosterone deprived rats. The acute administration of exogenous testosterone (0.2 and 2.0 μg/kg, IV), given to the testosterone deprived rats during cardiac ischemia was shown to reduce myocardial infarct size [22]. However, the precise molecular mechanism responsible for the beneficial effects in acute treatment was not determined [22]. In a chronic pretreatment model, exogenous testosterone (346 ng/kg and 2 mg/kg, 4 weeks) effectively reduced myocardial infarct size, LV dysfunction, and arrhythmias in testosterone deprived rats [10, 35, 42]. These studies proposed that testosterone could reduce cardiac inflammation [42], oxidative stress, apoptosis and cardiac mitochondrial dysfunction [10, 35] in testosterone deprived rats. These beneficial effects of exogenous testosterone were also observed in the obese–insulin resistant rats with testosterone deprivation subjected to cardiac I/R through a similar procedure [10]. A summary of these reports is shown in Table 3.3.

One study used a naturally aged rat model with age related low testosterone levels [43]. Exogenous testosterone was given to the rats at 1 mg/kg for 30 days prior to cardiac I/R, and the results showed that this intervention increased myocardial infarct size, when compared with the vehicle control group [43]. However, the degree of LV dysfunction was decreased. The molecular mechanism demonstrated that exogenous testosterone decreased p-Akt/Akt, which is a survival protein, however it did not affect apoptosis and ER stress levels [43] (Table 3.3). These data suggested that exogenous testosterone reduced LV dysfunction through the activation of this survival protein.

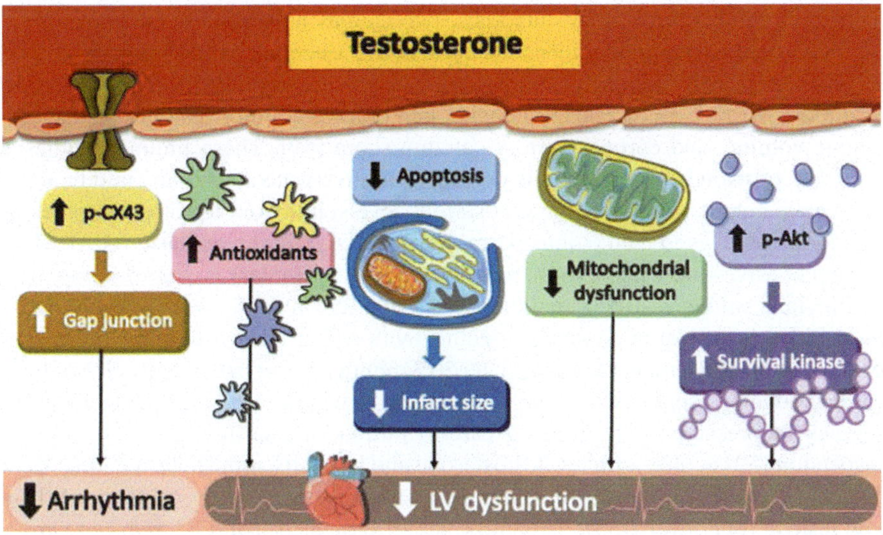

Fig. 3.1 This summary diagram demonstrates the potential mechanisms involved in the mediation of the action of testosterone during cardiac ischemia/reperfusion including the activation of Cx43, antioxidants, apoptosis, mitochondrial function, and survival kinase. Cx43: connexin 43; LV: left ventricle

From these reports, it could be concluded that exogenous testosterone application reduced cardiac mitochondrial dysfunction and led to a reduction in cardiac apoptosis, resulting in a reduction in myocardial infarct size and LV dysfunction. In addition, exogenous testosterone reduced matrix metalloproteinase activity, cellular infiltration, and inflammation in testosterone deprived rats. A summary regarding the roles of testosterone on the heart of cardiac I/R is depicted in Fig. 3.1.

The Effects of Testosterone Replacement Therapy (TRT) on Cardiovascular Outcome in Clinical Reports

A previous study has shown that testosterone plays an important role in regulating cardiac damage during MI, specifically that testosterone deprivation led to an increased incidence of microvascular obstruction in STEMI patients [7]. Therefore, TRT is used to prevent cardiac complications in patients with low circulating testosterone levels. There are large cohort studies showing that long–term TRT use (for more than 17 months) could increase cardiovascular event free survival rate, decrease all–cause mortality, and reduce major cardiac adverse events in middle to old aged men with low testosterone levels [44, 45, 47, 48]. However, TRT for 1 year has been shown neither to increase nor decrease MI events in middle to old aged men with low testosterone levels [46]. Although long–term TRT showed benefits as regards the cardiovascular outcomes, two studies have reported adverse effects of TRT in increasing the incidence of obstructive sleep apnea [44] and stroke [48]. In older aged men (the age of all participants was greater than 70 years old), TRT promoted MI risk by increasing noncalcified coronary artery plaque volume, total plaque volume, and coronary artery calcium score [49]. These findings suggested that TRT provided several benefits on cardiovascular outcomes in the middle to old aged male cohort who have low testosterone levels, however, testosterone prescription to the older aged men (subjects over 70 years old) might increase risk for MI. A comprehensive summary of these reports is shown in Table 3.4.

Previous studies also reported the effects of TRT on cardiovascular outcomes in men with low testosterone levels, together with MI or heart failure. In the case of both short-term (1 month) and longer-term (3 months-1 year) treatment, TRT reduced cardiac adverse events by delaying the time to ischemia [50] and time to 1 mm ST segment depression [51], reducing carotid intima–media thickness [50] and QT dispersion [52] in men with low testosterone and heart disease. However, long–term TRT did not affect MI events in aging men (subjects over 70 years old) with MI and low testosterone levels [53]. This information indicated that TRT reduced cardiac adverse events in middle to old aged men who have cardiac complications and low testosterone levels. A summary of these reports is shown in Table 3.4.

Table 3.4 Clinical reports concerning the effects of exogenous testosterone administration on cardiovascular outcomes

Study design	TRT regimen	Results	Interpretation	Refs
Men with low testosterone levels (0–60 years)	TRT group (n = 3,422) versus Control (n = 3,422) F/U = 17 month	–↑ CV event free survival –↓ CAD event –↔ CHF event Stroke event –↑ OSA	TRT increased survival rate and reduced risk of coronary artery disease, but it did not affect chronic heart failure and stroke event in men with low testosterone levels. However, TRT increased obstructive sleep apnea event	[44]
Men with low testosterone levels (Age ≥ 50 years)	TRT group (n = 8,137) versus Control (n = 4,418) Mean F/U = 6.6 years	↓ CHF event CAD event All–cause mortality MI event	TRT reduced mortality and cardiovascular events in men with low testosterone levels	[45]
Men with low testosterone levels (Age > 50 years)	TRT group (n = 207,176) versus Control (n = 207,176) F/U = 1 year	↔ MI event	TRT did not affect MI events in men with low testosterone levels	[46]

(continued)

Table 3.4 (continued)

Study design	TRT regimen	Results	Interpretation	Refs
Men with low testosterone levels (Age ≥ 50 years)	TRT group (n = 43,931) versus Control (n = 13,378) Mean F/U = 6.2 years	↓ All−cause mortality MI event Stroke event	TRT reduced mortality and cardiovascular events in men with low testosterone levels	[47]
Men with low testosterone levels (Age > 60 years)	TRT group (n = 2,241) versus low testosterone group (n = 801) F/U = 3 years	↓ MACE ↑ Stroke event	TRT reduced major cardiac adverse events, but it increased stroke events in men with low testosterone levels	[48]
Aging men with low testosterone levels (Age > 70 years)	TRT group (n = 73) versus Control (n = 65) F/U = 1 year	↑ non calcified coronary artery plaque volume Total plaque volume Coronary artery calcium score	TRT increased coronary plaques in aging men with low testosterone levels	[49]
Men with stable, chronic angina pectoris and low testosterone levels (Age > 20 years)	TRT group (n = 7) versus Control (n = 6) F/U = 1 year	↑ Time to ischemia ↓ Carotid intima−media thickness	TRT delayed time to ischemia and reduced carotid intima−media thickness in men with stable chronic angina pectoris and low testosterone levels	[50]
Men with ischemic heart disease and low testosterone levels (Age > 60 years)	TRT group (n = 11) versus Control (n = 10) F/U = 1 month	↑ Time to 1 mm ST segment depression	TRT delayed time to ischemia in men with ischemic heart disease and low testosterone levels	[51]
Men with heart failure and low testosterone levels (Age > 60 years)	TRT group (n = 44) versus their baseline F/U = 3 month	↓ QTd	TRT reduced QT dispersion in men with heart failure and low testosterone levels	[52]
Aging men with MI and low testosterone levels (Age > 70 years)	TRT group (n = 299) versus non−TRT group (n = 29,551) Mean F/U = 2.8 years	↔ MI events	TRT did not alter incidence of MI events in aging men with MI and low testosterone levels	[53]

TRT: testosterone replacement therapy; F/U: follow−up; MI: myocardial infarction; CV: cardiovascular; CAD: coronary artery disease; CHF: chronic heart failure; OSA: obstructive sleep apnea; MACE: major adverse cardiac event

Conclusion

Testosterone plays an important role in the regulation of myocardial injury during cardiac I/R, the extent of the impact being dependent on the duration of testosterone deprivation. Exogenous testosterone administration modulates myocardial infarction, LV function, and arrhythmias during cardiac I/R injury, the degree of modulation being dependent on dose and time of administration in both normal and testosterone deprived rats.

Acknowledgments This work was supported by grants from Thailand Research Fund (NA: TRG6280005), the Senior Research Scholar grant from the National Research Council of Thailand (SCC), a NSTDA research chair grant from the National Development Agency of Thailand (NC), and the Chiang Mai University Center of Excellence Award (NC).

References

1. Benjamin EJ, Virani SS, Callaway CW, Chamberlain AM et al (2018) Heart Disease and Stroke Statistics−2018 Update: A Report From the American Heart Association. Circulation 137:e67–e492
2. Collins P (2006) Risk factors for cardiovascular disease and hormone therapy in women. Heart 92(Suppl 3):iii24–8
3. Takawale A, Fan D, Basu R, Shen M et al (2014) Myocardial recovery from ischemia−reperfusion is compromised in the absence of tissue inhibitor of metalloproteinase 4. Circ Heart Fail 7(4):652–662
4. Ma RC, Tong PC (2010) Testosterone levels and cardiovascular disease. Heart 96(22):1787–1788
5. Nieschlag E, Nieschlag S (2019) ENDOCRINE HISTORY: The history of discovery, synthesis and development of testosterone for clinical use. Eur J Endocrinol 180(6):R201–R212
6. Farias JM, Tinetti M, Khoury M, Umpierrez GE (2019) Low testosterone concentration and atherosclerotic disease markers in male patients with type 2 diabetes. J Clin Endocrinol Metab 99(12):4698–4703
7. Niccoli G, Milardi D, D'Amario D, Fracassi F et al (2015) Hypotestosteronemia is frequent in ST−elevation myocardial infarction patients and is associated with coronary microvascular obstruction. Eur J Prev Cardiol 22(7):855–863
8. Apaiajai N, Chunchai T, Jaiwongkam T, Kerdphoo S et al (2018) Testosterone Deprivation Aggravates Left−Ventricular Dysfunction in Male Obese Insulin−Resistant Rats via Impairing Cardiac Mitochondrial Function and Dynamics Proteins. Gerontology 64(4):333–343
9. Pongkan W, Pintana H, Sivasinprasasn S, Jaiwongkam T et al (2016) Testosterone deprivation accelerates cardiac dysfunction in obese male rats. J Endocrinol 229(3):209–220
10. Pongkan W, Pintana H, Jaiwongkam T, Kredphoo S et al (2016) Vildagliptin reduces cardiac ischemic−reperfusion injury in obese orchiectomized rats. J Endocrinol 231(1):81–95
11. Arinno A, Apaijai N, Kaewtep P, Pratchayasakul W, et al (2019) Combined low−dose testosterone and vildagliptin confers cardioprotection in castrated obese rats. J Endocrinol (in press)
12. Osterberg EC, Bernie AM, Ramasamy R (2014) Risks of testosterone replacement therapy in men. Indian J Urol 30(1):2–7
13. Goodale T, Sadhu A, Petak S, Robbins R (2017) Testosterone and the Heart. Methodist Debakey Cardiovasc J 13(2):68–72

14. Seftel AD, Kathrins M, Niederberger C (2015) Critical Update of the 2010 Endocrine Society Clinical Practice Guidelines for Male Hypogonadism: A Systematic Analysis. Mayo Clin Proc 90(8):1104–1115
15. Palee S, McSweeney CM, Maneechote C, Moisescu DM, et al (2019) PCSK9 inhibitor improves cardiac function and reduces infarct size in rats with ischaemia/reperfusion injury: Benefits beyond lipid−lowering effects. J Cell Mol Med (in press)
16. Maneechote C, Palee S, Kerdphoo S, Jaiwongkam T et al (2019) Balancing mitochondrial dynamics via increasing mitochondrial fusion attenuates infarct size and left ventricular dysfunction in rats with cardiac ischemia/reperfusion injury. Clin Sci (Lond) 133(3):497–513
17. Apaijai N, Chinda K, Palee S, Chattipakorn S et al (2014) Combined vildagliptin and metformin exert better cardioprotection than monotherapy against ischemia−reperfusion injury in obese−insulin resistant rats. PLoS ONE 9(7):e102374
18. Lindsey ML, Bolli R, Canty JM Jr, Du XJ et al (2018) Guidelines for experimental models of myocardial ischemia and infarction. Am J Physiol Heart Circ Physiol 314(4):H812–H838
19. Borst SE, Quindry JC, Yarrow JF, Conover CF et al (2010) Testosterone administration induces protection against global myocardial ischemia. Horm Metab Res 42(2):122–129
20. Callies F, Stromer H, Schwinger RH, Bolck B et al (2003) Administration of testosterone is associated with a reduced susceptibility to myocardial ischemia. Endocrinology 144(10):4478–4483
21. Liu J, Tsang S, Wong TM (2006) Testosterone is required for delayed cardioprotection and enhanced heat shock protein 70 expression induced by preconditioning. Endocrinology 147(10):4569–4577
22. Rubio−Gayosso I, Ramirez−Sanchez I, Ita−Islas I, Ortiz−Vilchis P et al (2013) Testosterone metabolites mediate its effects on myocardial damage induced by ischemia/reperfusion in male Wistar rats. Steroids 78(3):362–369
23. Tsang S, Wu S, Liu J, Wong TM (2008) Testosterone protects rat hearts against ischaemic insults by enhancing the effects of alpha(1)−adrenoceptor stimulation. Br J Pharmacol 153(4):693–709
24. Ghimire A, Bisset ES, Howlett SE (2019) Ischemia and reperfusion injury following cardioplegic arrest is attenuated by age and testosterone deficiency in male but not female mice. Biol Sex Differ 10(1):42
25. Hadi NR, Yusif FG, Yousif M, Jaen KK (2014) Both castration and goserelin acetate ameliorate myocardial ischemia reperfusion injury and apoptosis in male rats. ISRN Pharmacol 2014:206951
26. Huang C, Gu H, Zhang W, Herrmann JL et al (2010) Testosterone−down−regulated Akt pathway during cardiac ischemia/reperfusion: a mechanism involving BAD, Bcl−2 and FOXO3a. J Surg Res 164(1):e1−11
27. Kohno H, Takahashi N, Shinohara T, Ooie T et al (2007) Receptor−mediated suppression of cardiac heat−shock protein 72 expression by testosterone in male rat heart. Endocrinology 148(7):3148–3155
28. Le TY, Ashton AW, Mardini M, Stanton PG et al (2014) Role of androgens in sex differences in cardiac damage during myocardial infarction. Endocrinology 155(2):568–575
29. Nam UH, Wang M, Crisostomo PR, Markel TA et al (2007) The effect of chronic exogenous androgen on myocardial function following acute ischemia−reperfusion in hosts with different baseline levels of sex steroids. J Surg Res 142(1):113–118
30. Pavon N, Martinez−Abundis E, Hernandez L, Gallardo−Perez JC, et al (2012) Sexual hormones: effects on cardiac and mitochondrial activity after ischemia−reperfusion in adult rats. Gender difference. J Steroid Biochem Mol Biol 132(1–2):135–46
31. Wang M, Tsai BM, Kher A, Baker LB et al (2005) Role of endogenous testosterone in myocardial proinflammatory and proapoptotic signaling after acute ischemia−reperfusion. Am J Physiol Heart Circ Physiol 288(1):H221–H226
32. Wang M, Wang Y, Abarbanell A, Tan J et al (2009) Both endogenous and exogenous testosterone decrease myocardial STAT3 activation and SOCS3 expression after acute ischemia and reperfusion. Surgery 146(2):138–144

33. O'Rourke B, Van Eyk JE, Foster DB (2011) Mitochondrial protein phosphorylation as a regulatory modality: implications for mitochondrial dysfunction in heart failure. Congest Heart Fail 17(6):269–282
34. Bertero E, Maack C (2018) Calcium Signaling and Reactive Oxygen Species in Mitochondria. Circ Res 122(10):1460–1478
35. Pongkan W, Chattipakorn SC, Chattipakorn N (2015) Chronic testosterone replacement exerts cardioprotection against cardiac ischemia–reperfusion injury by attenuating mitochondrial dysfunction in testosterone–deprived rats. PLoS ONE 10(3):e0122503
36. Svartberg J (2007) Epidemiology: testosterone and the metabolic syndrome. Int J Impot Res 19(2):124–128
37. Donner DG, Elliott GE, Beck BR, Bulmer AC et al (2015) Impact of Diet–Induced Obesity and Testosterone Deficiency on the Cardiovascular System: A Novel Rodent Model Representative of Males with Testosterone–Deficient Metabolic Syndrome (TDMetS). PLoS ONE 10(9):e0138019
38. Alexandre J, Milliez P, Rouet R, Manrique A et al (2015) Aldosterone and testosterone: two steroid hormones structurally related but with opposite electrophysiological properties during myocardial ischemia–reperfusion. Fundam Clin Pharmacol 29(4):341–351
39. Lauro FV, Francisco DC, Elodia GC, Eduardo PG et al (2014) Activity exerted by a testosterone derivative on myocardial injury using an ischemia/reperfusion model. Biomed Res Int 2014:217865
40. Kuhar P, Lunder M, Drevensek G (2007) The role of gender and sex hormones in ischemic–reperfusion injury in isolated rat hearts. Eur J Pharmacol 561(1–3):151–159
41. Seara FAC, Barbosa RAQ, de Oliveira DF, Gran da Silva DLS, et al (2017) Administration of anabolic steroid during adolescence induces long–term cardiac hypertrophy and increases susceptibility to ischemia/reperfusion injury in adult Wistar rats. J Steroid Biochem Mol Biol 171: 34–42
42. Maldonado O, Ramos A, Guapillo M, Rivera J, et al (2019) Effects of chronic inhibition of Testosterone metabolism on cardiac remodeling after ischemia/reperfusion–induced myocardial damage in gonadectomized rats. Biol Open 8(5)
43. Seara FAC, Barbosa RAQ, Santos MVN, Domingos AE et al (2019) Paradoxical effect of testosterone supplementation therapy on cardiac ischemia/reperfusion injury in aged rats. J Steroid Biochem Mol Biol 191:105335
44. Cole AP, Hanske J, Jiang W, Kwon NK et al (2018) Impact of testosterone replacement therapy on thromboembolism, heart disease and obstructive sleep apnoea in men. BJU Int 121(5):811–818
45. Oni OA, Sharma R, Chen G, Sharma M et al (2017) Normalization of testosterone levels after testosterone replacement therapy is not associated with reduced myocardial infarction in smokers. Mayo Clin Proc Innov Qual Outcomes 1(1):57–66
46. Li H, Mitchell L, Zhang X, Heiselman D et al (2017) Testosterone Therapy and Risk of Acute Myocardial Infarction in Hypogonadal Men: An Administrative Health Care Claims Study. J Sex Med 14(11):1307–1317
47. Sharma R, Oni OA, Gupta K, Chen G et al (2015) Normalization of testosterone level is associated with reduced incidence of myocardial infarction and mortality in men. Eur Heart J 36(40):2706–2715
48. Anderson JL, May HT, Lappé DL, Bair T et al (2016) Impact of testosterone replacement therapy on myocardial infarction, stroke, and death in men with low testosterone concentrations in an integrated health care system. Am J Cardiol 117(5):794–799
49. Budoff MJ, Ellenberg SS, Lewis CE, Mohler ER 3rd et al (2017) Testosterone treatment and coronary artery plaque volume in older men with low testosterone. JAMA 317(7):708–716
50. Mathur A, Malkin C, Saeed B, Muthusamy R et al (2009) Long–term benefits of testosterone replacement therapy on angina threshold and atheroma in men. Eur J Endocrinol 161(3):443–449
51. Malkin CJ, Pugh PJ, Morris PD, Kerry KE et al (2004) Testosterone replacement in hypogonadal men with angina improves ischaemic threshold and quality of life. Heart 90(8):871–876

52. Malkin CJ, Morris PD, Pugh PJ, English KM et al (2003) Effect of testosterone therapy on QT dispersion in men with heart failure. Am J Cardiol 92(10):1241–1243
53. Etminan M, Skeldon SC, Goldenberg SL, Carleton B, et al. Testosterone therapy and risk of myocardial infarction: a pharmacoepidemiologic study. Pharmacotherapy 35(1):72–8

Chapter 4
The Influence of Sex and Age on Responses of Isolated Ventricular Myocytes to Simulated Ischemia and Reperfusion

David G. Sapp and Susan E. Howlett

Abstract Myocardial ischemia occurs when the blood flow to the heart is disrupted because of conditions like atherosclerosis or during cardiac surgery. Ischemia ultimately can lead to cardiomyocyte cell death unless blood flow to the heart is restored by reperfusion, although reperfusion itself also can cause myocardial injury. Ischemia and reperfusion (IR) damages the myocardium and is a major cause of cardiovascular morbidity and mortality in our aging population. This has motivated interest in developing different preclinical models of IR injury to explore underlying mechanisms. This review focuses on various cellular models of IR that have been developed for use in isolated cardiac myocytes along with their strengths and weaknesses. In general, these cellular models permit tight control of the extracellular milieu and facilitate the application of substances directly onto the myocytes themselves. As these models eliminate the influence of other cell types and circulating factors on responses to IR, effects on cardiomyocyte contractile, electrophysiological and molecular mechanisms can readily be evaluated. This review highlights what is known about the influence of age and sex on cellular responses to IR and that most research has used only cells from young adult male animals. The need to use preclinical models of IR that more closely represent the older individuals of both sexes who are the most likely to develop these diseases is emphasized. This will create a stronger knowledge base that will help promote translation into clinical populations.

Keywords Aging · Ventricular myocytes · Ischemia · Reperfusion · Sex difference

D. G. Sapp · S. E. Howlett (✉)
Department of Pharmacology, Dalhousie University, Halifax, NS, Canada
e-mail: susan.howlett@dal.ca

S. E. Howlett
Department of Medicine (Division of Geriatric Medicine), Dalhousie University, Halifax, NS, Canada

B. Ostadal and N. S. Dhalla (eds.), *Sex Differences in Heart Disease*, Advances in Biochemistry in Health and Disease 21,
https://doi.org/10.1007/978-3-030-58677-5_4

Introduction

Myocardial ischemia occurs when blood flow to heart cells is restricted. It frequently arises because of atherosclerosis but can occur in other situations such as during cardiac bypass surgery and in positioning of stents. By preventing blood flow to the working myocardium, ischemia attenuates cardiac contractile function and disrupts the electrical activity of the heart [1]. Ischemia can lead to myocardial infarction, so it is important to restore blood flow to the heart, by reperfusion, as soon as possible. However, reperfusion itself can cause myocardial injury as it promotes intracellular calcium overload, oxidative stress and inflammation, all of which can further damage the heart [1]. The importance of ischemia and reperfusion (IR) injury in ischemic heart disease has motivated preclinical researchers to explore underlying mechanisms. These studies have utilised various established models of IR injury such as coronary occlusion models in intact animals, perfused hearts exposed to global ischemia and isolated heart muscle subjected to simulated IR [2, 3]. Cellular models of IR also have been developed for use in isolated cardiac myocytes. Together, these preclinical studies have begun to shed light on mechanisms involved in IR injury and identify new targets for interventions.

Clinical translation of knowledge about IR mechanisms requires basic research in areas that are relevant to the populations affected by ischemic heart disease. Ischemic heart disease occurs most frequently in older individuals and is prevalent in both sexes [4–6]. However, experimental studies of IR injury typically use young, mostly male animals. To facilitate translation of knowledge to humans it is important to use preclinical models that represent both sexes and to investigate responses to IR in the setting of aging. This review will evaluate available cellular models of IR injury in isolated ventricular myocytes, where cellular mechanisms that underpin IR damage have been investigated. We will highlight studies that use isolated cardiomyocytes to study sex and age differences in responses to IR. Knowledge gaps will be discussed and areas for further exploration will be identified. Additional research into IR injury in animal models that more closely reflect the older men and women who get these diseases will help create a stronger foundation of knowledge to develop novel clinical treatments.

Multicellular Models of Ischemia and Reperfusion

Preclinical investigations of IR have utilised various established models of IR injury. One established approach is to transiently ligate coronary arteries in intact animals [2]. Generally, such models are open chest, although newer methods to create closed chest IR injury in mice have been developed [7]. These models are useful in that they closely replicate IR damage that occurs clinically, provide information about left ventricular function and estimate infarct size as a function of the area at risk [2]. However, they can be complicated by effects of other cell types (e.g. fibroblasts, immune cells, etc.) or circulating factors and neuroendocrine signals from other organs. They are also technically challenging, especially in small animals where coronary vessels are more difficult to ligate [2].

Isolated perfused hearts exposed to global ischemia followed by reperfusion also have been used to model IR injury [2]. This model also can be used to measure left ventricular function and infarct size following exposure to IR. Isolated heart muscle preparations exposed to conditions that simulate IR can be used to investigate its effects on contraction and arrhythmic activity in coupled cardiac cells [3]. Isolated muscle and perfused heart models have the advantage that they minimize effects of circulating factors, neuroendocrine signals and other cells types. However, they do not replicate all aspects of IR injury. Advantages and limitations of these classic models of IR have been reviewed in detail recently [2].

Models of IR in Isolated Ventricular Myocytes

An alternative approach is to use isolated cardiac myocytes as an in vitro model of IR. This technique allows the extracellular environment to be both tightly controlled and systematically varied. The use of cell preparations eliminates the influence of extracellular factors, so the impact of IR on cardiac myocytes can be investigated without effects of other cell types (e.g. fibroblasts, immune cells etc.) or circulating factors (e.g. neurotransmitters, hormones). The impact of IR on cell survival and on contractile, electrophysiological and molecular mechanisms in the myocytes themselves can be evaluated [2]. This provides a unique insight into the responses of individual cells to IR and is ideal for gaining mechanistic insights.

Many different methods to investigate IR in isolated cardiomyocytes have been developed. These models mimic one or more components of true myocardial ischemia including acidosis, hypercapnia, hyperkalemia, hypoxia, lactate accumulation, substrate deprivation, metabolic inhibition and/or reactive oxygen species (ROS) accumulation [8, 9]. Table 4.1 describes models of simulated ischemia, hypoxia, acidosis, metabolic inhibition and ROS induction in cardiomyocytes. The table highlights the many features of myocardial ischemia that are simulated in each case and the references refer to seminal papers where these models were developed and characterized. In the "simulated ischemia" models in Table 4.1, myocytes are superfused with buffer that includes most components of true in vivo ischemia and then reperfused with regular buffer [10–15]. In many of these simulated ischemia models anoxic gas (e.g. nitrogen gas) is layered over the top of the superfusion chamber to exclude atmospheric oxygen. A schematic diagram of the first simulated ischemia model developed by Cordeiro et al. [12] for use in ventricular myocytes is represented in Fig. 4.1. In the "hypoxia" models, cells are transiently deprived of oxygen (and occasionally also substrate) and allowed to produce other components of ischemia naturally [16–21]. Others cellular models have focussed on specific conditions encountered in true ischemia such as acidosis [22], metabolic inhibition [23] or ROS induction [24], as summarized in Table 4.1.

Table 4.1 Components of ischemia mimicked in cellular models of ischemia/reperfusion

Model[a]	Acidosis	Hyper-capnia	Hyper-kalemia	Hypoxia	Lactate buildup	Lack of substrate	Lack of metabolism	ROS	Temp (°C)	References
Simulated ischemia	X	X	X	X	X	X			37	Cordeiro et al. [12]
	X	X	X	X	X	X			37	*Liu et al.* [14]
	X		X	X		X			37	Maddaford et al. [15]
	X		X	X		X			37	Heller et al. [13]
	X			X		X			37	Athias et al. [10]
	X	X	X	X	X	X			Unspecified	Brady et al. [11]
Hypoxia	X			X	X	X			37	Vander Heide et al. [19]
				X					37	Stern et al. [18]
				X	X				Unspecified	Rajs and Härm [17]
	X			X	X				37	Pitts and Toombs [16]
		X		X	X	X			37	Vemuri et al. [20]
				X	X	X			37	Yang et al. [21]
Acidosis	X								37	Fry et al. [22]
Metabolic inhibition							X		37	Priori et al. [23]
ROS induction		X				X		X	37	Cicconi et al. [24]

[a] Overview of components of true myocardial ischemia that are simulated in a variety of different models of ischemia and reperfusion (IR) for use in isolated cardiomyocytes. The symbol 'X' indicates that the model includes the listed component of IR. ROS = reactive oxygen species

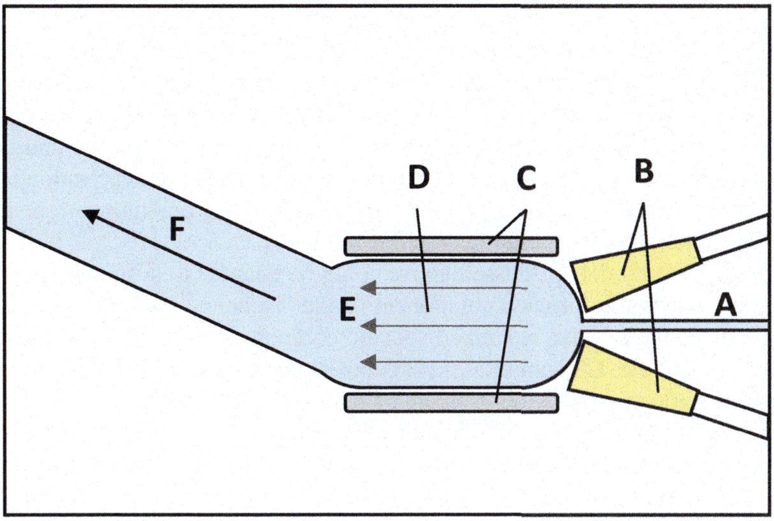

Fig. 4.1 Schematic diagram of an experimental set-up for simulated IR of isolated ventricular myocytes based on the model originally developed by Cordeiro and colleagues [12]. A. Inflow for both control and simulated "ischemic" buffer solutions. B. Tubing that allows the inflow of N_2 gas creates a layer of anoxic gas over the experimental chamber. C. Raised barriers to maintain laminar flow of anoxic gas. D. Cell bath containing ventricular myocytes adhered to the bottom and ischemic/control solution. E. Laminar flow of anoxic gas, creating a sealed hypoxic environment with aerial access. F. Outflow of ischemic/control solution at a steady flow rate

Simulated ischemia is a commonly used cellular model of IR (Table 4.1). Some approaches incorporate multiple features of true ischemia including acidosis, hypercapnia, hyperkalemia, hypoxia, lactate accumulation and substrate deprivation [11, 12, 14], while others focus on a subset of these [10, 13, 15]. Indeed, the ischemic solution composition varies between different simulated ischemia models. For example, aspects such as pH, gas type/concentrations, ion concentrations, and lactate concentrations differ between models. The simulated ischemia models established by Cordeiro et al. [12] and Liu et al. [14] use very similar solutions, although the latter has greater acidity and contains deoxyglucose. The model developed by Maddaford et al. [15] is a modification of the Cordeiro model, but the solution has no lactate and is not bubbled with high (10%) CO_2 to produce hypercapnia. Interestingly, more recent studies with the Maddaford model have supplemented the ischemic solution with lactate and used higher CO_2 to more closely resemble true myocardial ischemia [25, 26].

Cellular models that focus on hypoxia are also commonly used to investigate IR at the cellular level. As shown in Table 4.1, simulated ischemia models clearly reproduce more aspects of true myocardial ischemia than the hypoxia models. However, when cells are subjected to hypoxia some of the other ischemic components are induced indirectly, as a result of cellular responses to hypoxia [16, 17, 19–21]. For instance, an early model by Vander Heide and colleagues [19] used centrifugation to pellet the myocytes, then removed the extracellular fluid and covered the cells with oil to induce hypoxia. In consequence, the cells produced lactate and exhausted available ATP [19]. Other studies have induced hypoxia by exposing cells to between 95 and 100% N_2 [17, 20, 21], 100% argon gas [18] or by incubating them under a coverslip [16]. These investigators found that hypoxic myocytes produce lactate and exhaust ATP stores. In some cases, hypoxia is combined with substrate deprivation to more closely mimic ischemia in vivo [20, 21]. The ischemic factors known to be present in various hypoxia models are illustrated in Table 4.1.

In theory, the models discussed above could be adapted to systematically investigate the contribution of each component of the "ischemic solution" individually. Indeed, some studies have examined specific components that are implicated in IR injury. For example, Cicconi et al. [24] exposed myocytes (the HL-5 cardiac atrial cell line) to hydrogen peroxide to create oxidative stress and induce ROS. Oxidative stress and ROS generation are implicated in IR injury and are associated with opening of the mitochondrial permeability transition pore (mPTP) [27]. While the ROS induction model does not simulate IR per se, it does isolate a clinically important aspect of the disease. An entirely different approach to model IR in cardiac myocytes is to stress the cells by inhibiting metabolism. For example, severe metabolic inhibition of isolated ventricular myocytes with cyanide reversibly mimics effects of myocardial ischemia on cellular electrical activity [23]. While these models are a simple way to imitate ischemia, they do not accurately simulate the full range of cellular stresses associated with IR.

Cell type is a key factor in experimental models of IR injury, as illustrated in Table 4.2. Freshly isolated ventricular myocytes from adult animals have been used by many investigators [12–15, 17–19, 22, 23]. They are considered the gold standard because they are most like myocardial cells exposed to ischemia in vivo [2]. Other cell types are easier to prepare but have critical physiological differences when compared to adult cardiomyocytes [2]. For example, as shown in Table 4.2, neonatal myocytes and early post-natal cells have been used in some investigations [10, 16, 20]. These preparations create a monolayer of spontaneously beating myocytes, which allows the study of cell-to-cell connections [2]. However, as myocytes undergo marked physiological changes during maturation, results obtained in neonatal cells may not readily translate to the older population most likely to experience IR injury clinically. Other studies have used cardiac cell lines including the HL-5 [24], HL-1 [11] and H9c2 cell lines [21]. However, HL-1 and HL-5 cells are derived from a mouse atrial myocyte tumor lineage [28, 29] and the H9c2 cells are derived from embryonic rat heart [30]. Therefore, these cells lines do not model the working ventricular myocardium in adults. Both cardiac-derived cell lines and neonatal/embryonic cardiomyocytes do not respond to IR in the same way as adult ventricular myocytes [2, 11]. Cardiac cell types that have been used in IR experiments are summarized in Table 4.2.

Strengths and limitations of the various cellular IR models, along with the cell type and model of ischemia employed are highlighted in Table 4.2. Models that simulate ischemia in freshly isolated adult myocytes are preferred. They have the advantage that they mimic major aspects of ischemia in vivo [2]. The use of an open experimental chamber allows the investigator to record electrical activity (e.g. membrane

Table 4.2 Models of ischemia and reperfusion in isolated ventricular myocytes

Model[a]	Cell type	Strengths	Weaknesses	References
Simulated ischemia	Freshly isolated adult myocytes	Mimics major aspects of ischemia in vivo; open chamber permits access to myocytes	Requires specialised equipment	Cordeiro et al. [12]
		Mimics major aspects of ischemia in vivo; open chamber permits access to myocytes	Requires specialised equipment	Liu et al. [14]
		No lactate may reduce cellular necrosis; open chamber for cell access	Does not mimic all aspects of ischemia in vivo; requires specialised equipment	Maddaford et al. [15]
		Simultaneous experiments possible; can quickly alter bathing solution; practical to use a single myocyte more than once	Myocyte adherence to capillaries influences contractility measure; measurement error due to curvature; requires specialised equipment; no access to cells	Heller et al. [13]
	Neonatal myocytes	Ease of preparation; creates monolayer	Does not simulate all aspects of ischemia in vivo; does not isolate individual myocytes; myocytes are not identical to adult in vivo	Athias et al. [10]
	HL-1 cell line	Ease of preparation; long term survival in culture	Derived from mouse atrial cardiomyocyte tumor; not ventricle; cell line does not accurately mimic myocytes in vivo	Brady et al. [11]
Hypoxia	Freshly isolated adult myocytes	Spun cells into a pellet with no extracellular fluid to exchange metabolites; cellular suspension induces ischemia without substrate; open chamber for cell access	Substrate composition is not well controlled; requires specialised equipment	Vander Heide et al. [19]
		Argon gas exposure; gas concentration can be varied; isolates hypoxia/anoxia component; open chamber for cell access	Does not accurately mimic ischemia in vivo; requires specialised equipment	Stern et al. [18]

(continued)

Table 4.2 (continued)

Model[a]	Cell type	Strengths	Weaknesses	References
		N_2 gas exposure; N_2 concentrations can be varied; isolates hypoxia/anoxia component; open chamber for cell access	Does not accurately mimic ischemia in vivo; requires specialised equipment	Rajs and Härm [17]
	Neonatal myocytes	Ease of preparation using cover slip to induce hypoxia; contains ischemic and control myocytes in one monolayer; simulates regional ischemia	Does not accurately mimic ischemia in vivo; does not isolate individual myocytes; myocytes are not identical to in vivo; no cell access	Pitts and Toombs [16]
	Post-natal myocytes	Multicompartmental unit, ease of preparation; creates monolayer; open chamber for cell access	Does not accurately mimic ischemia in vivo; does not isolate individual myocytes; myocytes are not identical to in vivo; does not allow electrophysiological measurements	Vemuri et al. [20]
	H9c2 cell line	Ease of preparation; long term survival in culture; open chamber for cell access	Does not accurately mimic ischemia in vivo; derived from embryonic rat heart; does not mimic ventricular myocytes in vivo	Yang et al. [21]
Acidosis	Freshly isolated adult myocytes	Isolates low pH component; open chamber for cell access	Does not accurately mimic ischemia in vivo; requires specialised equipment	Fry et al. [22]
Metabolic inhibition	Freshly isolated adult myocytes	Expose cells to cyanide; simple method to stress myocytes as in ischemia; open chamber for cell access	Does not accurately mimic ischemia in vivo; requires specialised equipment	Priori et al. [23]
ROS induction	HL-5 cell line	Ease of preparation; long term survival in culture; isolates ROS component; open chamber for cell access	Derived from mouse atrial cardiomyocyte tumor; not ventricle; cell line does not accurately mimic myocytes in vivo	Cicconi et al. [24]

[a]Qualitative assessment of various cellular models of ischemia and reperfusion (IR) injury. ROS = reactive oxygen species

potentials, ionic currents) along with contractile activity in individual myocytes [12]. Cells also can be loaded with calcium-sensitive fluorescent dyes to allow intracellular calcium transients to be measured in conjunction with mechanical and/or electrical responses [31]. The open chamber also facilitates the use of a rapid solution switcher, so individual cells can be superfused rapidly with drugs of interest [31, 32]. On the other hand, these approaches do require specialised equipment and technical skill. Other models are more easily implemented and can have utility for specific investigations. For example, the approach developed by Pitts and Toombs [16] uses neonatal cardiomyocytes and simulates regional hypoxia by placing a coverslip over part of the cellular monolayer. This creates a model with cells in ischemic, semi-ischemic, and control conditions on one plate. As ischemia manifests regionally in vivo, this model uniquely simulates that in cardiomyocytes [16]. Heller et al. [13] have introduced myocytes into multiple tracks of capillaries rather than in a regular open chamber. This is beneficial in experiments where responses of individual cells to sequential solution changes are desired. However, neither the coverslip model nor the capillary model allows access to the bath and the cylindrical nature of the glass capillaries can introduce measurement error.

Models of IR in isolated adult ventricular myocytes, neonatal myocytes and cell lines are inherently different from in vivo cells because the cells are isolated from each other and the extracellular matrix. Still, there are advantages to these cellular models in that the extracellular environment can be controlled and effects of IR on individual cells can be investigated. The next sections will highlight how such models have been used to investigate the influence of sex and age on responses to IR.

Sex Differences in Cellular Responses to IR Injury

While sex differences in the electrical and contractile function of cardiomyocytes have been identified in cells studied under normoxic conditions [33], relatively little is known about sex differences in responses to IR. This is a critical knowledge gap because understanding both sexes' responses to IR is key to understanding male–female differences in mechanisms underlying the responses to IR and developing sex-specific interventions in the setting of IR. Major sex differences in the responses of cardiomyocytes to IR injury are summarised in Table 4.3.

A few studies have explored male–female differences in the recovery of cardiomyocyte contractile function in reperfusion in young adult rats. Ross and Howlett [34] showed that peak contractions were abolished in simulated ischemia in both sexes and remained depressed in cells from males (Table 4.3). By contrast, contractions recovered to exceed baseline levels in myocytes from females [34]. Exposure to ischemia increased resting calcium levels in both sexes and this returned to normal in reperfusion, but more cells died during reperfusion in males than in females. These observations suggest that young adult females are resistant to IR injury when compared to age-matched males. Bell et al. [26] found a female-specific increase in calcium transients during simulated ischemia with no change in contractions suggesting that

Table 4.3 Influence of sex on responses to ischemia and reperfusion in ventricular myocytes

Model[a]	Age	Sex	Species	Outcomes	References
Simulated ischemia	3 & 5–6 mos (Sham & OVX)	M & F	Rat	Recovery of contractile function and cell survival lower in young male than young female in reperfusion; OVX abolished female protection, increased abnormal contractile activity (alternans)	Ross and Howlett [34]
	~3–4 mos	M & F	Rat	Recovery of myocyte contractile function and cell survival in reperfusion similar in both sexes; diastolic calcium increased in IR more in males than females; suggests more effective control of internal calcium in females	Bell et al. [25, 26]
	~2 mos	F	Rat	Estrogen plus testosterone treatment abrogates harmful effects of OVX on lactate dehydrogenase release, contractile performance, and cell survival in reperfusion; individual hormones alone had fewer beneficial effects on the same variables	Liu et al. [14]
Hypoxia	Neonatal (1–3 days)	M	Rat	Cells exposed to hypoxia/reoxygenation show p53-dependent apoptosis after p53 is phosphorylated by p38α MAPK. 17β-estradiol protects cardiomyocytes by inhibiting p38α-p53 signalling; both estrogen receptors (ERα & ERβ) are involved	Liu et al. [35]

(continued)

Table 4.3 (continued)

Model[a]	Age	Sex	Species	Outcomes	References
Hypoxia plus isoproterenol	Adult	M & F	Rat	Male cells treated with isoproteronol then hypoxia had more cell death than females with isoproterenol or both sexes without drug; due to higher intracellular calcium loading in hypoxia in males; sex differences were reduced by nitric oxide synthase inhibition	Chen et al. [36]
Metabolic inhibition	Adult	M & F	Mouse	Cells exposed to sodium cyanide (0 glucose) show no sex difference in wildtype mice; male cells overexpressing sodium calcium exchanger show higher internal calcium than females; 17β-estradiol reduces this increase in calcium in male cells	Sugishita et al. [37]
	~4–5 mos	M	Rat	GDX increased lactate dehydrogenase release and reduced cell survival (cyanide model); testosterone attenuated these effects; α_1-adrenoreceptor stimulation enhanced beneficial effects and androgen receptor antagonists blocked them	Tsang et al. [38]

[a]Overview of male–female differences in cellular responses to ischemia and reperfusion (IR) demonstrated in cardiomyocyte models of IR. MAPK = mitogen-activated protein kinase; OVX = ovariectomy; GDX = gonadectomy; M = male; F = female

myofilament calcium sensitivity declined in females [26]. Even though they found no sex difference in cell survival in reperfusion [26], they did report that diastolic calcium increased more during IR in cells from males than females [25]. This suggests that cardiomyocytes from females are better able to regulate diastolic calcium than cells from males.

Studies of sex differences in responses to IR have used models that eliminate or add endogenous sex steroid hormones (e.g. estrogen, testosterone), as detailed in Table 4.3. The impact of low estrogen on responses to IR has been explored in ovariectomy (OVX) models. In OVX models, the ovaries are removed to model menopause, the time when the risk of ischemic heart disease rises in women [5]. Interestingly, myocytes from young adult females become highly susceptible to simulated IR injury following OVX. Unlike sham-operated controls, all myocytes from OVX animals developed arrhythmic activity and died in ischemia, so that none survived into reperfusion [34]. Liu et al. [14] used a simulated IR model and reported a reduction in cardioprotection following OVX, where contractions and myocyte survival were lower in OVX than sham controls and lactate dehodrogenase release was increased. Interestingly, physiological doses of estrogen and testosterone given together abolished these harmful effects of OVX [14]. However, these hormones were less effective when applied individually than when applied together (Table 4.3). Other work explored potential underlying mechanisms implicated in estrogen-mediated cardioprotection [35]. They subjected neonatal rat cardiomyocytes to 18 h of hypoxia followed by reoxygenation and showed that, when 17-β-estradiol was present prior to the onset of hypoxia, it was highly protective against apoptosis [35]. This beneficial effect is due to inhibition of p38α MAPK, which in turn inhibits p53-dependent apoptosis. This study also showed that both estrogen receptors (ERα and ERβ) are involved in this cardioprotective pathway [35]. Of note, selective agonists for each receptor were less cardioprotective than 17-beta-estradiol itself [35]. Sugishita et al. [37] showed that there was no sex difference in calcium accumulation by male and female cells exposed to sodium cyanide with no glucose. However, male cells that overexpressed the sodium calcium exchanger show higher internal calcium than females, an effect that was reversed by 17β-estradiol [37]. In combination, these findings suggest that estrogen has beneficial effects on cellular IR injury and that these benefits are at least partially attributable to effects on the myocytes themselves.

Interest in the role of circulating testosterone in IR injury has been motivated by the links between circulating testosterone levels and the risk of cardiovascular disease in both sexes [39, 40]. The effects of low testosterone on cellular responses to IR has been investigated with the help of gonadectomy (GDX) models, where animals are subjected to bilateral removal of the testes. Tsang et al. [38] used a metabolic inhibition model of IR to compare responses in ventricular myocytes from sham operated and GDX rats. They found that GDX increased lactate dehydrogenase release and reduced cell survival; these effects were attenuated by chronic testosterone supplementation [38]. Interestingly, exposure to adrenergic stress (e.g. an α_1-adrenergic agonist) augmented these beneficial effects while androgen receptor antagonists blocked these actions. These findings demonstrate that low testosterone exacerbates cellular IR injury and that supplementation with testosterone can reverse this.

Several investigators have used cellular models to explore male–female differences in responses to IR under pathophysiological conditions. Bell et al. [26] used a rat model of cardiac hypertrophy to investigate sex differences in cellular responses to simulated ischemia in the setting of disease. They found that the patterns of responses differed between the sexes in hypertrophied hearts. In females, calcium availability is reduced in hypertrophy whereas the sensitivity to calcium declines in males [26]. These findings suggest that novel sex differences can be revealed by modeling IR in a pathological setting. Another study used a ß-adrenergic agonist (e.g. isoproterenol) to induce a hyperadrenergic, hypercontractile state prior to exposing male and female cells to hypoxia followed by reoxygenation [36]. This hyperadrenergic state mimics that seen in many cardiovascular diseases [41]. Results showed that cell death was higher in cardiomyocytes from males exposed to isoproterenol prior to hypoxia compared with female cardiomyocytes [36]. This was attributable to higher intracellular calcium load in cells from males exposed to isoproterenol followed by hypoxia when compared to females. This female advantage was due to increased production of nitric oxide, which attenuated calcium loading following adrenergic challenge in females but not males [36]. These data may translate to females having greater resistance to IR damage when a hyperadrenergic state is present prior to the ischemic insult.

The results discussed above demonstrate that there are marked male–female differences in cellular responses to IR. These studies show that effects of estrogen and testosterone on susceptibility to IR injury are due, at least in part, to actions at the level of the cardiomyocytes. Furthermore, as both men and women suffer from ischemic heart disease, this highlights the need to conduct experiments in preclinical models using both sexes to identify underlying mechanisms that may well differ between the two. However, consideration of sex alone may not be enough, as advanced age is the major risk factor for cardiovascular disease in both sexes. The next section describes what is currently known about the influence of age on cellular responses to IR.

The Influence of Age on Cardiomyocyte Responses to IR

Age clearly increases the risk of ischemic heart disease in men and in women [4–6]. Still, as illustrated in Table 4.3, experimental studies of IR injury tend to use cardiomyocytes from young adult and even neonatal animals. Investigation of IR injury in the setting of aging is imperative to develop clinically relevant treatments, although few studies have explored cellular responses to IR injury in cardiomyocytes from older animals (Table 4.4).

O'Brien et al. [42] were the first to investigate the impact of age on cellular responses to simulated IR in ventricular myocytes from male rats (Table 4.4). They compared responses to IR in cells from young adult (3-month-old) and aged (24-month-old) animals and found that contractions were abolished in ischemia and only partially recovered in reperfusion regardless of age. Calcium transients declined in early reperfusion but then recovered and there was no age-dependent difference in cell survival following exposure to IR [42]. Despite these similarities, cells from aged males accumulated much more intracellular calcium in ischemia than cells from younger animals and this was associated with abnormal contractile activity, known as mechanical alternans [42]. These findings indicate that myocytes from older male animals are more susceptible to intracellular calcium overload and contractile dysfunction following IR than cells from younger hearts.

Table 4.4 Influence of age on responses to ischemia and reperfusion in isolated ventricular myocytes

Model[a]	Age	Sex	Species	Outcomes	References
Simulated ischemia	3 versus 24 mos	F	Rat	Young female cells protected from IR injury, but young male cells were not. Age and OVX abolished protection from IR injury in females and increased calcium loading plus abnormal contractile activity (alternans)	Ross and Howlett [34]
	3 versus 24 mos	M	Rat	Many responses similar in cells from young and old males; age enhanced diastolic calcium accumulation in ischemia and promoted mechanical alternans	O'Brien et al. [42]
	3 mos versus 18 mos	M	Rat	Exercise attenuated age-related loss of cardioprotection by IPC; IPC plus exercise reduced ROS generation and maintained mitochondrial membrane potential in old mice, but exercise or preconditioning alone did not	Wang et al. [43]
ROS induction	3–5 versus 20–24 mos	M	Rat	The cardioprotective glycogen synthase kinase-3β inhibitor increased ROS threshold for opening the mPTP in young, but not old cardiomyocytes; mPTP opened at lower ROS threshold in aged cells	Zhu et al. [44]

[a]Overview of the influence of age on cellular responses to ischemia and reperfusion (IR) injury. F = female; M = male; mPTP = mitochondrial permeability transition pore; OVX = ovariectomy; ROS = reactive oxygen species; IPC = ischemic preconditioning

Although no experimental studies have done a head-to-head comparison of responses to IR in myocytes from aged males and females, one study has compared data from young (3-month-old) and old (24-month-old) female rats [34]. This work has shown that myocytes from young female rats are much more resistant to contractile dysfunction in reperfusion than cells from young males, but this female advantage disappears with increasing age and fewer cells survive IR [34]. Interestingly, like the cells from older females, myocytes from females subjected to OVX exhibited reduced cell survival after exposure to simulated ischemia [34]. Together these observations suggest that the age-related decline in serum estrogen [45, 46] may help explain the enhanced sensitivity of aging female myocytes to IR injury. Results of these studies are summarized in Table 4.4.

Cells from older females also accumulate more intracellular calcium during ischemia and exhibit mechanical alternans when compared to younger females [34]. Thus, cells from females become more sensitive to IR injury with age and they appear to respond to simulated IR much the same as cells from older males [34, 42]. These effects of age on calcium homeostasis in males and females may be mediated by reduced sarco(endo)plasmic reticululm calcium-ATPase (SERCA) expression. Lower SERCA expression has been reported in the aging heart [47] and this would be expected to impair calcium sequestration, especially during cellular stress such as ischemia. Additionally, reduced SERCA expression is associated with cellular alternans [48] and this may help explain mechanical alternans in the setting of aging.

Cellular models of simulated IR also have been utilised to determine whether myocytes from aged hearts can respond to cardioprotective interventions (Table 4.4). Ischemic preconditioning (IPC) is a powerful cardioprotective intervention, but its efficacy is known to decline with age [49]. O'Brien and Howlett [50] used a cellular model of simulated IR to show that IPC improves cell survival in myocytes from young male rats but not in cells isolated from older male animals. This suggests that many of the beneficial effects of IPC may reside within the myocytes themselves. Wang et al. [43] extended these observations to demonstrate that IPC reduced ROS generation and increased mitochondrial membrane potential in young male rat myocytes exposed to simulated IR. However, ROS generation and mitochondrial membrane potentials were similar in older rat myocytes exposed to IR, regardless of whether they were exposed to IPC [43]. Interestingly, they showed this loss of cardioprotection at the cellular level is modulated by exercise. When older male rats were subjected to a 6-week exercise training program, the efficacy of IPC was restored in myocytes from older hearts and this was mediated by improved mitochondrial function during ischemia [43]. Whether IPC is affected by age in females and whether exercise can restore IPC in aging females has not yet been investigated.

A cellular model of IR mediated by ROS induction has been used to investigate responses to potential cardioprotective agents in the setting of aging. Glycogen synthase kinase-3β (GSK-3β) is a constitutively active Ser/Thr protein kinase that regulates opening of the mPTP, which leads to mitochondrial dysfunction and myocyte death in IR [51]. Zhu et al. [44] showed that a GSK-3ß inhibitor attenuated mPTP opening in young myocytes exposed to an increase in ROS, but this strategy was not effective in cells from older male rats [44]. These findings suggest that age modifies the regulation of the mPTP by GSK-3ß, although whether this applies in females has not yet been investigated.

The observations presented here provide at least preliminary evidence age can modify cellular responses to IR. In general, cardiomyocytes from older animals are more sensitive to the detrimental effects of IR. Importantly, there is emerging evidence that the influence of age on susceptibility to IR injury may differ between the sexes, although very few studies have addressed this critical issue at the cellular level. There is evidence that interventions such as IPC, exercise and mPTP inhibition may attenuate cellular IR injury, at least in males. Additional studies in aging animal models, and especially studies in both sexes, are now warranted.

Conclusions

Ischemic heart disease is a complex, multi faceted disease. Many preclinical studies of myocardial ischemia use isolated cardiac myocytes as an in vitro model of IR. This allows tight control of the extracellular environment and eliminates the impact of other cell types and circulating factors on cardiomyocyte responses. This review highlights the various cellular models of IR that have been used in the literature. From this analysis several important concepts emerge. First, even though ischemic heart disease occurs most often in older people and is prevalent in both men and women, most preclinical studies use young adult, mostly male, animals as cellular models of IR. Second, there are critical male–female differences in cellular responses to IR and these are mediated, at least in past, by the actions of sex steroid hormones on the myocytes themselves. Finally, very few studies have investigated the impact of IR on cardiomyocytes from older animals of either sex. This is an important knowledge gap. There is a critical need to utilise preclinical models of IR that more closely represent those who develop these diseases. The use of older animals of both sexes will help create a stronger foundation of knowledge that will facilitate translation and help develop novel clinical treatments.

Acknowledgments The authors are grateful to Peter Nicholl for assistance with the illustration and to Stefan Heinze-Milne for discussions about organization of the review and structuring the tables. This work is supported, in part, by grants from the Canadian Institutes of Health Research (MOP 97973; PJT 155961; PJT 162462;) and the Heart and Stroke Foundation of Canada (HSFC G-19-0026260).

References

1. Caricati-Neto A, Errante PR, Menezes-Rodrigues FS (2019) Recent advances in pharmacological and non-pharmacological strategies of cardioprotection. Int J Mol Sci 20:4002
2. Lindsey ML, Bolli R, Canty JM et al (2018) Guidelines for experimental models of myocardial ischemia and infarction. Am J Physiol Heart Circ Physiol 314:H812-838

3. Ferrier GR, Moffat MP, Lukas A (1985) Possible mechanisms of ventricular arrhythmias elicited by ischemia followed by reperfusion. Studies on isolated canine ventricular tissues. Circ Res 56:184–194
4. Maggioni AP, Maseri A, Fresco C et al (1993) Age-related increase in mortality among patients with first myocardial infarctions treated with thrombolysis. The Investigators of the Gruppo Italiano per lo Studio della Sopravvivenza nell'Infarto Miocardico (GISSI-2). N Engl J Med 329:1442–1448
5. Mozaffarian D, Benjamin EJ, Go AS et al (2016) American heart association statistics committee; stroke statistics subcommittee. Heart disease and stroke statistics-2016 update: a report from the American heart association. Circulation 133:e38-360
6. Shaw LJ, Bugiardini R, Merz CN (2009) Women and ischemia heart disease: evolving knowledge. J Am Coll Cardiol 54:1561–1575
7. Kim SC, Boehm O, Meyer R et al (2012) A murine closed-chest model of myocardial ischemia and reperfusion. J Vis Exp 65:e3896
8. Kalogeris T, Baines CP, Krenz M, Korthuis RJ (2012) Cell biology of ischemia/reperfusion injury. Int Rev Cell Mol Biol 298:229–317
9. Kazbanov IV, Clayton RH, Nash MP et al (2014) Effect of global cardiac ischemia on human ventricular fibrillation: insights from a multi-scale mechanistic model of the human heart. PLoS Comput Biol 10:e1003891
10. Athias P, Vandroux D, Tissier C, Rochette L (2006) Development of cardiac physiopathological models from cultured cardiomyocytes. Ann Cardiol Angeiol (Paris) 55:90–99
11. Brady NR, Hamacher-Brady A, Gottlieb RA (2006) Proapoptotic BCL-2 family members and mitochondrial dysfunction during ischemia/reperfusion injury, a study employing cardiac HL-1 cells and GFP biosensors. Biochim Biophys Acta 1757:667–678
12. Cordeiro JM, Howlett SE, Ferrier GR (1994) Simulated ischaemia and reperfusion in isolated guinea pig ventricular myocytes. Cardiovasc Res 28:1794–1802
13. Heller LJ, Mohrman DE, Smith JA, Wallace KB (2003) Multitrack system for superfusing isolated cardiac myocytes. Am J Physiol Heart Circ Physiol 284:H1872–H1878
14. Liu A, Gao L, Kang S et al (2012) Testosterone enhances estradiol's cardioprotection in ovariectomized rats. J Endocrinol 212:61–69
15. Maddaford TG, Hurtado C, Sobrattee S et al (1999) A model of low-flow ischemia and reperfusion, beating adult cardiomyocytes. Am J Physiol 277:H788–H798
16. Pitts KR, Toombs CF (2004) Coverslip hypoxia: a novel method for studying cardiac myocyte hypoxia and ischemia in vitro. Am J Physiol Heart Circ Physiol 287:H1801–H1812
17. Rajs J, Härm T (1980) Isolated rat cardiac myocytes as an experimental tool in the study of anoxic cell injury. Effect of reoxygenation--a preliminary report. Forensic Sci Int 16:185–190
18. Stern MD, Silverman HS, Houser SR et al (1988) Anoxic contractile failure in rat heart myocytes is caused by failure of intracellular calcium release due to alteration of the action potential. Proc Natl Acad Sci USA 85:6954–6958
19. Vander Heide RS, Rim D, Hohl CM, Ganote CE (1990) An in vitro model of myocardial ischemia utilizing isolated adult rat myocytes. J Mol Cell Cardiol 22:165–181
20. Vemuri R, Yagev S, Heller M, Pinson A (1985) Studies on oxygen and volume restrictions in cultured cardiac cells. I. A model for ischemia and anoxia with a new approach. Basic Res Cardiol 21:521–525
21. Yang GZ, Xue FS, Liu YY et al (2018) Feasibility analysis of oxygen-glucose deprivation-nutrition resumption on H9c2 cells in vitro models of myocardial ischemia-reperfusion injury. Chin Med J (Engl) 131:2277–2286
22. Fry CH, Powell T, Twist VW, Ward JP (1984) The effects of sodium, hydrogen and magnesium ions on mitochondrial calcium sequestration in adult rat ventricular myocytes. Proc R Soc Lond B Biol Sci 223:239–254
23. Priori SG, Yamada KA, Corr PB (1991) Influence of hypoxia on adrenergic modulation of triggered activity in isolated adult canine myocytes. Circulation 83:248–259
24. Cicconi S, Ventura N, Pastore D et al (2003) Characterization of apoptosis signal transduction pathways in HL-5 cardiomyocytes exposed to ischemia/reperfusion oxidative stress model. J Cell Physiol 195:27–37

25. Bell JR, Raaijmakers AJ, Curl CL et al (2015) Cardiac CaMKIIδ splice variants exhibit target signaling specificity and confer sex-selective arrhythmogenic actions in the ischemic-reperfused heart. Int J Cardiol 181:288–296
26. Bell JR, Curl CL, Harding TW et al (2016) Male and female hypertrophic rat cardiac myocyte functional responses to ischemic stress and β-adrenergic challenge are different. Biol Sex Differ 7:32
27. Kwong JQ, Molkentin JD (2015) Physiological and pathological roles of the mitochondrial permeability transition pore in the heart. Cell Metab 21:206–214
28. Claycomb WC, Lanson NA Jr, Stallworth BS et al (1998) HL-1 cells: a cardiac muscle cell line that contracts and retains phenotypic characteristics of the adult cardiomyocyte. Proc Natl Acad Sci USA 95:2979–2984
29. White SM, Constantin PE, Claycomb WC (2004) Cardiac physiology at the cellular level: use of cultured HL-1 cardiomyocytes for studies of cardiac muscle cell structure and function. Am J Physiol Heart Circ Physiol 286:H823–H829
30. Kimes BW, Brandt BL (1976) Properties of a clonal muscle cell line from rat heart. Exp Cell Res 98:367–381
31. Louch WE, Ferrier GR, Howlett SE (2002) Changes in excitation-contraction coupling in an isolated ventricular myocyte model of cardiac stunning. Am J Physiol Heart Circ Physiol 283:H800-810
32. Louch WE, Ferrier GR, Howlett SE (2005) Attenuation of cardiac stunning by losartan in a cellular model of ischemia and reperfusion is accompanied by increased sarcoplasmic reticulum calcium stores and prevention of cytosolic calcium elevation. J Pharmacol Exp Ther 312:238–247
33. Parks RJ, Howlett SE (2013) Sex differences in mechanisms of cardiac excitation-contraction coupling. Pflugers Arch 465:747–763
34. Ross JL, Howlett SE (2012) Age and ovariectomy abolish beneficial effects of female sex on rat ventricular myocytes exposed to simulated ischemia and reperfusion. PLoS ONE 7:e38425
35. Liu H, Pedram A, Kim JK (2011) Oestrogen prevents cardiomyocyte apoptosis by suppressing p38α-mediated activation of p53 and by down-regulating p53 inhibition on p38β. Cardiovasc Res 89:119–128
36. Chen J, Petranka J, Yamamura K et al (2003) Gender differences in sarcoplasmic reticulum calcium loading after isoproterenol. 285: H2657–H2662
37. Sugishita K, Su Z, Li F et al (2001) Gender influences $[Ca]_i$ during metabolic inhibition in myocytes overexpressing the Na-Ca exchanger. Circulation 104:2101–2106
38. Tsang S, Wu S, Liu J, Wong TM (2008) Testosterone protects rat hearts against ischaemic insults by enhancing the effects of $alpha_1$-adrenoceptor stimulation. Br J Pharmacol 153:693–709
39. Davis SR, Wahlin-Jacobsen S (2015) Testosterone in women–the clinical significance. Lancet Diabetes Endocrinol 3:980–992
40. Kloner RA, Carson C 3rd, Dobs A et al (2016) Testosterone and cardiovascular disease. J Am Coll Cardiol 67:545–557
41. Ali DC, Naveed M, Gordon A et al (2019) β-Adrenergic receptor, an essential target in cardiovascular diseases. Heart Fail Rev. Accessed 13 Aug 2019 [Epub ahead of print]
42. O'Brien JD, Ferguson JH, Howlett SE (2008) Effects of ischemia and reperfusion on isolated ventricular myocytes from young adult and aged Fischer 344 rat hearts. Am J Physiol Heart Circ Physiol 294:H2174–H2183
43. Wang W, Zhang H, Xue G et al (2014) Exercise training preserves ischemic preconditioning in aged rat hearts by restoring the myocardial polyamine pool. Oxid Med Cell Longev 2014:457429
44. Zhu J, Rebecchi MJ, Glass PS et al (2011) Cardioprotection of the aged rat heart by GSK-3beta inhibitor is attenuated: age-related changes in mitochondrial permeability transition pore modulation. Am J Physiol Heart Circ Physiol 300:H922–H930
45. Sze SC, Tong Y, Zhang YB et al (2009) A novel mechanism: erxian decoction, a Chinese medicine formula, for relieving menopausal syndrome. J Ethnopharmacol 123:27–33

46. Zhao W, Wen HX, Zheng HL et al (2011) Action mechanism of Zuo Gui Yin decoction's promotion on estradiol production in rats during the peri-menopausal period. J Ethnopharmacol 134:122–129
47. Dhalla NS, Rangi S, Babick AP et al (2012) Cardiac remodeling and subcellular defects in heart failure due to myocardial infarction and aging. Heart Fail Rev 17:671–681
48. Kulkarni K, Merchant FM, Kassab MB et al (2019) Cardiac alternans: mechanisms and clinical utility in arrhythmia prevention. J Am Heart Assoc 8:e013750
49. Randhawa PK, Bali A, Virdi JK, Jaggi AS (2018) Conditioning-induced cardioprotection: aging as a confounding factor. Korean J Physiol Pharmacol. 22:467–479
50. O'Brien JD, Howlett SE (2008) Simulated ischemia-induced preconditioning of isolated ventricular myocytes from young adult and aged Fischer-344 rat hearts. Am J Physiol Heart Circ Physiol 295:H768-777
51. Morciano G, Bonora M, Campo G et al (2017) Mechanistic role of mPTP in ischemia-reperfusion injury. Adv Exp Med Biol 982:169–189

Chapter 5
Non-atherosclerotic Acute Cardiac Events in Young Women

Fabiola B. Sozzi, Marco Schiavone, and Marta Belmonte

Abstract Acute coronary syndromes (ACS) represent a major cause of morbidity and mortality worldwide and are mostly related to coronary artery disease (CAD). Acute myocardial infarction (AMI) has historically been regarded as a men's disease and, for many years, women have been underdiagnosed and mistreated, especially at a young age. Relevant disparities still exist in the diagnosis, management and outcomes of ACS between men and women, especially during the pre-menopausal age. Moreover, although CAD is uncommon among young patients due to the low prevalence of atherosclerosis related risk factors, ACS may have destructive effects on their lifestyle and prognosis. Here we describe three frequent causes of ACS and AMI or conditions mimicking it, with a specific focus on women with absent or non-critical CAD: spontaneous coronary artery dissection (SCAD), AMI with non-obstructive coronary arteries (MINOCA) and Takotsubo cardiomyopathy (TTS). These conditions are rare but potentially life-threatening and are frequently underdiagnosed or misdiagnosed, since presenting symptoms and signs are atypical and not always overt and since most affected subjects have always been considered at low risk of having ACS. Due to difficulties in the diagnostic process and their potential devastating consequences, it is important to raise awareness on the prevalence and on clinical features of ACS in this class of patients.

Keywords Young women · Spontaneous coronary artery dissection (SCAD) · Microvascular obstruction · Takotsubo syndrome (TTS) · Myocardial infarction with non-obstructive coronary arteries (MINOCA)

Introduction

Acute myocardial infarction (AMI) has historically been regarded as a men's disease, and for many years, women have been underdiagnosed and undertreated. Relevant disparities still exist in the diagnosis, management and outcomes of acute coronary

F. B. Sozzi (✉) · M. Schiavone · M. Belmonte
Cardiology Unit, Ospedale Maggiore IRCCS Policlinico of Milan, Milan, Italy
e-mail: fabiola_sozzi@yahoo.it

B. Ostadal and N. S. Dhalla (eds.), *Sex Differences in Heart Disease*, Advances in Biochemistry in Health and Disease 21,
https://doi.org/10.1007/978-3-030-58677-5_5

syndromes (ACS) between women and men. A reason is related to less frequent medical investigations in symptomatic women, especially during the pre-menopausal age, with subsequent undertreatment of suspected cardiac events [1]. Recognizing and managing coronary artery diseases (CAD) and related acute ischemic events in women is crucial since, as a consequence of contemporary lifestyles and increasing prevalence of obesity, the number of ACS has increased in specific populations not commonly considered at high risk, such as young women. Both physicians and young people need to be aware of cardiovascular risk factors, as well as signs and symptoms linked to CAD, in order to avoid delays in presentation times to the emergency department and to optimize patient care.

Important differences between women and men have to be underlined in the pathophysiology of AMI, since they lead to different treatment strategies. According to the current classification of AMI, type 1 myocardial infarction (MI) is related to atherothrombotic CAD and usually is precipitated by atherosclerotic plaque disruption (rupture or erosion). As known, it is more frequent in men [2], though it is demonstrated that in the post-menopausal age the incidence difference flattens out. The mismatch between oxygen supply and demand leading to AMI has been classified as type 2 MI [3]. Many different factors may contribute to imbalance the ratio between myocardial oxygen supply and demand. Most of them are intrinsic coronary mechanism like: diminished myocardial perfusion due to coronary atherosclerosis without plaque rupture and thrombosis, microvascular dysfunction, spontaneous coronary artery dissection (SCAD) with or without intramural hematoma, coronary embolism, vasospastic angina. Other factors that may contribute or cause type 2 MI are: arrhythmias (severe bradyarrhythmias or tachyarrhythmias that may reduce blood flow or augment myocardial oxygen demand), respiratory failure (predominantly hypoxemia), severe anemia, acute heart failure with hypotension or shock [4]. Most studies showed a higher frequency of type 2 MI in women [5–7]. SCAD and myocardial infarction with non-obstructive coronary arteries (MINOCA) may be considered as specific subsets of type 2 MI in which gender differences are extremely relevant. Type 3 MI instead, is diagnosed when a patient with a typical presentation of acute myocardial event (including presumed new ischemic electrocardiographic changes or ventricular fibrillation) dies before it is achievable to sample cardiac biomarkers, or if the patient might have died soon after the onset of symptoms before elevation of biomarker values has occurred [8]. The incidence of type 3 AMI is variable, but a study showed an annual incidence below 10/100.000 person-years and a frequency of 3–4% among all types of MI (the percentage of male is around 66.6%) [9]. Type 4 AMI consists of 3 different subsets [8]: type 4a, coronary intervention-related; type 4b, stent/scaffold thrombosis related, as documented by angiography or autopsy; type 4c in-stent restenosis or restenosis following balloon angioplasty in the infarct territory related. Type 5 AMI is associated with coronary artery bypass grafting [8]. So far, no specific studies have been addressed to assess relevant gender-related difference in type 4 and type 5 AMI.

Based on the Global Registry of Acute Coronary Events (GRACE) [10], reviews [11, 12] and validation studies [13], the last universal definition of AMI published in 2018 states that significantly lower values of high-sensitivity cardiac troponin are observed among women compared with men, with the 99th centile in men being twice that in women [14, 15]. Therefore, sex-specific 99th percentile upper reference limits (URLs) are recommended for troponin assays [8]. Using sex-specific cut-off values have been reported to improve diagnostic and prognostic information in patients with suspected ACS, even if there is no definitive evidence [8, 10–13, 16–18]. One of the most recent study revealed that using sex-specific thresholds identifies 5-times more women than men with myocardial injury. Nevertheless, the women received approximately one-half the number of treatments for CAD as men, and outcomes were not improved [19]. In the clinical practice, the role of sex-specific cut-off values is underused and it is still controversial if this approach provides valuable additional diagnostic information.

Different electrocardiographic cut-off points are found for women in diagnosing ST-elevation MI (STEMI), since J-point in healthy women in leads V2 and V3 might be less elevated than in men, so that, to diagnose STEMI, new ST-elevation at the J-point in two contiguous leads requires these cut-points: ≥1 mm in all leads other than leads V2–V3 where elevation ≥2 mm in men ≥40 year, elevation ≥2.5 mm in men <40 years, or elevation ≥1.5 mm in women regardless of age is needed [8].

The most frequent clinical scenarios in women with suspected or diagnosed AMI secondary to non-obstructive CAD are represented by (1) spontaneous coronary artery dissection (SCAD), (2) AMI with non-obstructive coronary arteries (MINOCA), and (3) Takotsubo syndrome (TTS).

Spontaneous Coronary Artery Dissection (SCAD)

Introduction

Spontaneous coronary artery dissection (SCAD) has emerged as a relevant cause of acute coronary syndrome (ACS) and sudden cardiac death, especially in women.

The first case was described in a 42-years old woman by Pretty in 1931 at autopsy [20] and from then on, especially in the last decades, different case series has been reported during pregnancy, in the postpartum period and not related to pregnancy [21–23].

Moreover, position papers and scientific statements [24, 25] from the main American and European Societies of Cardiology have defined SCAD-associated conditions, diagnosis, management and treatment of SCAD. Advances in diagnostic techniques, such as high sensitivity biomarkers assays (e.g. high sensitivity troponin) and intravascular imaging (especially optical coherence tomography—OCT), increased knowledge about SCAD and allow easier diagnosis of this condition in acute settings.

Definition

SCAD is defined as a spontaneous (neither iatrogenic nor related to trauma) and usually abrupt formation of a tear along the inside wall of an artery, leading to a development of a false lumen within the coronary artery wall which might compress the true lumen and compromise coronary flow.

Contemporary usage of the term SCAD should be reserved for the non-atherosclerotic variant and most modern series exclude SCAD due to atherosclerotic CAD, when SCAD is a consequence of a penetrating ulcer or plaque rupture secondary to atherosclerotic disease or primary aortic dissection [26]. Therefore, it represents the result of a false lumen development, generally in the outer third of the tunica media (Fig. 5.1) [27–30]. Two different possible mechanisms have been proposed: (1) the "inside-out" model: a casual development of an intimal tear, allowing blood to cross the internal elastic lamina and accumulate in the media; (2) the "outside-in" model: a causal disruption of a vasa vasorum micro-vessel leading to hemorrhage directly into the tunica media [31, 32]. In both cases the true lumen is compressed by blood that propagates when the false lumen extends. It is still unclear if there is a single dominant mechanism in SCAD or if both causal events are possible, but recent OCT-based studies have shown cases in which there is no communication between false and true lumens, suggesting that "outside-in" mechanism is more likely, in at least a proportion of cases [33, 34]. A large number of histopathological reports have highlighted the presence of periadventitial mixed inflammatory infiltrate, frequently with a predominance of eosinophils [35–37]. However, the pathophysiological significance of this eosinophilic infiltrate is controversial, so that some authors regard it as pathognomonic for SCAD [35, 37] while others as a nonspecific response to vascular injury [27].

Epidemiology

The true prevalence of SCAD remains uncertain, primarily because it is an underdiagnosed condition. Since SCAD frequently affects young women (about 90% of SCAD patients are women) [25], typically with few or no traditional cardiovascular risk factors [38], diagnostic suspicion is often low, even in the presence of classic symptoms. Limitations of current coronary angiographic techniques and lack of physician familiarity with the condition contributes to misdiagnose SCAD [24]. Current angiographic series report SCAD diagnosis rates of 0.07–0.2% of all angiograms and 2–4% of angiograms performed for ACS [21, 39, 40], even if the study with the highest ACS rate did not exclude all atherosclerotic SCAD.

In young women SCAD is considered one of the main causes of ACS. The frequency of SCAD varies in the different case series from 36% of women <60 years with ACS and one or fewer conventional cardiovascular risk factors in a French series [41] to 24% of women <50 years with AMI in a Canadian series [21] and to 23% of

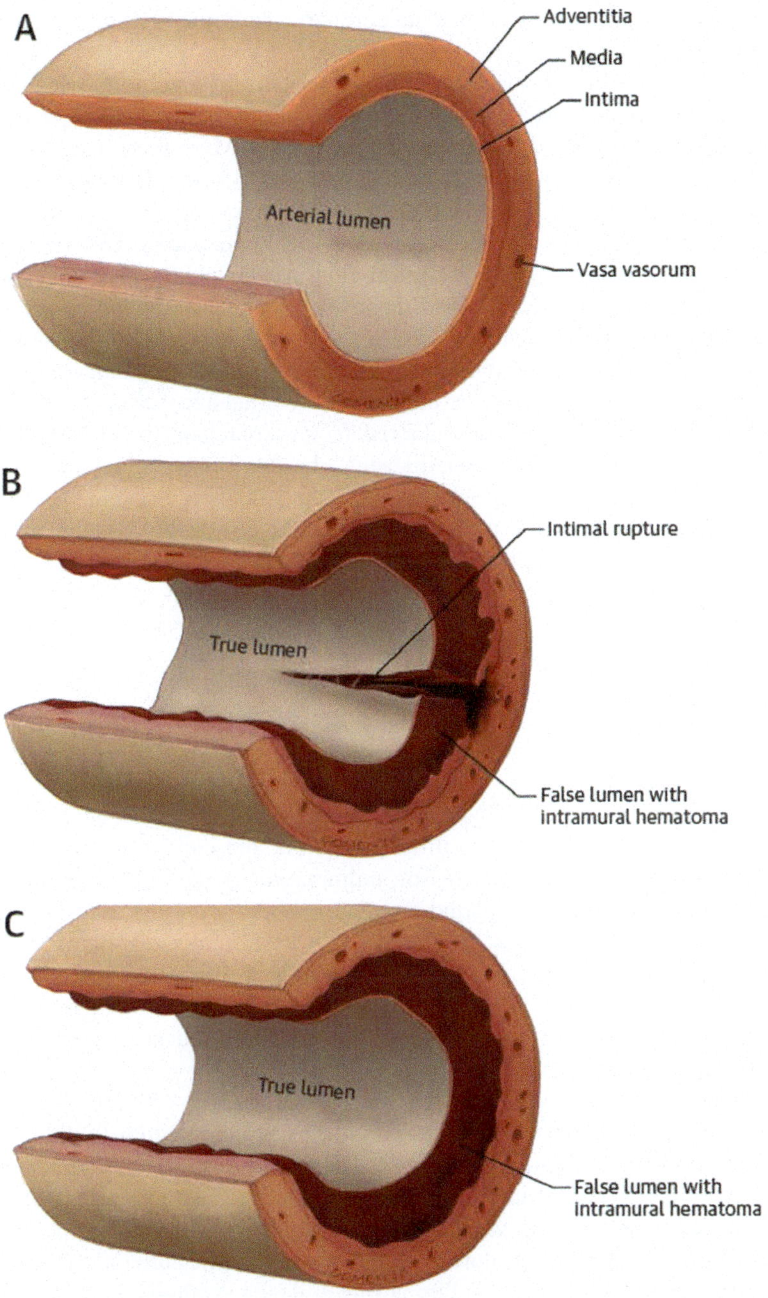

Fig. 5.1 Mechanisms of spontaneous coronary artery dissection (SCAD). **A** Normal coronary artery. **B** Intimal rupture initiating tear, with intramural hematoma formation. **C** Spontaneous bleeding into the arterial wall, creating a false lumen filled with intramural hematoma. Figure 5.1 from: Saw J, Mancini GBJ, Humphries KH. Contemporary Review on Spontaneous Coronary Artery Dissection. J Am Coll Cardiol. 2016; 68(3):297–312. Reproduced with permission

women <60 years presenting with acute coronary syndrome in a smaller Australian series [42].

SCAD has not to be considered primarily a peripartum condition since pregnancy and peripartum cases (P-SCAD) account for a minority of these cases (around 10% in most recent series) [42–44]. However, 21–27% of AMI in pregnancy and 50% of post-partum coronary events are likely due to SCAD [43–45]. The coronary disease in pregnancy generally has non-atherosclerotic features: pregnancy related-SCAD accounts for 43% of cases of acute coronary events in pregnancy, 18% of cases have normal coronary arteries and coronary thrombosis accounts for 17% of cases (often in the absence of atherosclerosis and most likely due to the hypercoagulability of pregnancy or from paradoxical embolization) [43, 46]. As the birth rate in women > 40 years increases, acute coronary syndrome complicating pregnancy will become more common, since for every year increase in maternal age there is a 20% increase in AMI risk [45]. Classic cardiovascular risk factors are related to CAD in pregnancy, such as smoking, age, hypertension, diabetes, obesity, and dyslipidemia [43, 47–50] but also additional risk factors such as pre-eclampsia, thrombophilia, transfusions, infections, multiparity, and post-partum hemorrhage have to be considered [48, 51].

SCAD has been described in patients aged from 18 to 84 years [21, 52] with the mean diagnostic age in most recent case series ranging from 44 to 53 years [25, 52–54].

Etiology and Pathophysiology

The underlying etiology of SCAD appears to be multifactorial. There is a predisposing arteriopathy, associated to different factors and conditions, which may be deteriorated by a precipitating stressor, culminating in the clinical expression of SCAD. The most commonly reported association is fibromuscular dysplasia, ranging from 25 to 86% in some cohort studies [52, 55, 56]. The association of SCAD and fibromuscular dysplasia was first described in 2005 and subsequently confirmed in several case series [57, 58]. It is a non-atherosclerotic, non-inflammatory vascular disease that can manifest as arterial stenosis, aneurysm, tortuosity, or dissection and that can affect coronary and extracoronary arteries.

Pregnancy related SCAD (P-SCAD) accounts for about 43% of cases of CAD in pregnancy. P-SCAD can occur antepartum, early post-partum (within 6 weeks of delivery), late post-partum (6 weeks to 12 months), and very late post-partum (12–24 months) [23]. The nature of this association needs to be clarified, but it is related to hormonal role on vascular connective tissue and on the microvasculature. High progesterone levels during pregnancy can weaken arterial media through their action on the elastic fiber and impairing collagen synthesis. Estrogen also can contribute to SCAD creating a hypercoagulable state. A weakened arterial wall plus a prothrombotic state are considered as major factors to increase the risk of SCAD and thrombosis [26, 59]. Hemodynamic changes during pregnancy can also set off P-SCAD. The increased cardiac output, the circulatory overload and the increasing in intra-abdominal pressures during pregnancy and labor can raise shear stress. Especially when an underlying vascular disease is present (e.g. fibromuscular dysplasia, connective tissues disorders, systemic lupus erythematosus), the established hemodynamic condition can trigger P-SCAD. It also has been postulated that multiparity might be a risk factor for SCAD, since the arterial wall integrity may be damaged due to the repeated exposures to hormonal changes [23, 26]. Breastfeeding may likewise be associated with late and very late postpartum P-SCAD, showing potential similarities in hormonal changes during pregnancy and lactation [26, 60]. Prolonged therapies with estrogen or progesterone are likely to undermine arterial architecture in the same way [26]. Indeed, 51 out of 187 (27.3%) SCAD patients were on hormone replacement therapy in a cohort described by Eng et al., and these patients had higher rate of recurrent SCAD at follow-up [61]. Furthermore, several connective tissues disorders, inherited arteriopathies and systemic inflammatory diseases are potentially associated with SCAD. Associated conditions and precipitating factors related to SCAD are summarized in Table 5.1.

The link between systemic inflammation and SCAD is suspected to be due to chronic inflammation from vasculitis [62], while in connective tissue diseases there is a weakness of the arterial wall that may lead to SCAD. Precipitating stressors may increase thoraco-abdominal pressure, cardiocirculatory shear stress and catecholamine release (increasing myocardial contractility or vasospasm, which can lead to intimal rupture or disruption of the vasa vasorum) that may trigger SCAD, especially in patients with underlying predisposing arteriopathies [26]. These factors include labor, delivery, Valsalva-like activities, intense emotional stress, physical activities (especially isometric) and use of sympathomimetic drugs [52].

Clinical Presentation

There is evidence that SCAD is still underdiagnosed [1, 63]. Some patients never present to hospital or cardiology services and sometimes the only manifestation of SCAD may be sudden cardiac death. Among the ones referred to hospitals, many patients are not referred for coronary investigations, primarily for their age, sex and because they do not present classical features of patients at high risk of obstructive atherosclerotic ACS and they are regarded having a low risk for ACS on the basis of traditional risk scores for ischemic heart disease [25].

Chest pain is the most frequently presenting symptom, ranging from 60 to 90% of SCAD patients, resulting even more frequent than in patients with atherosclerotic AMI. A possible explanation is that, besides pain generated from myocardial ischemia, dissection itself may be painful. In a Canadian series described by Luong et al., chest pain was associated with radiation to the arm in 49.5% of cases and to the neck in 22.1% of cases; nausea and vomiting was present in 23.4% of patients, diaphoresis in 20.9%, dyspnea in 19.3% and back pain in 12.2% [64]. In this study, most patients presented with typical chest pain but burning (9%), pleuritic (3.0%), tearing (1.0%), and positional (1%) pain was described by a minority of patients [64].

Table 5.1 Associated conditions and precipitating factors related to SCAD—adapted from Saw et al. [26]

Associated conditions or factors	Precipitating factors
Fibromuscular dysplasia (FMD)	Labor and delivery
Pregnancy related SCAD – antepartum – early post-partum – late postpartum – very late post-partum – recurrent pregnancies	Intense exercise – isometric – aerobic
Connective tissue disorders and inherited arteriopathies – Marfan syndrome – Loeys-Dietz syndrome – Ehler-Danlos syndrome – Cystic medial necrosis – Alpha-1-antitrypsin deficiency – Polycystic kidney disease	Intense Valsalva – Retching – Vomiting – Coughing – Bowel movement – Lifting heavy objects
Systemic inflammatory diseases: – Autoimmune diseases (systemic lupus erythematosus, rheumatoid arthritis, celiac disease) – Inflammatory bowel diseases (Crohn's disease, ulcerative colitis) – Sarcoidosis – Vasculitis (Polyarteritis nodosa, Churg-Strauss syndrome, Wegener's granulomatosis, Kawasaki disease, giant cell arteritis)	Intense emotional stress
Hormone therapy – Corticosteroids – Oral contraceptive – Estrogen, progesteron, β-chorionic gonadotropin and testosterone	Hormone therapy – β-chorionic gonadotropin injections – Corticosteroid injections
Coronary artery spasm	Drugs – Cocaine – Methamphetamines

In a low percentage of cases, SCAD presentation was with ventricular arrhythmia (2.8–10%) [52–54, 65], cardiogenic shock (<3%) [52–54] or sudden cardiac death (<1%), although this presentation may be underestimated [28]. The typical clinical presentation of patients with SCAD is ACS with elevation of cardiac markers (e.g. cardiac troponin) in about 90–100% of cases, with different percentages in STEMI (26.1–87%) versus non-STEMI (NSTEMI) (13–65.4%) presentations, as reported in the larger published series [52–54, 65].

Diagnosis

Misdiagnosis patients with SCAD is not unusual, since clinical features of SCAD do not resemble the classic phenotype of patients with atherosclerotic CAD, mainly due to their young age and the absence of classical cardiovascular risk factors [24].Correct diagnosis of SCAD in the early stages of ACS presentation is crucial to avoid wrong management and early discharge of these patients. SCAD should be suspected on the basis of young age, female sex, and few or no conventional cardiovascular risk factors.

Once SCAD is suspected, coronary angiography represents the principal tool for the diagnosis in clinical practice [63, 66]. Most SCAD cases are diagnosed only on angiographic basis and intracoronary imaging is reserved for cases where diagnosis is uncertain [66], such as when coronary arteries appear normal, without clear evidence of obstruction to flow, or when the findings resemble atherosclerosis.

The most commonly used angiographic classification of SCAD has been adapted from Saw et al. [67, 68]. From the Saw et. al classification: type 1 is a radiolucent flap and linear double lumen, associated with contrast staining (Fig. 5.2A). This pattern is present in 29– 48% of cases [52, 65, 69]. Type 2 pattern (52–67%) [52, 65, 69] is a long (>20 mm) diffuse stenosis typically located in mid or distal segments. This pattern has been divided into Type 2a (Fig. 5.2B) where there is resumption of a normal caliber distal vessel and Type 2b (Fig. 5.2C) where the stenosis extends to the end of the vessel. Type 3 lesions (Fig. 5.2D) are indistinguishable from a focal atherosclerotic stenosis [65, 67–69]. They require diagnostic confirmation by intracoronary imaging, using intravascular ultrasound (IVUS) or optical computed tomography (OCT). These account for a small minority of cases (0–3.9%) [52, 65, 69]. IVUS and OCT techniques are able to distinguish atherosclerotic plaques from SCAD, demonstrating true and false coronary lumens. The principal advantage of IVUS over OCT is that blood clearance and subsequent contrast injection is not required and then that it allows complete visualization of the vessel wall to the external elastic lamina. The main advantage of OCT is the greater spatial resolution which can help characterize small structures associated with SCAD such as the intimal-medial membrane and localized fenestrations connecting true and false lumens. Type 4 lesions are described as a total occlusion, usually of a distal vessel (Fig. 5.2E) [66]. The diagnosis is particularly challenging and frequently they are established once coronary flow is re-established or after subsequent vessel healing and after the exclusion of an embolic cause. Intermediate type 1/2 SCAD appearance has also been described (Fig. 5.2F) [25].

Management

No-randomized trial has compared medical therapies versus revascularization strategies (percutaneous coronary intervention or coronary artery bypass grafting), so the

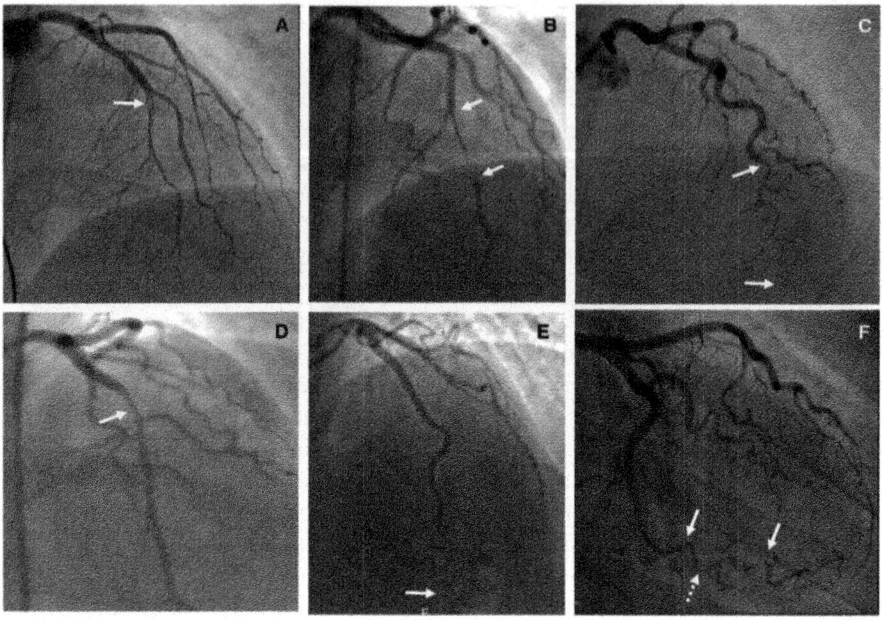

Fig. 5.2 Angiographic classification of spontaneous coronary artery dissection. **A** Type 1 spontaneous coronary artery dissection (SCAD) of distal left anterior descending (LAD) artery with staining of artery wall (asterisk). **B** Type 2A SCAD of mid-distal LAD (between arrows). **C** Type 2B SCAD of diagonal branch (asterisk), which healed 1 year later (asterisk in **D**). **E** Type 3 SCAD of mid-circumflex artery (asterisk), with corresponding optical coherence tomography showing intramural hematoma in **F**. Figure 5.2 from: Saw J, Mancini GBJ, Humphries KH. Contemporary Review on Spontaneous Coronary Artery Dissection. J Am Coll Cardiol. 2016; 68(3):297–312. Reproduced with permission

optimal management of SCAD is still undetermined. Current recommendations on management of SCAD, based on expert opinions and endorsed by the main American and European societies [24, 25], state that the majority of SCAD will first stabilize and then heal completely over time if managed conservatively. Moreover, it has to be remarked that revascularization in patients with SCAD is very challenging due to the presence of underlying vessel frailty and that it is associated with high failure rates. It is reported that revascularization lead to worse outcomes for coronary angioplasty than in atherosclerotic CAD [52–66, 69]. When there is no need for urgent revascularization (e.g. in hemodynamically stable patients), a conservative strategy is generally preferred [24–26, 63, 66]. When revascularization is mandatory due to hemodynamic conditions or when it is taken into account due to high risk features, such as ongoing or recurrent ischemia/chest pain or left main dissection, specific additional risks associated with SCAD interventions (e.g. increased risk of iatrogenic catheter-induced coronary dissection, difficulty in advancing coronary wire into distal lumen or guidewire passage into the false lumen or false lumen propagation during stent deployment) have to be considered [25, 26, 39, 53, 70, 71].

Revascularization with coronary artery bypass should be considered for patients with prohibitive risk for percutaneous angioplasty (dissection involving the left main or extensive dissections involving proximal arteries), or in patients in whom coronary angioplasty failed or who are not anatomically suitable for it, as a bail-out strategy [25, 26]. Limited data on coronary artery bypass are reported in literature. Short term data are good, but a Mayo Clinic series reported high rates of graft failure at follow-up, maybe due to healing of the native coronary leading to competitive flow and conduit thrombosis [25, 53].

Medical therapy consists in beta-blockers, that reduce coronary arterial wall stress and are usually administered acutely and long-term after SCAD [26], while the use of antiplatelet therapies and the duration of treatment remains controversial in clinical practice (aspirin seems reasonable in the acute setting of MI, reducing false lumen thrombus burden, and for the long-term SCAD management) [24–26].

Outcomes and Follow-Up

Acute in-hospital mortality was <5% in modern series, and need for urgent revascularization in conservatively managed patients, in-hospital recurrent AMI or other major adverse cardiac events were 5–10% [38, 52, 54].

Considering the association between SCAD, fibromuscular dysplasia and other extra-coronary vascular abnormalities, affecting multiple vascular territories and the implications for patient management and follow-up, imaging of extra-coronary vascular beds in patients with SCAD is advised [72–74].

At discharge, cardiac rehabilitation should be highly encouraged, especially with SCAD-patients dedicated programs.

Long term mortality of SCAD is low: 92% and 94.4% of 10 and 6 years-survival rates has been reported in a Mayo Clinic series and in an Italian series respectively [38, 54]; 1.2% mortality has been described in a Canadian series (median follow-up 3.1 years) [75], while a Japanese series and a Swiss series reported one death and no deaths among 63 patients each (median follow-up 34 months and 4.5 years respectively) [65, 69]. SCAD recurrences have been extensively described [26, 38, 52–54, 65, 69]: they often affects new territories and stenting does not offer protection. No medical therapy has been shown to reduce rates of recurrence, apart from the potential benefit of beta-blockers and control of hypertension. Among women of childbearing age, further pregnancies should also be avoided considering the risk of recurrent SCAD [25].

Following SCAD, as in the post-AMI setting for other causes, an assessment of left ventricular systolic function is mandatory to optimize medical treatment (e.g. with ACE-inhibitors and mineralocorticoid receptor antagonist) and evaluate the benefit of device therapy [76, 77].

Myocardial Infarction with Nonobstructive Coronary Arteries (Minoca)

Last universal definition of MI introduced the concept of MI with non-obstructive coronary arteries (MINOCA) [8].

The prevalence of MINOCA is estimated to be around 6–8% among patients diagnosed with MI (especially in patients with NSTEMI) and it is more common in women than men (10.5% vs. 3.4%) [78]. Women represent almost 50% of the MINOCA population, suggesting that the hormonal influences may play a role as it is in SCAD population, with or without a clear coronary flow obstruction. On the other hand, women represent only 25% of the population with AMI due to CAD [78]. Women with MINOCA show higher mortality than men overall (3.6% versus 2.4%) [79].

As the definition of MINOCA implies, diagnosis requires the presence of criteria of AMI (according to the current 4th definition of AMI [8]), the evidence of non-obstructive coronary arteries following coronary angiography (no lesions ≥50% in a major epicardial vessel) and the absence of other overt specific cause that can explain otherwise the acute clinical presentation (such as myocarditis). The angiographic criteria for non-obstructive CAD in the MINOCA definition consider the conventional cut-off of 50% stenosis, which is consistent with contemporary angiographic guidelines [80–82].

The finding of non-critical coronary stenosis does not exclude a role of thrombotic disease in MINOCA, so the most used approach worldwide is to define MINOCA on the basis of the absence of a potentially obstructive stenosis on coronary angiography rather than on the presence of any coronary atherosclerosis. Atherosclerotic plaque disruption with thrombosis and/or thromboembolism may be a cause of MINOCA (type 1 AMI, 5–20% of all type-1 AMI cases), likely followed by spontaneous thrombolysis or autolysis of a coronary thrombosis [80]. However, coronary spasm (vasospastic angina), coronary thromboembolism (in the setting of hereditary/acquired thrombotic disorders or of coronary/systemic arterial thrombosis), coronary microvascular dysfunction (even if its role is not conclusively related to an acute presentation) and SCAD, may be involved as well in the pathogenesis of MINOCA (type 2 MI) [8, 80, 81].

In the clinical setting of SCAD, accurate review of coronary angiogram has to be made: despite most patients with SCAD have some obstruction to coronary flow (so that a diagnosis of MINOCA is excluded), occasionally arteries can appear normal or near normal and these cases should be classified as MINOCA [81]. With the increasing use of intracoronary imaging in assessing SCAD, all cases that are not regarded as "obstructive", will increasingly be recognized as a relevant cause of MI related to SCAD [81].

Other forms of type-2 AMI may be related to MINOCA, when myocardial damage and infarction are due to supply–demand imbalance in different conditions such as arrhythmia, anemia, respiratory failure, shock (e.g. hypotensive shock or septic shock), hypertensive crisis, severe aortic valve disease, acute heart failure or end-stage cardiomyopathy [4]. It has to be underlined that all these triggers may also precipitate underlying obstructive CAD.

Frequently, imaging and functional tests are useful to understand etiology and pathophysiology of MINOCA. Echocardiography should be performed in the acute setting to assess wall motion and left ventricular ejection fraction and to diagnose specific clinical entity, such as Takotsubo syndrome (TTS). Cardiac magnetic resonance imaging is the key diagnostic tool to be performed in MINOCA patients since it can provide confirmation of AMI and exclude TTS, myocarditis and cardiomyopathies. Late gadolinium enhancement allows localizing the area of myocardial damage and may suggest the mechanism of injury: ischemic if subendocardic (although the late gadolinium enhancement pattern does not identify the cause at the base of the ischemia), non-ischemic if subepicardic or absent. Intracoronary imaging with intravascular ultrasound or optical coherence tomography provide clues as to how to understand pathophysiology of MINOCA, in order to identify atherosclerotic plaque disruption, plaque erosion, coronary dissection or thrombosis, which may not have been clear during coronary angiography.

Characterization of MINOCA was essential to clarify a so-called "epidemiological paradox" described by Smilowitz et al. [79]: despite young women with AMI present without atherosclerotic CAD at coronary angiography more often than young men and although MINOCA have a better prognosis than type 1 MI, in-hospital mortality rate related to MI is higher in females than in men in the same age groups [2, 83–86]. Indeed, the ACTION Registry–GWTG (Acute Coronary Treatment and Intervention Outcomes Network Registry–Get With the Guidelines) [79] clarified that women have a higher risk of death after type 1 MI, highlighting that women with CAD are more vulnerable than men, while MINOCA can still be considered as a lower-risk MI type. This corroborates the theory that women, especially at a young age, take more advantage from a thorough assessment in case of suspected ACS in order to improve outcomes and avoid misdiagnosis.

Takotsubo Syndrome (TTS)

Takotsubo syndrome (TTS), also known as stress cardiomyopathy, is an acute cardiac condition that clinically mimics acute coronary syndrome (chest pain, heart failure, electrocardiographic changes, biomarkers), with reversible ventricular motion abnormalities not related to epicardial coronary obstruction. Left ventricular dysfunction generally extends beyond the territory subtended by a single coronary artery and recovers within weeks [87, 88].

Based on "International Expert Consensus Document on Takotsubo Syndrome", there are different diagnostic criteria (InterTAK Diagnostic Criteria) that can be summarized as follows [89]:

(1) Transient left ventricular dysfunction (hypokinesia, akinesia, or dyskinesia) presenting as apical ballooning or midventricular, basal, or focal wall motion abnormalities, with or without right ventricular involvement (usually, regional wall motion abnormality in TTS do not have coronary distribution, even if epicardial vascular distribution pattern may be present as a focal syndrome).
(2) Emotional, physical or combined factors as triggers.
(3) Neurologic diseases (e.g. subarachnoid hemorrhage, seizures or stroke/transient ischemic attack) or pheochromocytoma as triggers.
(4) New ECG abnormalities: ST segment-elevation or depression, T-wave inversion, QTc prolongation (rarely, TTS can present without ECG changes)
(5) Cardiac biomarkers elevation (preferably hs-Troponin or creatine kinase) and significant elevation of brain natriuretic peptide.
(6) Presence of significant atherosclerotic CAD at coronary angiography does not exclude TTS.
(7) No evidence of acute myocarditis.
(8) Occurrence in postmenopausal women.

Takotsubo syndrome may occur in up to 2% of patients presenting with suspected ACS and in about 1–2% of patients presenting with STEMI. Out-of-hospital cardiac arrest is also reported at presentation, especially in men [90]. If only women are considered, up to 10% of ACS patients are ultimately diagnosed as having TTS [90]. About 90% of patients with TTS in the NIS cohort (the largest reported cohort of affected patients from the Nationwide Inpatient Sample in the USA) were aged ≥50 years and 90% were women. Recurrence rate of this syndrome is estimated to be 1.8% per-patient year [90, 91].

Both emotional and pathological stressors can trigger and predispose to TTS: as shown by Patel et al., the former, such as chronic psychological disorders, are more common in women, while the latter, such as major non-cardiac diseases or previous procedures, in men [92]. Those findings were confirmed by a German registry that reported that the absence of pathological triggers or emotional stress alone were more common in women [93]. Acute precipitating factors, such as thyrotoxicosis, pheochromocytoma and subarachnoid hemorrhage, may increase cortisol and catecholamines levels by activating sympathetic nervous system, and may therefore trigger TTS [94]. It is postulated that those factors can induce myocardial damage during an adrenergic storm with several mechanisms, including direct or adrenoreceptor-mediated toxicity, epicardial and microvascular dysfunction and increased cardiac workload [95]. Notably, endomyocardial biopsies have shown apoptosis of microvascular endothelial cells that might highlight the role of microcirculatory vasoconstriction related to sympathetic stimulation [96]. Although little is known about gender differences in this topic, evolutionary theories suggest that men may be more protected from acute and massive catecholamine release due to a larger expression of adrenergic receptors in cardiomyocytes that could be associated with a greater exposure to physical stress developed over the centuries [97]. Also hormonal mechanisms may have a role in the gender differences, since TTS mostly occurs in postmenopausal women (about 90% of patients) [88], due to the lower estrogen levels that might expose women to a greater renin–angiotensin–aldosterone system activity (RAAS), sympathetic hyperactivity and subsequent loss of antioxidant and cardioprotective effects [98]. Based on these data and on findings on rats [99], it is supposed that hormone replacement therapy with estrogens may have benefits during menopause, mitigating the excessive adrenergic effects on the cardiovascular system in response to stress, although clinical trial are lacking. Despite TTS is more likely to occur in women, male gender is associated with higher acute mortality rate, in hospital complications (i.e. cardiac arrhythmias) and with worse disease severity (i.e. use of mechanical ventilation or inotropes) [100, 101]. Also long-term follow-up corroborate those findings, as reported by Templin et al., that showed that men have increased rates of death from any cause (12.9% vs. 5.0% per patient-year, $P < 0.001$) and of major adverse cardiac and cerebrovascular events (16.0% vs. 8.7% per patient-year, $P = 0.002$) beyond 30-days.

No randomized clinical trials on the specific treatment of TTS have been performed so far and treatment of TTS remains mostly supportive, especially during the acute phase and if cardiogenic shock, acute heart failure or ventricular arrhythmias occur. In the non-acute phase, β-blockers are supposed to be the most useful therapy for prevention of TTS recurrence, modulating sympathetic nervous system and preventing catecholamine surges, but they have failed to demonstrate beneficial effect on mortality after discharge during 1 year of follow-up [101, 102]. However, RAAS inhibitors such as angiotensin-converting–enzyme inhibitors (ACEi) and angiotensin-receptor blockers (ARBs) have shown to reduce TTS recurrence and improve survival at 1 and 4-year follow-up, due to the inhibition of adrenergic effects and/or having a direct anti-inflammatory effect of cardiomyocytes [101, 102].

Psychological counselling or antianxiety drugs may modulate emotional stressors and be useful in prevent TTS recurrence, especially in young women or in patients affected by neuropsychiatric disorders, in which emotional stressors play a major role in triggering TTS [102].

Conclusions

CAD is an uncommon clinical entity in young female patients but emerging clinical entity such as SCAD, MINOCA and stress cardiomyopathy (TTS) are extremely relevant in current female population.

Women have different risk factor profiles, clinical presentations, and prognoses compared to men, especially in the young age. All risk factors and symptoms should be taken into account in the different clinical scenarios, when physicians assess female patients with suspected AMI, since these conditions are potentially life-threatening when underdiagnosed or misdiagnosed.

When MINOCA is suspected, after the exclusion of other specific causes for symptoms and troponin elevation, the diagnostic process should be focused on recognizing specific causes, so that tailored therapies can be used.

Despite progresses in the characterization of SCAD and TTS, pathophysiology of these conditions remains poorly understood. Although it is known that there is a significant recurrence risk, to date, there is no specific therapy to reduce this risk.

Therefore, it is crucial to enhance awareness of these clinical entities related to AMI or mimicking AMI in women in order to improve the patient management in the acute setting and its prognosis.

Prospective studies and randomized control trials, evaluating optimal diagnostic and management strategies, along with the best medical therapies and the optimal coronary intervention strategy are needed.

References

1. Wenger NK (2012) Women and coronary heart disease: a century after Herrick: understudied, underdiagnosed, and undertreated. Circulation 126(5):604–611
2. Gehrie ER, Reynolds HR, Chen AY et al (2009) Characterization and outcomes of women and men with non-ST-segment elevation myocardial infarction and nonobstructive coronary artery disease: results from the Can Rapid Risk stratification of unstable angina Patients Suppress Adverse Outcomes with Early Implementation of the ACC/AHA Guidelines (CRUSADE) quality improvement initiative. Am Heart J 158(4):688–694
3. Saaby L, Poulsen TS, Hosbond S et al (2013) Classification of myocardial infarction: frequency and features of type 2 myocardial infarction. Am J Med 126(9):789–797
4. Januzzi JL, Sandoval Y (2017) The many faces of type 2 myocardial Infarction. J Am Coll Cardiol 70(13):1569–1572
5. Gupta S, Vaidya SR, Arora S et al (2017) Type 2 versus type 1 myocardial infarction: a comparison of clinical characteristics and outcomes with a meta-analysis of observational studies. Cardiovasc Diagn Ther 7(4):348–358
6. Safdar B, Spatz ES, Dreyer RP et al (2018) Presentation, clinical profile, and prognosis of young patients with myocardial infarction with nonobstructive coronary arteries (MINOCA): results from the VIRGO Study. J Am Heart Assoc 7(13)
7. Baron T, Hambraeus K, Sundström J et al (2015) Type 2 myocardial infarction in clinical practice. Heart 101(2):101–106
8. Thygesen K, Alpert JS, Jaffe AS et al (2019) Fourth universal definition of myocardial infarction (2018). Eur Heart J 40(3):237–269
9. Jangaard N, Sarkisian L, Saaby L et al (2017) Incidence, frequency, and clinical characteristics of type 3 myocardial infarction in clinical practice. Am J Med 130(7):862.e9-862.e14
10. Goodman SG, Steg PG, Eagle KA et al (2006) The diagnostic and prognostic impact of the redefinition of acute myocardial infarction: lessons from the Global Registry of Acute Coronary Events (GRACE). Am Heart J 151(3):654–660
11. Sandoval Y, Apple FS (2014) The global need to define normality: the 99th percentile value of cardiac troponin. Clin Chem 60(3):455–462
12. Apple FS, Sandoval Y, Jaffe AS, Ordonez-llanos J (2017) Cardiac troponin assays: guide to understanding analytical characteristics and their impact on clinical care. Clin Chem 63(1):73–81
13. Giannitsis E, Kurz K, Hallermayer K et al (2010) Analytical validation of a high-sensitivity cardiac troponin T assay. Clin Chem 56(2):254–261
14. Apple FS, Collinson PO (2012) Analytical characteristics of high-sensitivity cardiac troponin assays. Clin Chem 58(1):54–61
15. Kimenai DM, Janssen EBNJ, Eggers KM et al (2018) Sex-specific versus overall clinical decision limits for cardiac Troponin I and T for the diagnosis of acute myocardial infarction: A systematic review. Clin Chem 64(7):1034–1043

16. Shah AS, Griffiths M, Lee KK et al (2015) High sensitivity cardiac troponin and the underdiagnosis of myocardial infarction in women: prospective cohort study. BMJ 350:g7873
17. Eggers KM, Johnston N, James S et al (2014) Cardiac troponin I levels in patients with non-ST-elevation acute coronary syndrome-the importance of gender. Am Heart J 168(3):317-324.e1
18. Romiti GF, Cangemi R, Toriello F et al (2019) Sex-specific cut-offs for high-sensitivity cardiac troponin: Is less more? Cardiovasc Ther 2019:9546931
19. Lee KK, Ferry AV, Anand A et al (2019) Sex-specific thresholds of high-sensitivity troponin in patients with suspected acute coronary syndrome. J Am Coll Cardiol. 74(16):2032–2043
20. Pretty HC (1931) Dissecting aneurysm of coronary artery in a woman aged 42: rupture. BMJ 1:667
21. Saw J, Aymong E, Mancini GB et al (2014) Nonatherosclerotic coronary artery disease in young women. Can J Cardiol 30(7):814–819
22. Ito H, Taylor L, Bowman M et al (2011) Presentation and therapy of spontaneous coronary artery dissection and comparisons of postpartum versus nonpostpartum cases. Am J Cardiol 107(11):1590–1596
23. Vijayaraghavan R, Verma S, Gupta N, Saw J (2014) Pregnancy-related spontaneous coronary artery dissection. Circulation 130(21):1915–1920
24. Hayes SN, Kim ESH, Saw J et al (2018) Spontaneous coronary artery dissection: current state of the science: a scientific statement from the American Heart Association. Circulation 137(19):e523–e557
25. Adlam D, Alfonso F, Maas A, Vrints C (2018) European Society of Cardiology, acute cardiovascular care association, SCAD study group: a position paper on spontaneous coronary artery dissection. Eur Heart J 39(36):3353–3368
26. Saw J, Mancini GBJ, Humphries KH (2016) Contemporary review on spontaneous coronary artery dissection. J Am Coll Cardiol 68(3):297–312
27. Desai S, Sheppard MN (2012) Sudden cardiac death: look closely at the coronaries for spontaneous dissection which can be missed. A study of 9 cases. Am J Forensic Med Pathol 33(1):26–29
28. Hill SF, Sheppard MN (2010) Non-atherosclerotic coronary artery disease associated with sudden cardiac death. Heart 96(14):1119–1125
29. Melez İE, Arslan MN, Melez DO et al (2015) Spontaneous coronary artery dissection: Report of 3 cases and literature review hormonal, autoimmune, morphological factors. Am J Forensic Med Pathol 36(3):188–192
30. Basso C, Morgagni GL, Thiene G (1996) Spontaneous coronary artery dissection: a neglected cause of acute myocardial ischaemia and sudden death. Heart 75(5):451–454
31. Alfonso F, Bastante T, García-guimaraes M et al (2016) Spontaneous coronary artery dissection: new insights into diagnosis and treatment. Coron Artery Dis 27(8):696–706
32. Vrints CJ (2010) Spontaneous coronary artery dissection. Heart 96(10):801–808
33. Alfonso F, Paulo M, Gonzalo N et al (2012) Diagnosis of spontaneous coronary artery dissection by optical coherence tomography. J Am Coll Cardiol 59(12):1073–1079
34. Paulo M, Sandoval J, Lennie V et al (2013) Combined use of OCT and IVUS in spontaneous coronary artery dissection. JACC Cardiovasc Imaging 6(7):830–832
35. Mandal R, Brooks EG, Corliss RF (2015) Eosinophilic coronary periarteritis with arterial dissection: the mast cell hypothesis J. Forensic Sci 60(4):1088–1092
36. Pitliya A, Datta S, Kalayci A et al (2018) Eosinophilic inflammation in spontaneous coronary artery dissection: a potential therapeutic target? Med Hypotheses 121:91–94
37. Stoukas V, Dragovic LJ (2009) Sudden deaths from eosinophilic coronary monoarteritis: a subset of spontaneous coronary artery dissection. Am J Forensic Med Pathol 30(3):268–269
38. Tweet MS, Hayes SN, Pitta SR et al (2012) Clinical features, management, and prognosis of spontaneous coronary artery dissection. Circulation 126(5):579–588
39. Mortensen KH, Thuesen L, Kristensen IB, Christiansen EH (2009) Spontaneous coronary artery dissection: a Western Denmark Heart Registry study. Catheter Cardiovasc Interv 74(5):710–717

40. Nishiguchi T, Tanaka A, Ozaki Y et al (2016) Prevalence of spontaneous coronary artery dissection in patients with acute coronary syndrome. Eur Heart J Acute Cardiovasc Care 5(3):263–270
41. Motreff P, Malcles G, Combaret N et al (2017) How and when to suspect spontaneous coronary artery dissection: novel insights from a single-centre series on prevalence and angiographic appearance. EuroIntervention 12(18):e2236–e2243
42. Rashid HN, Wong DT, Wijesekera H et al (2016) Incidence and characterisation of spontaneous coronary artery dissection as a cause of acute coronary syndrome–a single-centre Australian experience. Int J Cardiol 202:336–338
43. Elkayam U, Jalnapurkar S, Barakkat MN et al (2014) Pregnancy-associated acute myocardial infarction: a review of contemporary experience in 150 cases between 2006 and 2011. Circulation 129(16):1695–1702
44. Faden MS, Bottega N, Benjamin A, Brown RN (2016) A nationwide evaluation of spontaneous coronary artery dissection in pregnancy and the puerperium. Heart 102(24):1974–1979
45. Bush N, Nelson-piercy C, Spark P et al (2013) Myocardial infarction in pregnancy and postpartum in the UK. Eur J Prev Cardiol 20(1):12–20
46. Tweet MS, Hayes SN, Codsi E et al (2017) Spontaneous coronary artery dissection associated with pregnancy. J Am Coll Cardiol 70(4):426–435
47. Lameijer H, Kampman MA, Oudijk MA, Pieper PG (2015) Ischaemic heart disease during pregnancy or post-partum: systematic review and case series. Neth Heart J 23(5):249–257
48. James AH, Jamison MG, Biswas MS et al (2006) Acute myocardial infarction in pregnancy: a United States population-based study. Circulation 113(12):1564–1571
49. Roth A, Elkayam U (2008) Acute myocardial infarction associated with pregnancy. J Am Coll Cardiol 52(3):171–180
50. Elkayam U, Goland S, Pieper PG, Silverside CK (2016) High-risk cardiac disease in pregnancy: part I. J Am Coll Cardiol 68(4):396–410
51. Ladner HE, Danielsen B, Gilbert WM (2005) Acute myocardial infarction in pregnancy and the puerperium: a population-based study. Obstet Gynecol 105(3):480–484
52. Saw J, Aymong E, Sedlak T et al (2014) Spontaneous coronary artery dissection: association with predisposing arteriopathies and precipitating stressors and cardiovascular outcomes. Circ Cardiovasc Interv 7(5):645–655
53. Tweet MS, Eleid MF, Best PJ et al (2014) Spontaneous coronary artery dissection: revascularization versus conservative therapy. Circ Cardiovasc Interv 7(6):777–786
54. Lettieri C, Zavalloni D, Rossini R et al (2015) Management and long-term prognosis of spontaneous coronary artery dissection. Am J Cardiol 116(1):66–73
55. Saw J, Ricci D, Starovoytov A et al (2013) Spontaneous coronary artery dissection: prevalence of predisposing conditions including fibromuscular dysplasia in a tertiary center cohort. JACC Cardiovasc Interv 6(1):44–52
56. Prasad M, Tweet MS, Hayes SN et al (2015) Prevalence of extracoronary vascular abnormalities and fibromuscular dysplasia in patients with spontaneous coronary artery dissection. Am J Cardiol 115(12):1672–1677
57. Saw J, Poulter R, Fung A et al (2012) Spontaneous coronary artery dissection in patients with fibromuscular dysplasia: a case series. Circ Cardiovasc Interv 5(1):134–137
58. Pate GE, Lowe R, Buller CE (2005) Fibromuscular dysplasia of the coronary and renal arteries? Catheter Cardiovasc Interv 64(2):138–145
59. Sozzi FB, Schiavone M, Maalhouf J et al (2019) Acute cardiac disease in the postpartum: what to think. Echocardiography 36:2099–2102
60. Labbok MH (2001) Effects of breastfeeding on the mother. Pediatr Clin North Am. 48(1):143–158
61. Eng L, Starovoytov A, Heydari M et al (2015) Abstract 18913: spontaneous coronary artery dissection in women and association with hormonal stressors (abstr). Circulation 132:A18913
62. Kamran M, Guptan A, Bogal M (2008) Spontaneous coronary artery dissection: case series and review. J Invasive Cardiol 20(10):553–559

63. Tweet MS, Gulati R, Hayes SN (2016) Spontaneous Coronary Artery Dissection. Curr Cardiol Rep 18(7):60
64. Luong C, Starovoytov A, Heydari M et al (2017) Clinical presentation of patients with spontaneous coronary artery dissection. Catheter Cardiovasc Interv 89(7):1149–1154
65. Nakashima T, Noguchi T, Haruta S et al (2016) Prognostic impact of spontaneous coronary artery dissection in young female patients with acute myocardial infarction: A report from the Angina Pectoris-Myocardial Infarction Multicenter Investigators in Japan. Int J Cardiol 207:341–348
66. Al-hussaini A, Adlam D (2017) Spontaneous coronary artery dissection. Heart 103(13):1043–1051
67. Saw J (2014) Coronary angiogram classification of spontaneous coronary artery dissection. Catheter Cardiovasc Interv 84(7):1115–1122
68. Saw J, Mancini GB, Humphries K et al (2016) Angiographic appearance of spontaneous coronary artery dissection with intramural hematoma proven on intracoronary imaging. Catheter Cardiovasc Interv 87(2):E54-61
69. Rogowski S, Maeder MT, Weilenmann D et al (2017) Spontaneous coronary artery dissection: angiographic follow-up and long-term clinical outcome in a predominantly medically treated population. Catheter Cardiovasc Interv 89(1):59–68
70. Conraads VM, Vorlat A, Colpaert CG et al (1999) Spontaneous dissection of three major coronary arteries subsequent to cystic medial necrosis. Chest 116(5):1473–1475
71. Walsh SJ, Jokhi PP, Saw J (2008) Successful percutaneous management of coronary dissection and extensive intramural haematoma associated with ST elevation MI. Acute Card Care 10(4):231–233
72. Persu A, Giavarini A, Touze E et al (2014) European consensus on the diagnosis and management of fibromuscular dysplasia. J Hypertens 32:1367–1378
73. Persu A, Van der Niepen P, Touze E et al (2016) Revisiting fibromuscular dysplasia: rationale of the european fibromuscular dysplasia initiative. Hypertension 68:832–839
74. Liang JJ, Prasad M, Tweet MS et al (2014) A novel application of CT angiography to detect extracoronary vascular abnormalities in patients with spontaneous coronary artery dissection. J Cardiovasc Comput Tomogr 8:189–197
75. Saw J, Humphries K, Aymong E et al (2017) Spontaneous coronary artery dissection: clinical outcomes and risk of recurrence. J Am Coll Cardiol 70(9):1148–1158
76. Roffi M, Patrono C, Collet JP et al (2016) 2015 ESC guidelines for the management of acute coronary syndromes in patients presenting without persistent ST-segment elevation: task force for the management of acute coronary syndromes in patients presenting without persistent ST-Segment elevation of the European Society of Cardiology (ESC). Eur Heart J 37(3):267–315
77. Ibanez B, James S, Agewall S et al (2018) 2017 ESC guidelines for the management of acute myocardial infarction in patients presenting with ST-segment elevation: The Task Force for the management of acute myocardial infarction in patients presenting with ST-segment elevation of the European Society of Cardiology (ESC). Eur Heart J 39(2):119–177
78. Pasupathy S, Air T, Dreyer RP et al (2015) Systematic review of patients presenting with suspected myocardial infarction and nonobstructive coronary arteries. Circulation 131(10):861–870
79. Smilowitz NR, Mahajan AM, Roe MT et al (2017) Mortality of myocardial infarction by sex, age, and obstructive coronary artery disease status in the ACTION Registry-GWTG (Acute Coronary Treatment and Intervention Outcomes Network Registry-Get With the Guidelines). Circ Cardiovasc Qual Outcomes 10(12):e003443
80. Agewall S, Beltrame JF, Reynolds HR et al (2017) ESC working group position paper on myocardial infarction with non-obstructive coronary arteries. Eur Heart J 38(3):143–153
81. Tamis-holland JE, Jneid H, Reynolds HR et al (2019) Contemporary diagnosis and management of patients with myocardial infarction in the absence of obstructive coronary artery disease: a scientific statement from the American Heart Association. Circulation 139(18):e891–e908

82. Scanlon PJ, Faxon DP, Audet AM et al (1999) ACC/AHA guidelines for coronary angiography. A report of the American College of Cardiology/American Heart Association task force on practice guidelines (committee on coronary angiography). Developed in collaboration with the Society for Cardiac Angiography and Interventions. J Am Coll Cardiol 33:1756–1824
83. Vaccarino V, Parsons L, Every NR et al (1999) Sex-based differences in early mortality after myocardial infarction. National registry of myocardial infarction 2 participants. N Engl J Med 341(4):217–225
84. Shaw LJ, Shaw RE, Merz CN et al (2008) Impact of ethnicity and gender differences on angiographic coronary artery disease prevalence and in-hospital mortality in the American College of Cardiology-National Cardiovascular Data Registry. Circulation 117(14):1787–1801
85. Ramanath VS, Armstrong DF, Grzybowski M et al (2010) Receipt of cardiac medications upon discharge among men and women with acute coronary syndrome and nonobstructive coronary artery disease. Clin Cardiol 33(1):36–41
86. Patel MR, Chen AY, Peterson ED et al (2006) Prevalence, predictors, and outcomes of patients with non-ST-segment elevation myocardial infarction and insignificant coronary artery disease: results from the Can Rapid risk stratification of unstable angina patients. Suppress adverse outcomes with early implementation of the ACC/AHA Guidelines (CRUSADE) initiative. Am Heart J 152(4):641–647
87. Sharkey SW, Lesser JR, Zenovich AG et al (2005) Acute and reversible cardiomyopathy provoked by stress in women from the United States. Circulation 1; 111(4):472–479
88. Lyon AR, Bossone E, Schneider B et al (2016) Current state of knowledge on Takotsubo syndrome: a position statement from the taskforce on Takotsubo syndrome of the Heart Failure Association of the European Society of Cardiology. Eur J Heart Fail 18(1):8–27
89. Ghadri JR, Wittstein IS, Prasad A et al (2018) International expert consensus document on Takotsubo syndrome (Part I): clinical characteristics, diagnostic criteria, and pathophysiology. Eur Heart J 39:2032–2046
90. Akashi YJ, Nef HM, Lyon AR (2015) Epidemiology and pathophysiology of Takotsubo syndrome. Nat Rev Cardiol. 12(7):387–397
91. Deshmukh A, Kumar G, Pant S et al (2012) Prevalence of Takotsubo cardiomyopathy in the United States. Am. Heart J. 164:66–71
92. Patel SM, Chokka RG, Prasad K et al (2013) Distinctive clinical characteristics according to age and gender in apical ballooning syndrome (Takotsubo/stress cardiomyopathy): an analysis focusing on men and young women. J Card Fail. 19(5):306–310
93. Schneider B, Athanasiadis A, Stöllberger C et al (2013) Gender differences in the manifestation of Tako-tsubo cardiomyopathy. Int J Cardiol 166(3):584–588
94. Lyon AR, Rees PS, Prasad S et al (2008) Stress (Takotsubo) cardiomyopathy-a novel pathophysiological hypothesis to explain catecholamine-induced acute myocardial stunning. Nat Clin Pract Cardiovasc Med 5:22–29
95. Pelliccia F, Kaski JC, Crea F, Camici PG (2017) Pathophysiology of Takotsubo syndrome. Circulation 135(24):2426–2441
96. Uchida Y, Egami H, Uchida Y et al (2010) Possible participation of endothelial cell apoptosis of coronary microvessels in the genesis of Takotsubo cardiomyopathy. Clin Cardiol 33:371–377
97. Stöllberger C, Finsterer J (2011) Why does Takotsubo ("broken heart syndrome") affect more females than males? Int J Cardiol 147:175–176
98. Fischer M, Baessler A, Schunkert H (2002) Renin angiotensin system and gender differences in the cardiovascular system. Cardiovasc Res. 53(3):672–677
99. Ueyama T, Ishikura F, Matsuda A et al (2007) Chronic estrogen supplementation following ovariectomy improves the emotional stress-induced cardiovascular responses by indirect action on the nervous system and by direct action on the heart. Circ J. 71(4):565–573
100. Singh K, Carson K, Dip LM et al (2014) Meta-analysis of clinical correlates of acute mortality in Takotsubo cardiomyopathy. Am J Cardiol 113:1420–1428

101. Templin C, Ghadri JR, Diekmann J et al (2015) Clinical features and outcomes of Takotsubo (stress) cardiomyopathy. N Engl J Med 373(10):929–938
102. Kato K, Lyon AR, Ghadri JR, Templin C (2017) Takotsubo syndrome: aetiology, presentation and treatment. Heart 103(18):1461–1469

Part II
Sex Differences in Heart Failure

Chapter 6
Sex Differences in Contractile Function in Cardiac Hypertrophy and Heart Failure Subsequent to Volume Overload

Paramjit S. Tappia, Anureet K. Shah, Mohamad Nusier, and Naranjan S. Dhalla

Abstract Cardiac hypertrophy is an adaptive response of the heart to hemodynamic overload that is initially designed to maintain cardiac functioning, but prolonged hypertrophy becomes detrimental and results in cardiac dysfunction and heart failure (HF). Population studies have indicated that men and women are different in their risk and etiology in developing HF. In fact, female sex hormones are believed to have a cardioprotective role during premenopause. Although cardiac remodeling in experimental volume overload has also been shown to occur in males, there is still a paucity of information regarding sex differences in the pattern of ventricular remodeling and contractile function in response to volume overload. Therefore, this article discusses the sex differences in changes of ventricular dimensions and contractile function in cardiac hypertrophy and HF due to volume overload. Furthermore, it describes cardiomyocyte Ca^{2+}-handling properties that could potentially contribute to sex differences in cardiac remodeling and contractile function. The influence of sex hormones that are present during these processes is also highlighted. Since sex differences in the development of cardiac hypertrophy have been reported to occur in pressure overload and myocardial infarction, some of these changes are also described for a comparative perspective. Understanding the sex differences in the pathophysiology of heart disease may form the foundations for the development of new approaches for sex-specific treatment and prevention of HF.

P. S. Tappia
Asper Clinical Research Institute, St. Boniface Hospital, Winnipeg, Canada

A. K. Shah
Department of Kinesiology, Nutrition and Food Science, California State University, Los Angeles, USA

M. Nusier
School of Medicine, Jordan University of Science and Technology, Irbid, Jordan

N. S. Dhalla (✉)
Institute of Cardiovascular Sciences, St. Boniface Hospital Albrechtsen Research Centre & Department of Physiology and Pathophysiology, College of Medicine, Rady Faculty of Health Sciences, University of Manitoba, Winnipeg, Canada
e-mail: nsdhalla@sbrc.ca

B. Ostadal and N. S. Dhalla (eds.), *Sex Differences in Heart Disease*,
Advances in Biochemistry in Health and Disease 21,
https://doi.org/10.1007/978-3-030-58677-5_6

Keywords Sex differences · Cardiac remodeling · Cardiac hypertrophy · Heart failure · Ca^{2+}-handling · Volume overload

Introduction

Globally cardiovascular disease (CVD) is the leading cause of death, however; significant differences exist between males and females in the development of CVD [1–12]. Premenopausal women have a reduced risk of CVD, in fact, the incidence and severity of CVD increases markedly after menopause [13]. Women tend to develop heart disease 10–15 years later than men, pointing to the existence of specialized mechanisms in females that attenuate their risk of developing heart disease [14]. Females have also been reported to be less susceptible to ischemic insult and have improved survival following myocardial ischemic-reperfusion injury [15, 16].

The incidence of heart failure (HF) due to elevated blood pressure and diabetes is greater in women than in men [17]. However, as compared to men, the development of HF in women is more benign and is predominantly characterized by preserved systolic function and diastolic dysfunction [17]. Interestingly, older women tend to be more hypertensive and are less likely to present with coronary heart disease [18]. Postmenopausal women with diabetes are at a 3- to 6-fold higher risk of a myocardial infarction (MI), in comparison to a 2- to 4-fold risk of MI in men [19]. Women seem to be better protected and show a later onset of cardiac decompensation than men with HF [20]. In terms of survival in older adults with HF, women fare significantly better than men [21]. It is interesting to note that since 1984, more women die of HF each year, and yet more men are diagnosed with HF [22].

In women, an increased left ventricle (LV) mass or degree of hypertrophy has been suggested to be a stronger predictor of mortality than traditional measures of LVsize and function [23]. Thus, changes in the size, shape, structure, and function (so-called ventricular remodeling or cardiac remodeling) of the heart, may occur in a sex-dependent manner. Accordingly, it is planned to discuss sex differences in the changes in LVdimensions and contractile function in cardiac hypertrophy and heart failure due to volume overload as well as to describe the potential of cardiomyocyte Ca^{2+}-handling properties that may contribute to sex differences in cardiac remodeling and contractile function (Fig. 6.1). Also, the role of sex hormones in these processes will be highlighted. Reference to cardiac remodeling and cardiac function in response to pressure overload and MI will also be made for the purpose of comparison.

General Characteristics of Cardiac Remodeling and Function in Hypertrophy and Heart Failure

Geometrical and structural alterations in the heart are the hallmarks of cardiac remodeling. Such cardiac changes occur in a range of different cardiac pathophysiological conditions including hypertrophic cardiomyopathy, dilated cardiomyopathy, diabetic

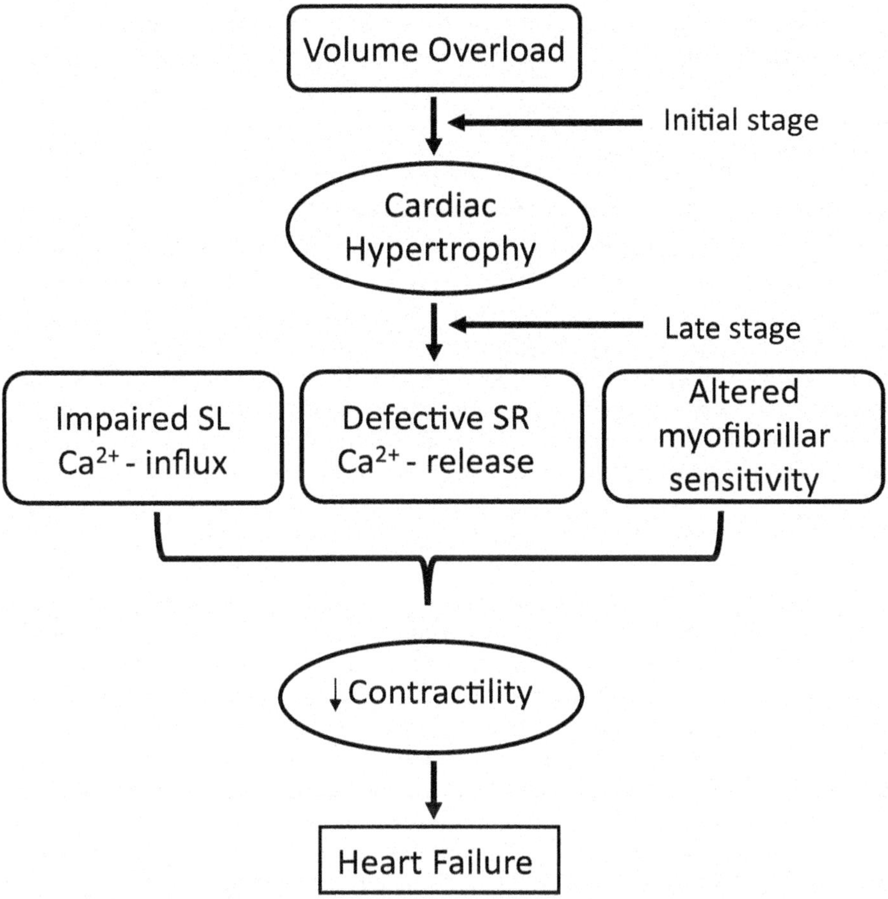

Fig. 6.1 Development of cardiac hypertrophy and heart failure at early stages and late stages of inducing volume overload, respectively. Alterations in subcellular Ca^{2+}-handling and myofibrillar Ca^{2+}-sensitivity are suggested to be associated with heart failure

cardiomyopathy, ischemic cardiomyopathy as well as hypertensive cardiomyopathy, which can progressively lead to HF. With an excessive workload or sustained hemodynamic overload, ventricular myocytes grow in response to a complex cascade of events [24] such as mechanical stress, ischemia, and the activation of multiple intracellular signaling pathways by neurohormones. In the initial stage, cardiac hypertrophy may be seen as a beneficial response to normalize LV wall stress and preserve normal cardiac contractile function, whereas the prolongation of hypertrophy leads to HF and sudden death [25–27]. This transition of cardiac hypertrophy to HF is generated by the occurrence of marked ventricular dilatation once the myocardial hypertrophic response is exhausted [27]. The progressive deterioration of LV function is an established characteristic of HF in men; however, it appears that a different pattern of ventricular remodeling occurs in women, which is designed to preserve cardiac contractility.

Sex Differences in Cardiac Remodeling in Hypertrophy and Heart Failure Due to Volume Overload

Experimental and clinical studies have revealed sex differences in myocardial remodeling in aging, pressure overload, volume overload and MI. The outcome of the remodeling processes are that women more often present HF with preserved systolic function, but exhibit an increased risk for acute low output syndrome [28]. Experimental studies have also shown that males experience 10-times higher mortality due to MI than females [14]. Female rats showed a different pattern of LV remodeling than males with less increase in the thickness of the noninfarcted portions of LV than males, but comparable LV cavity enlargement and systolic dysfunction due to MI [17, 18]. Although, we have earlier reviewed the mechanisms of sex differences in cardiac dysfunction due to heart disease [29–31], there is still a relative paucity of information regarding cardiac remodeling in hypertrophy and heart failure due to volume overload. Some studies have been conducted to examine differences in male and female rats during both the hypertrophic and failing stages following induction of volume overload. Gardner et al. [32] have identified that after 8 wks post AV fistula, mortality was tenfold less in female rats as compared to male rats. Although females had increased LV and right ventricle (RV) weights, these animals showed no signs of CHF whereas the males had increased lung weight, marked ventricular dilatation and increased compliance thus indicating the presence of HF. It should be pointed out that this study was carried out at 8 wks post fistula and the LV volume and function studies were carried out under in vitro conditions.

We have also observed sex differences in the cardiac remodeling process in volume overload induced by an AV-shunt [33]. Both males and females developed significant cardiac hypertrophy. Furthermore, echocardiographic data revealed increases in both diastolic and systolic left ventricle internal diameters (LVID) in male rats, whereas females showed no significant change in these parameters. It seems that the principle difference between males and females at this time point is that females show an increase in LV posterior wall thickness (PWT) indicating that concentric hypertrophic remodeling has taken place (Fig. 6.2). On the other hand, in males it appears that eccentric cardiac hypertrophy leading to ventricular dilatation occurs. Our study also examined remodeling at 16 wks post-AV shunt. In this regard, significant cardiac hypertrophy was seen to remain 16 wks post AV shunt in both males and female rats (Fig. 6.3). Such sex differences in cardiac remodeling have been shown in other models of heart failure, including pressure overload [34], MI [35], and spontaneously hypertension [36]. For example, it was shown that following pressure overload; males were unable to maintain concentric hypertrophy and transitioned to HF much faster than their female counterparts who were able to maintain a concentrically remodeled

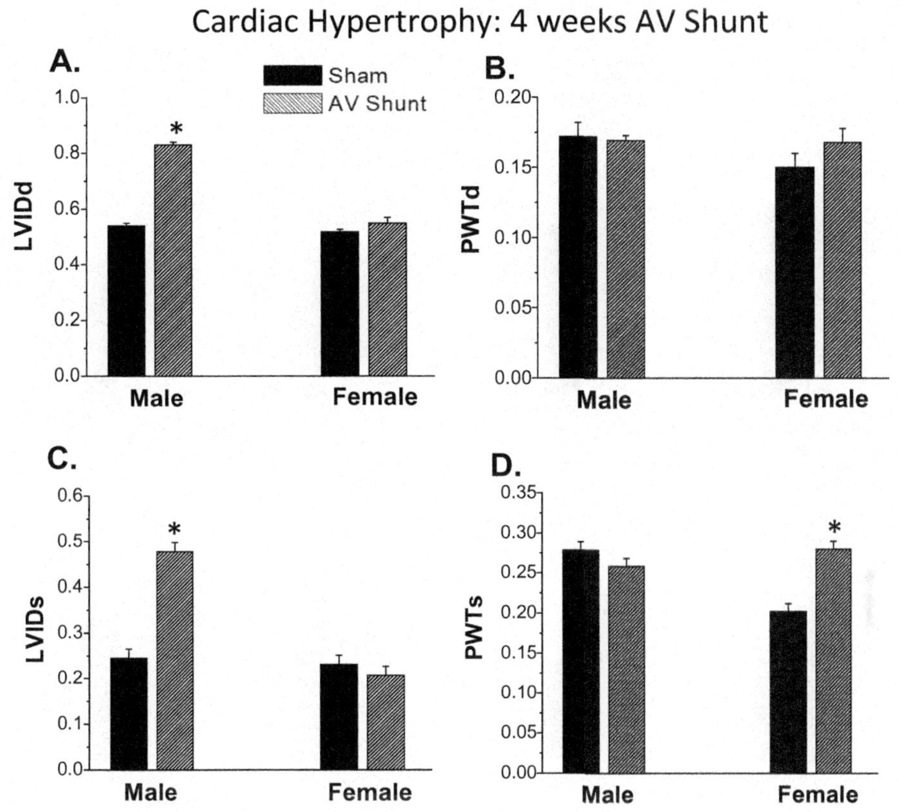

Fig. 6.2 Echocardiographic analysis of male and female rats 4 weeks post-AV shunt for inducing cardiac hypertrophy due to volume overload. Data are taken from our paper—Dent MR, Tappia PS, Dhalla NS. J Card Fail 16: 439–449, 2010. Erratum in: J Card Fail. 17:179, 2011 [33]. LVIDs, systolic left ventricular internal diameter; LVIDd, diastolic left ventricular internal diameter; PWTs, systolic posterior wall thickness; PWTd, diastolic posterior wall thickness; $^{*}P < 0.05$ versus sham values

heart, therefore delaying the transition to HF [34]. It is interesting to note that in aortic stenosis, women tolerate pressure overload with less concentric remodeling and myocardial fibrosis, but are more likely to develop symptoms, which may be related to higher wall stress and filling pressures in women [37]. Experimentally, female rats also demonstrated a different pattern of LV remodeling than males following MI, the LV noninfarcted region was not as thick in females as it was in male animals. Although both males and females exhibited similar infarct size, females were able to preserve LVdiastolic filling capacity [35].

It appears that the major sex difference that occurs in many of models of HF is that females are able to develop cardiac hypertrophy in such a way that they maintained cardiac function and delay the transition to HF. It has been suggested by Grossman et al. [38] that increased LVEDP is the trigger for developing eccentric

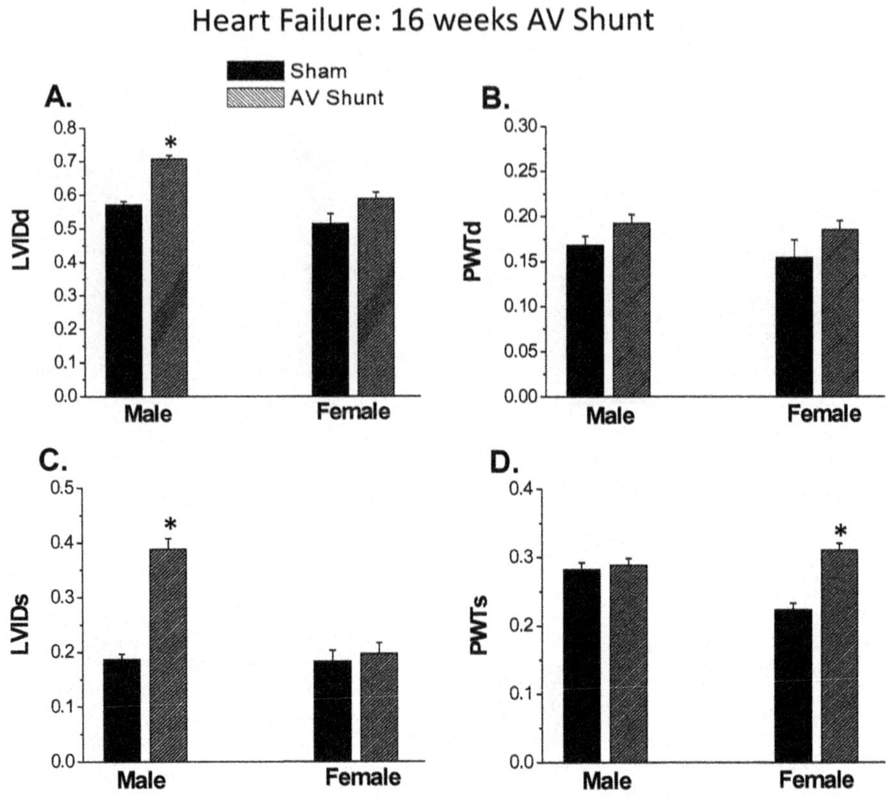

Fig. 6.3 Echocardiographic analysis of female and male sham and arteriovenous shunt groups at 16 weeks. Data are taken from our paper—Dent MR, Tappia PS, Dhalla NS. J Card Fail 16: 439–449, 2010. Erratum in: J Card Fail. 17:179, 2011 [33]. LVIDs, systolic left ventricular internal diameter; LVIDd, diastolic left ventricular internal diameter; PWTs, systolic posterior wall thickness; PWTd, diastolic posterior wall thickness; *$P < 0.05$ versus sham values.

hypertrophy or inappropriate hypertrophy that is unable to maintain cardiac function. However, females subjected to volume overload were able to increase LV wall thickness, thereby developing concentric hypertrophy to compensate for the increased workload and maintain relatively normal cardiac function.

The development of cardiac hypertrophy in response to volume overload subsequent to chronic aortic valve regurgitation (AVR) in both male and female rats has also been evaluated [39]. In this regard, AVR-inducd volume overload resulted in comparable increases in LV dilation and heart weight mass in both male and female rats, as compared to respective sham controls. However, the relative LV wall thickness as measured by the ratio between wall thickness to end-diastolic diameter, was diminished in male rats, but remained unchanged in females as compared to respective sham controls. Treatment of the experimental animals with the angiotensin II receptor blocker (ARB), valsartan, was observed to reverse LV thickening in female rats only, with no effect on LV dilation in both sexes. In addition, valsartan treatment normalized left atrial mass and E-wave slope, but only in female AVR rats [39].

Earlier we have reported differences in basal hemodynamic parameters between male and female rats [33]. In this regard, it was observed that both the rate of contraction (+dP/dt) and rate of relaxation (-dP/dt) were significantly lower in the female control rats as compared to the male counterparts at 4 (Fig. 6.4) and 16 wks (Fig. 6.5) time points, except –dP/dt at 4 wks. While LVEDP was seen to be higher at 4 wks, no changes in baseline left ventricular systolic pressure (LVSP) and left ventricular end-diastolic pressure (LVEDP) values were seen at 16 wks in both female and male rats (Figs. 6.4 and 6.5). On the other hand, hemodynamic assessment at the same time points i.e. at 4 and 16 wks post-AV shunt revealed that LVEDP, unlike other parameters, was significantly increased only in males at 4 wks post-AV shunt (Fig. 6.4). However, at 16 wks post-AV shunt, cardiac function was significantly depressed in males as evidenced by decreases in + dP/dt and -dP/dt as well as a significant increase in LVEDP; these parameters were unaltered in the 16 wk post-AV shunt female group (Fig. 6.5). No significant differences in heart rate (beats/min) were observed in both male and female rats at 4 wks and 16 wks post-AV shunt as compared to respective sham control values. In this study, an echocardiographic assessment of normal and post-AV male and female hearts was also conducted. In this regard, at 4 wks post-AV shunt (hypertrophy stage), an increase in the cardiac output, but decrease in the fractional shortening were seen in male rats (Fig. 6.4). At 16 wks following AV shunt (HF stage), while cardiac output was increased, a decrease in fractional shortening were seen in male rats. On the other hand, a small, but significant increase in the cardiac output, was observed at 16 wks post-AV shunt, but no changes occurred in fractional shortening (Fig. 6.5).

Role of Estrogen in Cardiac Hypertrophy and Heart Failure

Since there is an elevated risk of CVD in postmenopausal women, the role of female sex hormones has generally been considered to be a major factor in providing protection against CVD in premenopausal women. Observations from several investigations such as Women's Health Initiative Studies, Cochrane Review Studies, Early Versus Late Intervention Trial with Estradiol Study, and Kronos Early Estrogen Prevention Study have indicated that hormone replacement therapy (HRT) may exert some beneficial action in women. However, it was also observed that the beneficial action occurs in those women that were under 60 yrs of age, and if HRT was started within 10 years of menopause. On the other hand, there was no benefit to be found in women who were of more than 60 years, and in particular if their HRT was initiated at more than 10 years of menopause [40].

Estrogen has been thought to have a cardioprotective effect. However, the recent outcomes of the Heart and Estrogen Replacement Study (HERS) and The WHI have indicated that estrogen or a combination estrogen-progestin therapy increased the coronary heart disease events and therefore physicians were advised to use hormone

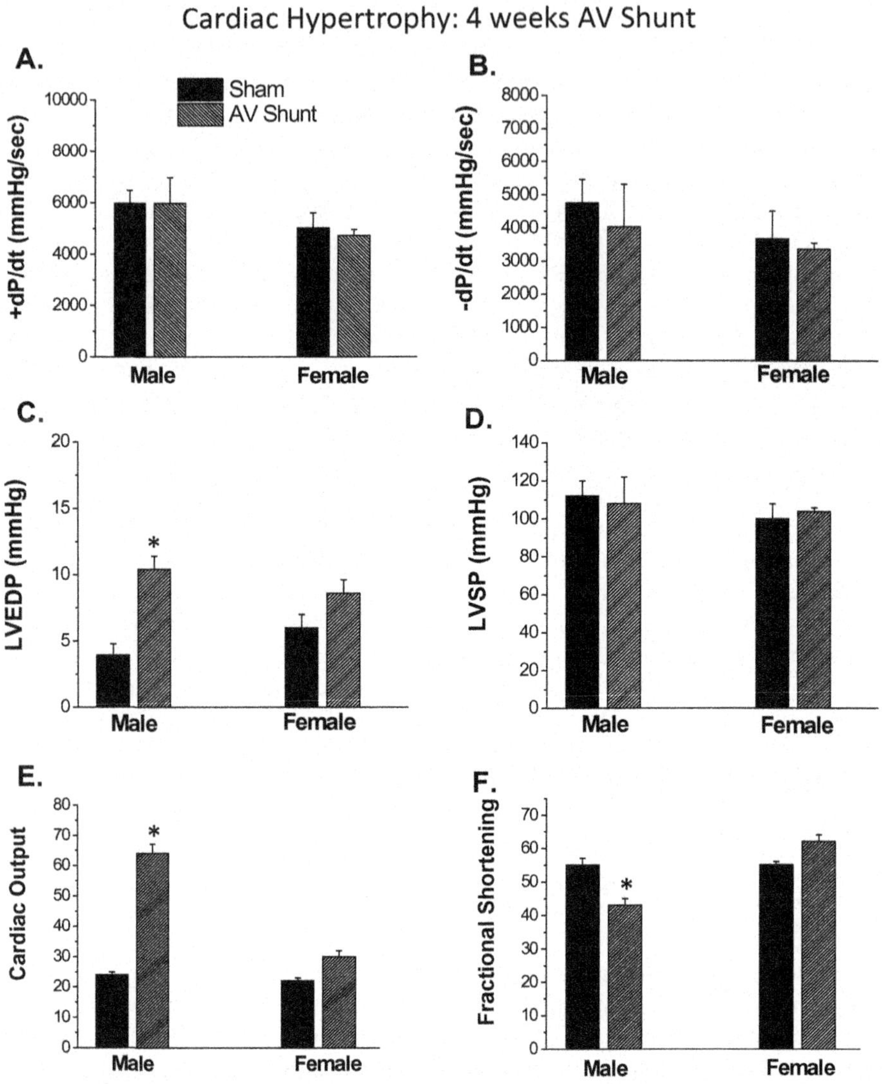

Fig. 6.4 Hemodynamic changes in male and female rats at 4 weeks post AV shunt. Values are taken from our paper—Dent MR, Tappia PS, Dhalla NS. J Card Fail 16: 439–449, 2010. Erratum in: J Card Fail. 17:179, 2011 [33]. Sham, age-matched controls; + dP/d*t*, rate of contraction; -dP/d*t*, rate of relaxation. LVEDP, LVend-diastolic pressure; LVSP, LVsystolic pressure. * $P < 0.05$ versus sham-operated values

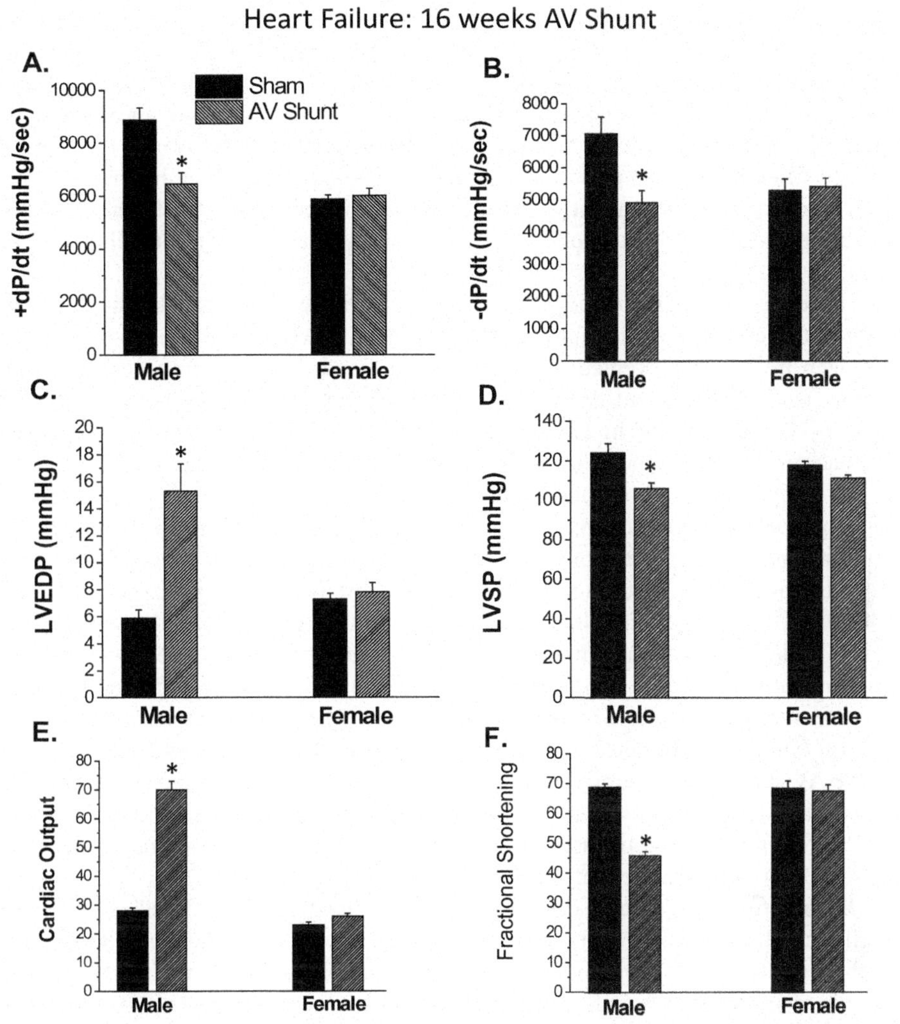

Fig. 6.5 Hemodynamic changes in male and female rats at 16 weeks post AV shunt. Values are taken from our paper—Dent MR, Tappia PS, Dhalla NS. J Card Fail 16: 439–449, 2010. Erratum in: J Card Fail. 17:179, 2011 [33]. Sham, age-matched controls; + dp/d*t*, rate of contraction; -dP/d*t*, rate of relaxation. LVEDP, LVend-diastolic pressure; LVSP, LVsystolic pressure. * $P < 0.05$ versus sham-operated control values

replacement therapy strictly for menopausal symptom relief [41]. Another publication reported on the CV data from two arms of WHI and concluded that the age of initiation of therapy may play a role on the effect on coronary heart disease outcomes. It appears that younger women who are closer to menopause have a lower event rate than women where therapy was initiated over the age of 60 [42]. In different rat models of CHF estrogen has been suggested to play a major role in the underlying mechanisms for the gender differences in the cardiac remodeling process. For example, in the in vitro study by Brower et al. [43] it was found that ovariectomy lead to the development of eccentric hypertrophy and low contractility measurements as opposed to concentric hypertrophy and maintained contractility in the control females. In addition, the prevention of adverse cardiac remodeling in volume overload induced by aortocaval fistula in the female rat has been suggested to be the result of an estrogen-altered mast cell phenotype and/or prevention of mast cell activation [44].

Subsequent studies conducted by this group showed that the administration of phytoestrogens added to the female advantage against volume overload [45] and the estrogen treatment prevented pulmonary edema and clearly attenuated LV hypertrophy and dilatation but did not maintain contractility [46]. In contrast, Drolet et al. [47], observed that in an aortic regurgitation model, female rats had increased cardiac remodeling than males, which imposed a greater workload on their hearts; it was concluded that hormone status did not have any implication on the remodeling process [47]. Following MI, it was shown that 17β-estradiol increased LV and cardiomyocyte hypertrophy [48] and prevented deterioration of cardiac function [49]. In contrast, 17β-estradiol attenuated hypertrophy following pressure overload [48, 50]. These authors concluded that the effect of estrogen on cardiac hypertrophy might be dependent on the initial stimulus [49]. Interestingly, we have shown that the administration of estrogen in ovariectomized female rats following the induction of volume overload induces a higher degree of hypertrophy than that seen in the intact control female rats. Echocardiographic analysis has also revealed that the treatment of the AV shunt ovariectomized female rat with estrogen results in attenuation in the ventricular dilatation, as well as increased the wall thickness. These findings indicate that estrogen may be implicated in the signaling mechanism that initiate concentric hypertrophy as opposed to inappropriate eccentric hypertrophy.

To investigate the role of estrogen in cardiac function in heart failure, we have earlier conducted experiments in sham-operated as well as ovariectomized animals with or without 17-β estradiol treatment. Figure 6.6 shows that in contrast to intact female rats, AV shunt produced significant depressions in + dP/dt, -dP/dt and LVSP and a marked increase in LVEDP in ovariectomized rats; these alterations were either fully or partially prevented by estrogen treatment. Although ovariectomy reduced LVEDP, unlike +dP/dt, -dP/dt and LVSP, this change was not affected by estrogen treatment (Fig. 6.6). Furthermore, echocardiographic assessment of cardiac performance of females and ovariectomized female rats treated with or without estrogen, revealed a preservation of fractional shortening and cardiac output subsequent to induction of volume overload by an AV shunt [33]. Since cardiac function as represented by ±dP/dt was maintained and a partial attenuation of the increase in the LVEDP was observed upon administration of estrogen to ovariectomized animals, estrogen likely plays an important role in the cardiac remodeling observed in females subjected to volume overload. Indeed, alterations in intraventricular dimensions of intact female rats and ovariectomized females treated with or without estrogen, in sham and AV animals can be prevented by estrogen treatment [33].

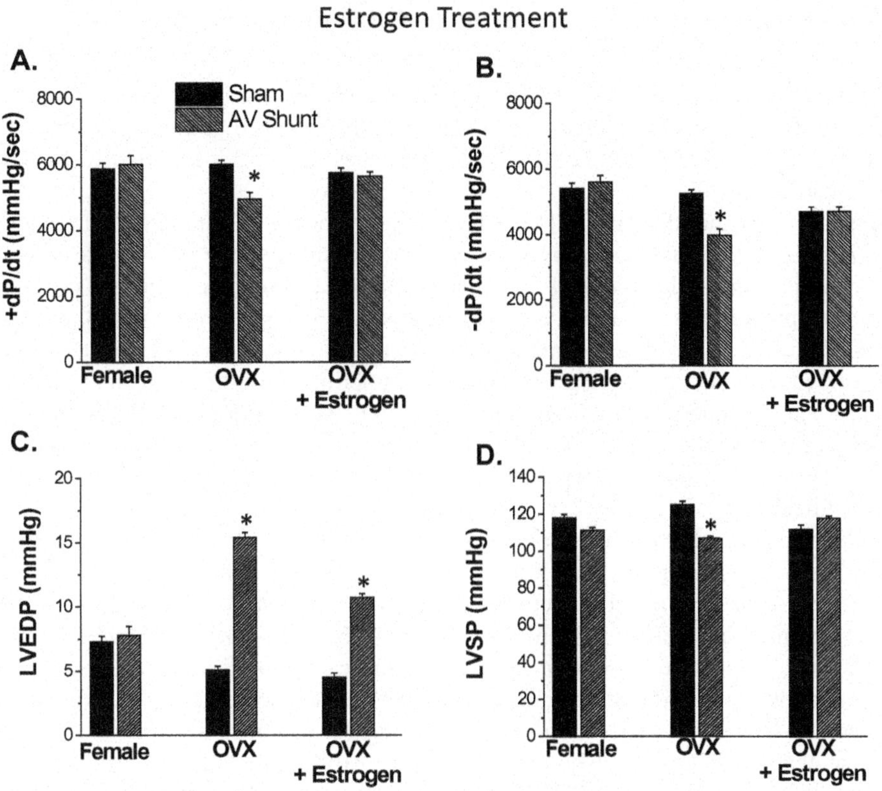

Fig. 6.6 Hemodynamic changes in female, ovariectomized treated with or without estrogen rats at 16 weeks post AV shunt. Data are taken from our paper—Dent MR, Tappia PS, Dhalla NS. J Card Fail 16: 439–449, 2010. Erratum in: J Card Fail. 17:179, 2011 [33]. Sham, age-matched controls; + dP/d*t*, rate of contraction; -dP/d*t*, rate of relaxation. LVEDP, LVend-diastolic pressure; LVSP, LVsystolic pressure; OVX, ovariectomized. * $P < 0.05$ versus sham-operated control values

Interestingly, sex differences in LV function at different stages of LV hypertrophy due to pressure overload induced by the banding of the abdominal aorta have also been reported [51]. In early stage of LV hypertrophy, increased LV mass index, heart weight-to-tibial length, cardiomyocyte diameter, concentric LV geometry, and moderate interstitial fibrosis were detected in both male and female rats subjected to aortic banding. These changes were associated with impaired relaxation, increased contractility, and preserved ventricular-arterial coupling in both sexes. However, the late stage was associated with eccentric remodeling, increased fibrosis, and enhanced chamber stiffness in male rats only. Furthermore, augmented contractility declined in male rats, but not in the female animals. It was concluded that contractile augmentation, preserved ventricular-arterial coupling, and improved myocardial compliance in female rats contribute to sex differences in LV function during the progression of pressure overload-induced LV hypertrophy [51].

Sex Differences in Cardiomyocyte Ca^{2+}-handling Proteins

Contractile abnormalities in HF have been linked to ventricular remodeling and progressive defects in cardiomyocyte Ca^{2+}-handling as well as changes in myofibrillar sensitivity to Ca^{2+} [52–56]. Sex differences have been reported to exist in cardiac contractility [57], as well as the inotropic responses to Ca^{2+} [58], myocardial Ca^{2+}-channel density [59] and the cardiac response to adrenergic stimulation [57, 60]. Thus, it has been postulated that sex hormones may have a role to play for changes in cardiac remodeling in males and females under a variety of different cardiac pathologies including volume overload and pressure overload [33, 45, 61–64]. Since the sarcoplasmic reticulum (SR) and sarcolemma (SL) are intimately involved in the Ca^{2+} handling and cardiac contractility, differences in abnormalities in SR and SL functions would be seen to contribute to sex differences in the development of cardiac dysfunction. Furthermore, alterations in the sensitivity of myofibrils (MF) to Ca^{2+} in male and female hearts would also be a factor in determining the sex difference.

The functionality of sex hormones has been examined in ovariectomized female rats as well as castrated male rats. Treatment of the ovariectomized female rats with estrogen eliminated defective cardiac function as well as alterations in SR Ca^{2+}-pump, Ca^{2+}-release channels (RyR_2), and SL Na^+-Ca^{2+}-exchange activities [65, 66]. While ovariectomy in female rats and castration in male rats have both been reported to produce a shift in myosin enzymes and decrease cardiac contractile function, treatment with estrogen and testosterone was demonstrated to attenuate these changes [67]. It is also pointed out that castration in male rats has also been reported to depress SL Na^+-Ca^{2+}-exchanger and L-type Ca^{2+}-channels mRNA levels, with partial normalization upon testosterone treatment [68–70].

It should be noted that 17-β estradiol and estrogen receptors have a sex-specific role in mitochondrial function and Ca^{2+} ion channel activity in the heart [71] as well as for sustaining cardiac contractile function. These observations provide further evidence that sex hormones are an important factor in determining differences in the regulation of cardiac function in male and females. The relation between sex and contractile properties at the actin-myosin level in patients with chronic volume overload due to mitral regurgitation (leakage of blood backwards through the mitral valve during each contraction from LV) has recently been examined [72]. This study demonstrated that female fibers from patients exposed to chronic volume overload developed higher force values at a given Ca^{2+} concentration compared to fibers from male patients. The Ca^{2+} sensitivity among the male and female patients was significantly different, and it was suggested that males have higher Ca^{2+} sensitivity and might compensate for lower force values at maximal Ca^{2+} concentrations by a higher affinity for this cation. Thus, female patients with mitral regurgitation would appear to work in a more energy efficient manner [72].

It should be mentioned that MI represents the major form of HF and a large amount of the experimental data have been obtained by employing male hearts. Consequently, the effects of female sex hormones in modifying MI-induced changes in Ca^{2+}-cycling proteins need to be examined in females with and without ovariectomy. On the other hand, the deleterious effects of testosterone and castration in the MI-induced changes in Ca^{2+}-cycling proteins should also be examined. It is possible that a divergent pattern of sex differences in the Ca^{2+}-cycling proteins may exist depending on the etiology of HF. In this regard, some of our work on MI-induced HF has indicated that changes in the SR Ca^{2+}-cycling proteins are more marked as compared to the changes at the level of the SL and myofibrils.

It is interesting to note that temporal sex differences in Ca^{2+}-signaling have been reported in pressure overload-induced LV hypertrophy [73]. An initial hypertrophy of the LV due to pressure overload-induced by thoracic aortic constriction was seen in female mice and was associated with a concomitant down-regulation of SERCA2a, CaMKII activation, and GSK3β inactivation. Although both male and females showed systolic dysfunction, which may be associated with the down-regulation of RyR2, only males exhibited preserved diastolic LV function. Thus, HF caused by different etiology may exert different types of changes for Ca^{2+}-handling in a sex-dependent manner. These possibilities, in volume overload-induced cardiac hypertrophy and heart failure, therefore warrant further investigation. In addition, since MI is the most prevalent cause of HF when compared to other etiologies such as pressure overload, volume overload, valvular defects and other type of cardiomyopathies [74–76], sex-differences in MI-induced HF should also be further examined.

Conclusions

It is evident that echocardiographic and hemodynamic studies have revealed significant sex differences in cardiac remodeling and contractile function due to volume overload. Females undergo concentric remodeling and maintain an appropriate level of cardiac hypertrophy that is sufficient to delay the transition to HF. In contrast, males develop eccentric hypertrophy and transition to HF. Estrogen plays a key role in cardiac remodeling following the induction of volume overload. While the mechanisms involved in sex differences in cardiac remodeling and contractile function are still a matter for extensive research, understanding sex differences in the pathophysiology of heart disease may help in developing novel sex-specific interventions for the management and prevention of HF.

Acknowledgments Infrastructural support was provided by the St. Boniface Hospital Research Foundation. Dr. Nusier is a Visiting Professor from Jordan University of Science and Technology, School of Medicine, Department of Physiology and Biochemistry, Jordan.

References

1. Zhang Y, Liu B, Zhao R et al (2020) The influence of sex on cardiac physiology and cardiovascular diseases. J Cardiovasc Transl Res 13:3–13
2. Lundberg G, Walsh MN, Mehta LS (2019) Sex specific differences in risk factors for development of heart failure in women. Heart Fail Clin 15:1–8
3. Mentzer G, Hsich EM (2019) Heart failure with reduced ejection fraction in women: epidemiology, outcomes, and treatment. Heart Fail Clin 15:19–27
4. Madan N, Itchhaporia D, Albert CM et al (2019) Atrial fibrillation and heart failure in women. Heart Fail Clin 15:55–64
5. Kessler EL, Rivaud MR, Vos MA, van Veen TAB (2019) Sex-specific influence on cardiac structural remodeling and therapy in cardiovascular disease. Biol Sex Differ 10:7
6. Stanhewicz AE, Wenner MM, Stachenfeld NS (2018) Sex differences in endothelial function important to vascular health and overall cardiovascular disease risk across the lifespan. Am J Physiol Heart Circ Physiol 315:H1569–H1588
7. Marra AM, Salzano A, Arcopinto M et al (2018) The impact of gender in cardiovascular medicine: lessons from the gender/sex-issue in heart failure. Monaldi Arch Chest Dis 88:998
8. Cauwenberghs N, Kuznetsova T (2018) Sex-specific differences in cardiac maladaptation to hypertension and arterial stiffening. Kardiol Pol 76:1303–1311
9. Kerkhof PLM, Peace RA, Macfarlane PW (2018) Sex- and Age-related reference values in cardiology, with annotations and guidelines for interpretation. Adv Exp Med Biol 1065:677–706
10. Savarese G, D'Amario (2018) Sex differences in heart failure. Adv Exp Med Biol 1065:529–544
11. Shufelt CL, Pacheco C, Tweet MS, Miller VM (2018) Sex-specific physiology and cardiovascular disease. Adv Exp Med Biol 1065:433–454
12. Kerkhof PLM, Peace RA, Heyndrickx GR et al (2018) Heart function analysis in cardiac patients with focus on sex-specific aspects. Adv Exp Med Biol 1065:361–377
13. Schonfelder G (2005) The biological impact of estrogens on gender differences in congestive heart failure. Cardiovasc Res 67:573–574
14. Rossouw JE (2002) Hormones, genetic factors, and gender differences in cardiovascular disease. Cardiovasc Res 53:550–557
15. Bae S, Zhang L (2005) Gender differences in cardioprotection against ischemia/reperfusion injury in adult rat hearts: focus on Akt and protein kinase C signaling. J Pharmacol Exp Ther 315:1125–1135
16. Wang M, Wang Y, Weil B et al (2009) Estrogen receptor beta mediates increased activation of PI3K/Akt signaling and improved myocardial function in female hearts following acute ischemia. Am J Physiol Regul Integr Comp Physiol 296:R972–R978
17. Stromberg A, Martensson J (2003) Gender differences in patients with heart failure. Eur J Cardiovasc Nurs 2:7–18
18. Rosengren A, Hauptman P (2008) Women, men and heart failure: a review. Heart Fail Monit 6:34–40
19. Regitz-Zagrosek V, Lehmkuhl E (2005) Heart failure and its treatment in women: role of hypertension, diabetes and estrogen. Herz 30:356–367
20. Guerra S, Leri A, Wang X et al (1999) Myocyte death in the failing human heart is gender dependent. Circ Res 85:856–866
21. Parashar S, Katz R, Smith NL et al (2009) Race, gender, and mortality in adults > or = 65 years of age with incident heart failure (from the Cardiovascular Health Study). Am J Cardiol 103:1120–1127
22. Sobhani K, Nieves Castro DK, Fu Q et al (2018) Sex differences in ischemic heart disease and heart failure biomarkers. Biol Sex Differ 9:43
23. Mejhert M, Kahan T, Edner M, Persson HE (2008) Sex differences in systolic heart failure in the elderly: the prognostic importance of left ventricular mass in women. J Womens Health 17:373–381

24. Eizema K, Van Heugton H, Bezstarosti K et al (2000) SERCA2 and ANF promoter-activity studies in hypertrophic cardiomyocytes using liposome-, gene gun-, and adenovirus-mediated gene transfer. In: Takeda N, Nagano M, Dhalla NS (eds) The hypertrophied heart. Kluwer Academic Publishers, Boston, pp 52–66
25. Sakurai K, Sugawara H, Watanabe T, et al (2000) Ca^{2+} transients, contractility, and inotropic responses in rabbit volume-overloaded cardiomyocytes. In: Takeda N, Nagano M, Dhalla NS (eds) The hypertrophied heart. Kluwer Academic Publishers, Boston, pp 67–81
26. Van Bilsen M, Chien KR (1993) Growth and hypertrophy of the heart: towards an understanding of cardiac specific and inducible gene expression. Cardiovasc Res 27:1140–1149
27. Takano H, Zou Y, Akazawa H et al (2003) Ca^{2+}-dependent signaling pathways through calcineurin and Ca^{2+}/calmodulin-dependent protein kinase in development of cardiac hypertrophy. In: Dhalla NS, Hryshko L, Kardami E, Singal PK (eds) Signal transduction and cardiac hypertrophy. Kluwer Academic Publishers, Boston, pp 85–103
28. Piro M, della Bona R, Abbate A et al (2010) Sex-related differences in myocardial remodeling. J Am Coll Cardiol 55:1057–1065
29. Dhalla NS, Dent MR, Elimban V, Zieroth S (2006) Mechanisms of gender differences in cardiac dysfunction due to heart disease. In: Kimchi A (ed) Proceedings of the16th world congress on heart disease. International Academy of Cardiology, Monduzzi Editore S.p.A-Medimond Inc. Bologna, pp 239–245
30. Dent MR, Elimban V, Malik A et al (2013) Functional adaptation during the development of cardiac hypertrophy and heart failure in females. In: Ostadal B, Dhalla NS (eds) Cardiac adaptation: molecular mechanisms. Springer Science+Business Media, LLC, New York, pp 201–212
31. Dhalla NS, Malik A, Zieroth S, Tappia PS (2013) Role of gender in Ca^{2+} cycling and cardiac remodeling due to heart failure. In: Jugdutt BI, Dhalla NS (eds) Cardiac remodeling - molecular mechanisms. Springer Science+Business Media, LLC, New York
32. Gardner JD, Brower GL, Janicki JS (2002) Gender differences in cardiac remodeling secondary to chronic volume overload. J Card Fail 8:101–107
33. Dent MR, Tappia PS, Dhalla NS (2010) Gender Differences in cardiac dysfunction and remodeling due to volume overload. J Card Fail 16: 439–449; Erratum in: J Card Fail. 17:179, 2011
34. Douglas PS, Katz SE, Weinberg EO et al (1998) Hypertrophic remodeling: gender differences in the early response to left ventricular pressure overload. J Am Coll Cardiol 32:1118–1125
35. Litwin SE, Katz SE, Litwin CM et al (1999) Gender differences in postinfarction left ventricular remodeling. Cardiology 91:173–183
36. Tamura T, Said S, Gerdes AM (1999) Gender-related differences in myocyte remodeling in progression to heart failure. Hypertension 33:676–680
37. Singh A, Chan DCS, Greenwood JP et al (2019) Symptom onset in aortic stenosis: relation to sex differences in left ventricular remodeling. JACC Cardiovasc Imaging 12:96–105
38. Grossman W, Jones D, McLaurin LP (1975) Wall stress and patterns of hypertrophy in human left ventricle. J Clin Invest 56:56–64
39. Walsh-Wilkinson E, Drolet MC, Le Houillier C et al (2019) Sex differences in the response to angiotensin II receptor blockade in a rat model of eccentric cardiac hypertrophy. PeerJ 7:e7461
40. Pabbidi MR, Kuppusamy M, Didion SP et al (2018) Sex differences in the vascular function and related mechanisms: role of 17β-estradiol. Am J Physiol Heart Circ Physiol 315:H1499–H1518
41. Rossouw JE, Anderson GL, Prentice RL et al (2002) Risks and benefits of estrogen plus progestin in healthy postmenopausal women: principal results from the Women's Health Initiative randomized controlled trial. JAMA 288:321–333
42. Rossouw JE, Prentice RL, Manson JE et al (2007) Postemenopausal hormone therapy and risk of cardiovascular disease by age and years since menopause. JAMA 297:1465–1477
43. Brower GL, Gardner JD, Janicki JS (2003) Gender mediated cardiac protection from adverse ventricular remodeling is abolished by ovariectomy. Mol Cell Biochem 251:89–95
44. Lu H, Meléndez GC, Levick SP, Janicki JS (2012) Prevention of adverse cardiac remodeling to volume overload in female rats is the result of an estrogen-altered mast cell phenotype. Am J Physiol Heart Circ Physiol 302:H811–H817

45. Gardner JD, Brower GL, Janicki JS (2005) Effects of dietary phytoestrogens on cardiac remodeling secondary to chronic volume overload in female rats. J Appl Physiol 99:1378–1383
46. Gardner JD, Brower GL, Voloshenyuk TG, Janicki JS (2008) Cardioprotection in female rats subjected to volume overload: synergistic interaction of estrogen and phytoestrogens. Am J Physiol Heart Circ Physiol 294:H198–H204
47. Drolet MC, Lachance D, Plante E et al (2006) Gender-related differences in left ventricular remodeling in chronic severe aortic valve regurgitation in rats. J Heart Valve Dis 15:345–351
48. Patten RD, Pourati I, Aronovitz MJ et al (2008) 17 Beta-estradiol differentially affects left ventricular and cardiomyocyte hypertrophy following myocardial infarction and pressure overload. J Card Fail 14:245–253
49. Cavasin MA, Sankey SS, Yu AL et al (2003) Estrogen and testosterone have opposing effects on chronic cardiac remodeling and function in mice with myocardial infarction. Am J Physiol Heart Circ Physiol 284:H1560–H1569
50. Van Eickels M, Grohe C, Cleutjens JP et al (2001) 17β-estradiol attenuates the development of pressure-overload hypertrophy. Circulation 104:1419–1423
51. Ruppert M, Korkmaz-icöz S, Loganathan S et al (2018) Pressure-volume analysis reveals characteristic sex-related differences in cardiac function in a rat model of aortic banding-induced myocardial hypertrophy. Am J Physiol Heart Circ Physiol 315:H502–H511
52. Dhalla NS, Das PK, Sharma GP (1978) Subcellular basis of cardiac contractile failure. J Mol Cell Cardiol 10:363–385
53. Dhalla NS, Afzal N, Beamish RE et al (1993) Pathophysiology of cardiac dysfunction in congestive heart failure. Can J Cardiol 9:873–887
54. Dhalla NS, Liu X, Panagia V, Takeda N (1998) Subcellular remodeling and heart dysfunction in chronic diabetes. Cardiovasc Res 40:239–247
55. Dhalla NS, Saini HK, Tappia PS et al (2007) Potential role and mechanisms of subcellular remodeling in cardiac dysfunction due to ischemic heart disease. J Cardiovasc Med 8:238–250
56. Dhalla NS, Rangi S, Babick AP et al (2012) cardiac remodeling and subcellular defects in heart failure due to myocardial infarction and aging. Heart Fail Rev 17:671–681
57. Capasso JM, Remily RM, Smith RH, Sonnenblick EH (1983) Sex differences in myocardial contractility in the rat. Basic Res Cardiol 78:156–171
58. Wang SN, Wyeth RP, Kennedy RH (1998) Effects of gender on the sensitivity of rat cardiac muscle to extracellular Ca^{2+} Eur J Pharmacol 361:73–77
59. Ishii K, Kano T, Ando J (1988) Sex differences in ^{3}H nitrendipine binding and effects of sex steroid hormones in rat cardiac and cerebral membranes. Jpn J Pharmacol 46:117–125
60. Schwertz DW, Vizgirda V, Solaro RJ et al (1999) Sexual dimorphism in rat left atrial function and response to adrenergic stimulation. Mol Cell Biochem 200:143–153
61. Gardner JD, Murray DB, Voloshenyuk TG et al (2010) Estrogen attenuates chronic volume overload induced structural and functional remodeling in male rat hearts. Am J Physiol Heart Circ Physiol. 298:H497–H504
62. Wankerl M, Böhm M, Morani I et al (1990) Calcium sensitivity and myosin light chain pattern of atrial and ventricular skinned cardiac fibers from patients with various kinds of cardiac disease. J Moll Cell Cardiol 22:1425–1438
63. Ayaz O, Howlett SE (2015) Testosterone modulates cardiac contraction and calcium homeostasis: cellular and molecular mechanisms. Biol Sex Differ 6:9
64. Dent MR, Tappia PS, Dhalla NS (2010) Gender differences in apoptotic signaling in heart failure due to volume overload. Apoptosis 15:499–510
65. Bupha-Intr T, Wattanapermpool J (2006) Regulatory role of ovarian sex hormones in calcium uptake activity of cardiac sarcoplasmic reticulum. Am J Physiol Heart Circ Physiol 291:H1101–H1108
66. Kravtsov GM, Kam KW, Liu J et al (2007) Altered Ca^{2+} handling by ryanodine receptor and Na^{+}-Ca^{2+} exchange in the heart from ovariectomized rats: role of protein kinase A. Am J Physiol Cell Physiol 292:C1625–C1635
67. Scheuer J, Malhotra A, Schaible TF et al (1987) Effects of gonadectomy and hormonal replacement on rat hearts. Circ Res 61:12–19

68. Golden KL, Marsh JD, Jaing Y (2002) Castration reduces mRNA levels for calcium regulatory proteins in rat heart. Endocrine 19:339–344
69. Golden KL, Marsh JD, Jaing Y et al (2003) Gonadectomy of adult male rats reduces contractility of isolated cardiac myocytes. Am J Physiol Endocrinol Metab 285:E449–E453
70. Golden KL, Marsh JD, Jaing Y (2004) Testosterone regulates mRNA levels of calcium regulatory proteins in cardiac myocytes. Endocrine 36:197–202
71. Mahmoodzadeh S, Dworatzek E (2019) The role of 17β-estradiol and estrogen receptors in regulation of Ca^{2+} channels and mitochondrial function in cardiomyocytes. Front Endocrinol 10:310
72. Bening C, Hamouda K, Leyh R (2016) Sex differences in volume overload in skinned fibers. BMC Cardiovasc Disord 16:197
73. Prévilon M, Pezet M, Semprez F et al (s2011) FKBP12.6 mice display temporal gender differences in cardiac Ca^{2+}-signaling phenotype upon chronic pressure overload. Can J Physiol Pharmacol 89:769–782
74. Müller AL, Dhalla NS (2012) Role of various proteases in cardiac remodeling and progression of heart failure. Heart Fail Rev 17:395–409
75. Yousef ZR, Redwood SR, Marber MS (2000) Postinfarction left ventricular remodeling: where are the theries and trials leading us? Heart 83:76–80
76. Duhamel TA, Dhalla NS (2007) New insights into the causes of heart failure. Drug Discov Today Dis Mech 4:175–184

Chapter 7
Sex-Specific Differences of Apoptosis in Heart Failure Due to Volume-Overload

Sukhwinder K. Bhullar, Anureet K. Shah, and Naranjan S. Dhalla

Abstract Volume overload induced by arteriovenous shunt for 4 and 16 weeks has been shown to produce sex-specific alterations in cardiac hypertrophy and heart failure, respectively. These changes were accompanied by sex-dependent alterations in the pro- and anti-apoptotic protein content and cardiomyocyte apoptosis in the heart. Cardiac hypertrophy in both male and female hearts produced a small depression in the extent of apoptosis without any change in mRNA levels for caspases 3 and 9. On the other hand, heart failure in males, unlike females, showed a marked increase in apoptosis and elevated mRNA levels for both caspase isoforms. Content for unphosphorylated and phosphorylated Bad proteins as well as Bax protein content in failing male hearts were higher than those in female hearts. Phosphorylated Bcl-2 protein content in male failing hearts were lower and that for females were higher in comparison to the respective sham control values. Increased apoptosis as well as the protein content for caspase 3, caspase 9, phosphorylated Bad and Bax in 16 weeks AV-shunt ovariectomized animals were attenuated by treatment with estrogen. AV-shunt induced alterations in Bcl-2 and phosphorylated Bcl-2 protein content in ovariectomized hearts were also prevented by estrogen treatment. These alterations in cardiomyocyte apoptosis as well as, pro-and anti-apoptotic factors in the heart may provide a possible mechanism to explain the sex-specific differences in the cardiac remodelling and cardiac function induced by volume overload.

Keywords Volume-overload · Heart failure · Cardiac hypertrophy · Cardiomyocyte apoptosis · Anti-apoptotic factors · Pro-apoptotic factors

---

S. K. Bhullar · N. S. Dhalla (✉)
Institute of Cardiovascular Sciences, St. Boniface Hospital Albrechtsen Research Centre, Department of Physiology and Pathophysiology, Max Rady College of Medicine, University of Manitoba, Winnipeg, Manitoba R2H 2A6, Canada
e-mail: nsdhalla@sbrc.ca

A. K. Shah
Department of Kinesiology, Nutrition and Food Science, California State University, Los Angeles, CA, USA

B. Ostadal and N. S. Dhalla (eds.), *Sex Differences in Heart Disease*, Advances in Biochemistry in Health and Disease 21,
https://doi.org/10.1007/978-3-030-58677-5_7

Introduction

Apoptosis is a tightly regulated and energy-requiring mechanism in which cell death follows a programmed sequence of structural changes including cell shrinkage, plasma membrane blebbing, nuclear condensation, DNA fragmentation, formation of apoptotic bodies in the nucleus and inflammation [1–5]. Various cellular stress signals including intracellular Ca^{2+}-overload, alterations in cytokines, loss of growth factors, oxidative stress, functional hypoxia and nitric oxide are considered to participate in the occurrence of apoptosis [6–16]. Apoptosis, in the cell is initiated by two major pathways: (i) extrinsic/death receptor pathway via ligand binding to external death receptors and (ii) intrinsic/mitochondrial pathway near the surface of the mitochondria. These are regulated by pro- and anti-apoptotic protein content including caspases 3 and 9, Bcl-2, Bad as well as BAX, which play a critical role in the development of apoptotic cell death in the myocardium [16–22].

Since apoptosis causes structural deformities in the myocardium, which has a limited ability for self-renewal, an increase in the rate of cardiomyocyte apoptosis has been shown to profoundly affect cardiac contractility and induce cardiac dysfunction [23–27]. Contribution of this mechanism is now well known in the development of contractile heart failure due to different etiologies such as myocardial infarction [28–34], ischemia-reperfusion injury [35–38], pressure overload [21, 39–44], volume overload [45–54], cardiomyopathies (dilated, ischemic, toxic and hypertrophic) [24–26, 55–70], hypertension [34, 71, 72], as well as atherosclerosis and restenosis [73–77]. Although cardiac hypertrophy is considered to be an adaptive mechanism for maintaining heart function, the ongoing loss of cardiomyocytes by apoptosis may result in the transition of cardiac hypertrophy to heart failure. In view of the well known differences in the responses of male and female hearts to various stressful situations, it appears that the development of cardiac hypertrophy and heart failure are gender-dependent events. Accordingly, the present article is intended to focus discussion on whether the occurrence of cardiomyocyte apoptosis and alterations of the associated pro-and anti-apoptotic proteins in different types of heart diseases are gender-dependent. In particular, it is planned to discuss the role and mechanisms of cardiomyocyte apoptosis in male and female hearts during the development of cardiac hypertrophy and heart failure upon the induction of volume overload.

Gender Differences of Cardiomyocyte Apoptosis in Heart Disease

It is now well known that the status of cardiac function and the extent of cardiomyocyte apoptosis in healthy males are essentially not different from those in the normal female hearts. However, cardiomyocyte apoptosis and cardiac dysfunction in diverse heart diseases have been found more prevalent in men compared to women

[21, 25, 31–34, 44–46, 78–88]. This could be due to the difference in the presence of apoptosis-related genes on the Y chromosomes or the role of estrogen and estrogen receptors in female population [25, 89–94]. Also, with aging the prevalence of cardiomyocyte apoptosis activity in human hearts has been found more pronounced in males than females [95–98]. In this regard, it is noteworthy that, unlike the females, a 4-fold increase in cardiomyocyte apoptosis has been described in the senescent male monkey hearts; this observation has signified the role of aging in gender differences [98].

Cardiomyocyte loss, which is invariably associated with heart failure, has been reported to exhibit gender difference [31, 99]. The extent of cardiomyocyte apoptosis was lower in female failing hearts in various heart diseases such as myocardial infarction [100], volume overload [25], pulmonary hypertension [101], and different cardiomyopathies [25, 88, 102, 103]. Improved cardiac function, and a greater reversibility of heart failure were observed in women with dilated cardiomyopathy [104–106]. However, the occurrence of heart failure as well as morbidity and mortality in patients with ≥20 years of age, were found to be equal in both genders [107]. Furthermore, the treatment approaches for heart failure in both males and females are similar except for using angiotensin-converting enzyme inhibitors and angiotensin receptor antagonists in females during pregnancy due to the risk of birth defects [108–111]. Nonetheless, there are dramatic sex differences with respect to heart dysfunction during the development of heart failure [112]. In comparison to males, females with heart failure have shown improved survival and were found to maintain normal left ventricular size and preserve ejection fraction when challenged with myocardial infarction [55, 93, 113–118]; atherosclerotic burden [93, 119]; and ischemia/reperfusion injury [114, 120–126]. In addition to, common causes including aging, pregnancy, environmental factors and lifestyle changes, there are other pathological influences such as hypertension which have shown sex differences with respect to hormonal imbalances, as well as cardiac remodeling processes, and signal transduction mechanisms for apoptosis [88, 89, 108, 127–132].

A growing number of studies have supported the view regarding gender differences for cardiomyocyte apoptosis in heart failure. Compared to men, a markedly lower extent of cardiomyocytes apoptosis was seen in the women's failing hearts [133, 134]. In hearts of patients undergoing cardiac transplantation, a significantly lesser degree of apoptosis in female heart was observed in comparison to males, and it has been pointed out that higher cardiomyocyte death in men was associated with earlier onset of heart failure [55]. It was also reported that due to increased extent of cardiomyocyte apoptosis after myocardial infarction [100], men were at much higher risk of worsened myocardial function and cardiac rupture while women showed greater myocardial retrieve [135]. Furthermore, females were found relatively more protected than males in response to prolonged acute coronary ischemia [55, 93, 110]. It may also be noted that there is a rapid progress in the transition from cardiac hypertrophy to heart failure in males compared to females, and cardiomyocyte apoptosis may play a role at the late stage of this transition [58, 63, 136–142]. The development of heart failure in pathological conditions such as pressure overload [143–149] and volume overload [25, 150–152] have been observed to be slower in females compared to males. These observations indicate sex-specific differences concerning apoptosis in a wide variety of cardiovascular diseases and provide evidence that cardiomyocyte apoptosis may play an important role in determining the gender-dependent differences in the development of heart failure.

Gender Differences of Cardiomyocyte Apoptosis in Heart Failure Due to Volume Overload

Since cardiomyocytes apoptosis plays a significant role in transition from cardiac hypertrophy to heart failure, this section is concerned with discussion on differences in cardiomyocyte apoptosis during the development of cardiac hypertrophy and heart failure induced by volume overload as a consequence of arteriovenous (AV) shunt in male and female rats [25]. Upon induction of volume overload, the degree of cardiac hypertrophy was increased and the cardiac output was unaltered, but the extent of apoptosis was depressed in both male and female hearts at early stage (4 weeks) (Fig. 7.1a). These events were not accompanied by any changes in mRNA levels for both caspase 3 or caspase 9 (Fig. 7.1b, c). On the other hand, at late stage of heart failure (16 weeks), the parameters of cardiac performance were considerably reduced in males hearts, while, these were maintained in female hearts [25]. Compared to the females, the levels of mRNA for caspase 3 and caspase 9 were higher (Fig. 7.1d, e) and extent of apoptosis was significantly greater in the male hearts (Fig. 7.1a). Furthermore, the data for different signaling transduction pathways involved in the regulation of apoptosis at 16 weeks volume overload are shown in Fig. 7.2a–e. The pro-apoptotic Bad protein content was higher in both phosphorylated and non- phosphorylated forms in male failing hearts whereas no change was seen in female volume overload hearts (Fig. 7.2a, b). The anti-apoptotic Bcl-2 protein content was remained unaltered in the non-phosphorylated form in both sexes (Fig. 7.2c) but the level of phosphorylated form was higher in female hearts and lower in male hearts upon the induction of volume overload for 16 weeks (Fig. 7.2d). Furthermore, protein content of pro-apoptotic factor BAX was higher in male failing hearts but was unaltered in the female volume overload heart (Fig. 7.2e). These results have demonstrated that the potential for the occurrence of apoptosis in the female heart is less than the male heart at the hypertrophic as well as the failing stages after the induction of volume overload. It appears that the observed decrease in the extent of apoptosis in both males and females may be due to an adaptive response to volume overload at the early stage of cardiac hypertrophy. However, the male failing heart showed a marked increase in the extent of apoptosis whereas female volume overload heart exhibited a lesser extent of apoptosis 16 weeks post AV shunt. These observations are consistent with the view that there are sex-specific differences in volume overload induced heart failure.

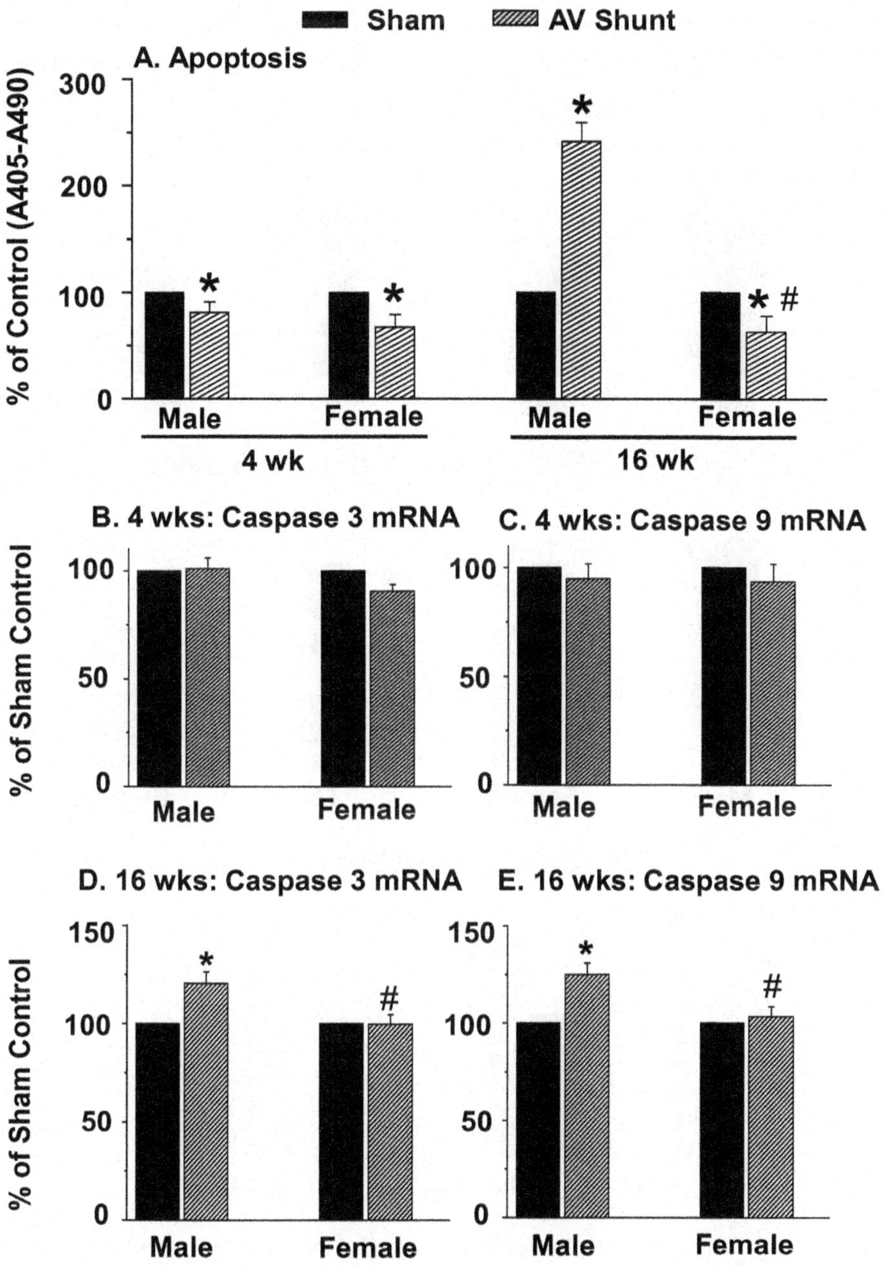

Fig. 7.1 Sex-specific changes in apoptosis as well as mRNA levels for caspases 3 and 9 in sham and AV shunt failing (4 weeks and 16 weeks) hearts due to volume overload in male and female rats. Data are taken from our paper—Dent MR, Tappia PS, Dhalla NS. Apoptosis 15: 499–510, 2010. Erratum in: Apoptosis. 16, 757–758, 2011. *_ $P < 0.05$ versus sham; #_ $P < 0.05$ versus corresponding value for male

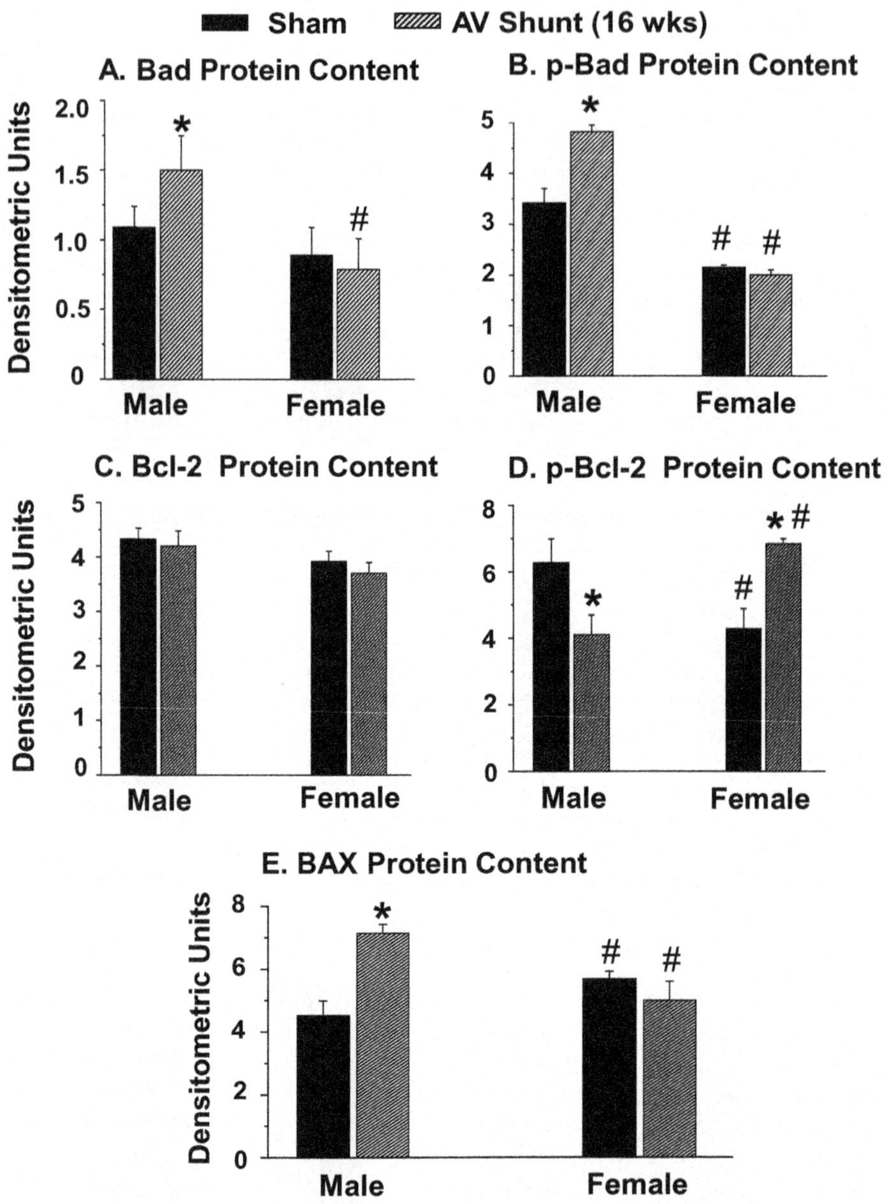

Fig. 7.2 Sex-specific changes in Bad, phosphorylated Bad (p-Bad), Bcl-2, phosphorylated Bcl-2 (p- Bcl-2) and BAX protein content in sham and 16 weeks AV shunt hearts due to volume overload in male and female rats. Data are taken from our paper—Dent MR, Tappia PS, Dhalla NS. Apoptosis 15: 499–510, 2010. Erratum in: Apoptosis. 16, 757–758, 2011. *_ $P < 0.05$ versus sham; #_ $P < 0.05$ versus corresponding value for male

Role of Estrogen in Cardiomyocyte Apoptosis Due to Volume Overload

Several studies have been reported that cardiomyocyte apoptosis is affected by sex hormones such as estrogen, testosterone and progesterone, which exert different actions during the development of heart failure in males and females [25, 92–94, 120, 153–163]. The effects of estrogens in the females include the delay in heart failure and improved survival during the pre-menopausal period in addition to slowing down the development of left ventricular hypertrophy [112, 144, 164–167]. Moreover, in comparison to male heart, the loss of viable cardiomyocytes was found to be markedly low in the healthy aging female heart [141, 145, 167, 168]. These effects in female hearts are believed to be due to estrogens and estrogen receptors signaling, particularly mediated by the estrogen receptor-β [90, 152–154, 169–172]. Since sex-specific differences in cardiovascular systems have been attributed to the consequence of higher levels of estrogen, as well as the presence or absence of other hormones, which modify apoptosis and reduce the risk of heart failure in women [144], estrogen has been suggested to prevent cardiomyocyte apoptosis due to volume overload in females [25].

The extent of cardiomyocyte apoptosis was increased markedly in ovariectomized female rats upon induction of volume overload at 16 weeks (Fig. 7.3a), but was depressed upon treatment of these animals with estrogen (Fig. 7.3a). It was seen that the level of protein content for caspase 3 was increased, whereas, that for caspase 9 remained unchanged due to volume overload in ovariectomized animals. Treatment of ovariectomized volume overload rats with estrogen, protein content for both caspase 3 and caspase 9 was observed to be depressed (Fig. 7.3b, c). Alterations in different signaling transduction pathways involved in the regulation of apoptosis after inducing AV shunt in the ovariectomized rats are depicted in Fig. 7.4a–e. Induction of volume overload in ovariectomized rats were found to reduce protein content for Bcl-2 and phosphorylated Bcl-2 but increased phosphorylated Bad and BAX protein content without any changes in bad protein content (Fig. 7.4). It can also be seen from Fig. 7.4d that protein content for phosphorylated Bcl-2 was increased in sham control ovariectomized rats. Treatment of ovariectomized volume overload rats with estrogen was found to attenuate the ovariectomy-induced changes in these parameters fully or partially (Fig. 7.4). Thus, estrogen may modify both pro-apoptotic and anti-apoptotic regulatory mechanisms and prevent the occurrence of cardiomyocyte apoptosis in females. Accordingly, these observations are consistent with the view that estrogen may play an important role in determining the reduced susceptibility of females to different cardiovascular risk factor.

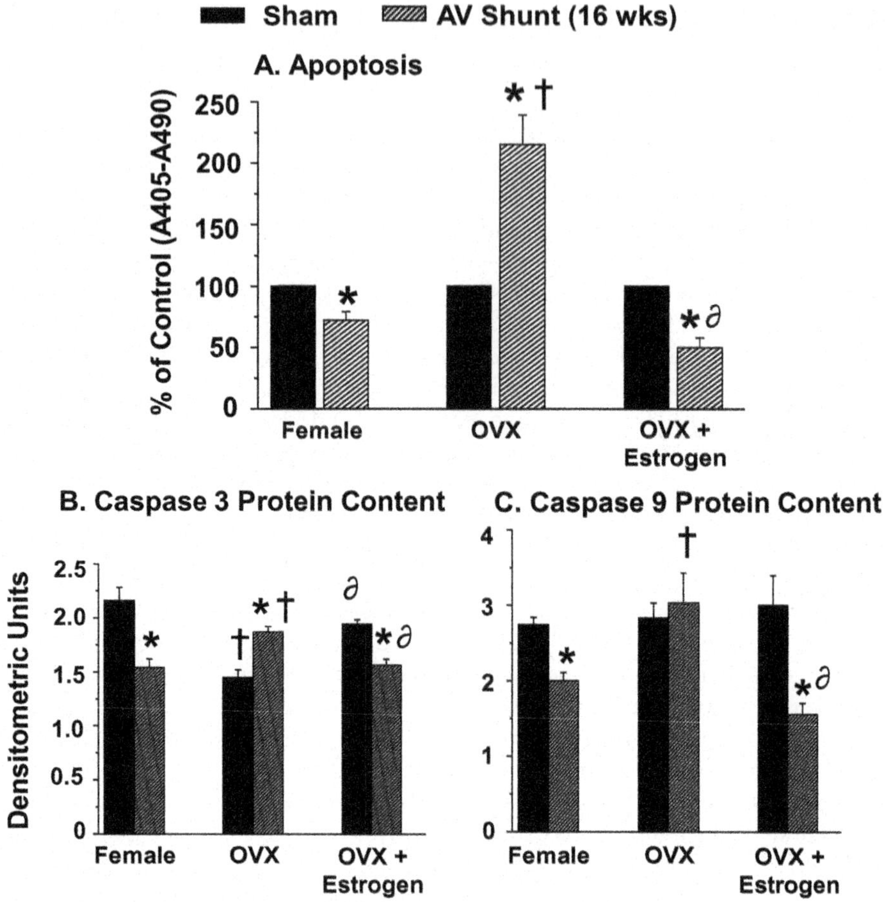

Fig. 7.3 Changes in apoptosis as well as protein content for caspases 3 and 9 in sham and 16 weeks post AV-shunt female rats with or without ovariectomy (OVX) as well as OVX animals treated with estrogen. Data are taken from our paper – Dent MR, Tappia PS, Dhalla NS. Apoptosis 15: 499–510, 2010. Erratum in: Apoptosis. 16, 757–758, 2011. *_ $P < 0.05$ vs sham; †_ $P < 0.05$ versus corresponding value for control female; ∂_ $P < 0.05$ versus corresponding value for OVX animals without 17-β estradiol treatment

Conclusions

This study has provided evidence that cardiac hypertrophy due to induction of volume overload at early stages in both male and female rats was associated with an equal extent of depressions in cardiomyocyte apoptosis without any changes in the mRNA levels of caspase 3 and 9. However, a marked degree of apoptosis in male hearts, unlike female hearts, was evident in heart failure at late stage of volume overload. These sex-dependent differences in the cardiomyocyte apoptosis in the failing heart

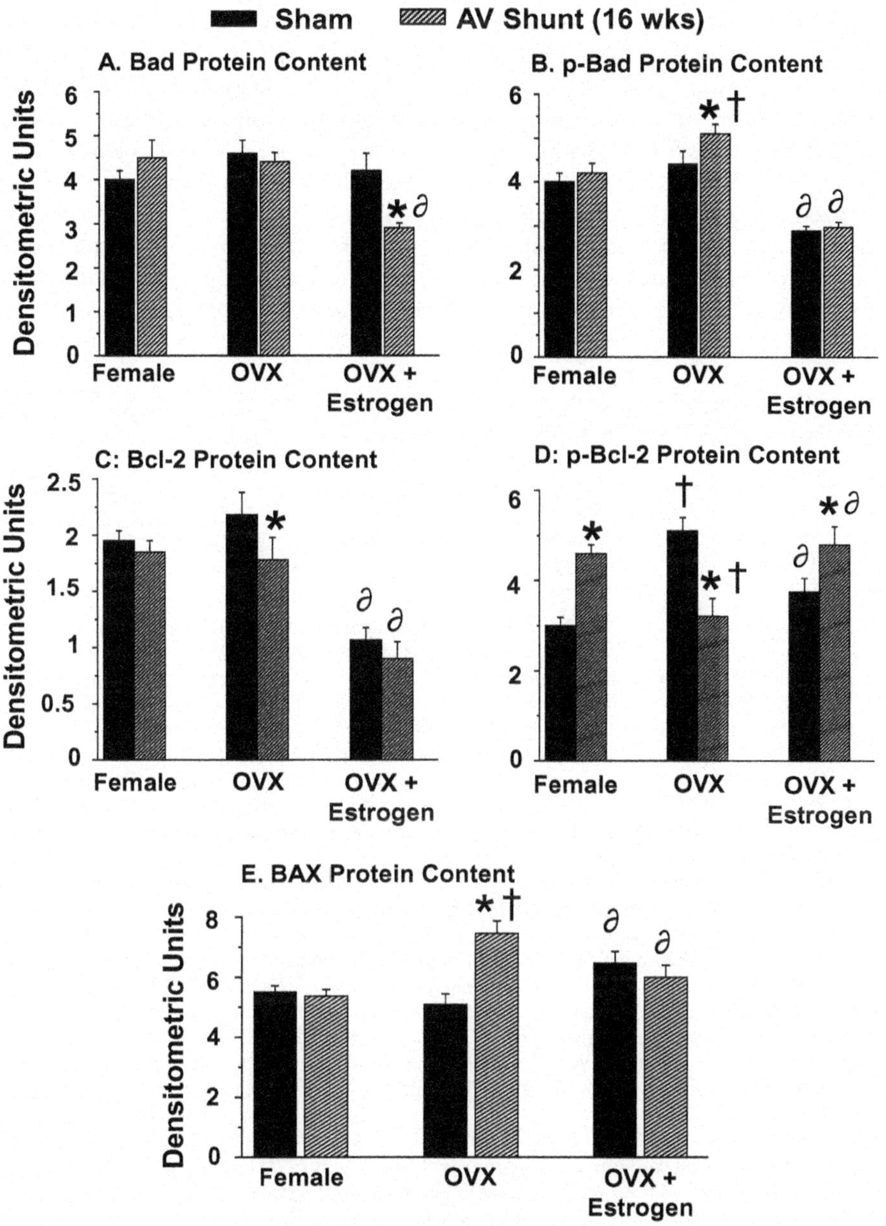

Fig. 7.4 Changes in Bad, phosphorylated Bad (p-Bad), Bcl-2, phosphorylated Bcl-2 (p- Bcl-2), and BAX protein content in sham and 16 weeks post AV-shunt female rats with or without ovariectomy (OVX) as well as OVX animals treated with estrogen (16 weeks post AV-shunt). Data are taken from our paper – Dent MR, Tappia PS, Dhalla NS. Apoptosis 15: 499–510, 2010. Erratum in: Apoptosis. 16, 757–758, 2011. *_ $P < 0.05$ vs sham; †_ $P < 0.05$ versus corresponding value for control female; ∂_ $P < 0.05$ versus corresponding value for OVX animals without 17-β estradiol treatment

were associated elevated levels of mRNA for both caspase 3 and 9. Sex-specific differences in some pro-apoptotic and anti apoptotic parameters such as Bad, BAX and Bcl-2 protein content were also seen in the failing hearts due to volume overload. Volume overload in ovariectomized animals was found to increase the extent of apoptosis in association with increased levels of caspase 3 as well as BAX protein content and decreased protein of phosphorylated Bcl-2. These alterations in ovariectomized volume overload animals were partially or fully prevented by treatment with estrogen. These observations support the view that cardiomyocyte apoptosis and associated heart failure is sex-specific. Furthermore, this study provides evidence for the role of estrogen in reducing the susceptibility of female population to different cardiovascular risk factor.

Acknowledgments The infrastructure support for this project was provided by the St. Boniface Hospital Research Foundation, Winnipeg, Canada.

References

1. Kerr JFR, Wyllie AH, Currie AR (1972) Apoptosis: a basic biological phenomenon with wide-ranging implications in tissue kinetics. Br J Cancer 26:239–257
2. Wyllie AH, Kerr JF, Currie AR (1980) Cell death: the significance of apoptosis. Int Rev Cytol 68:251–306
3. Searle J, Kerr JFR, Bishop CJ (1982) Necrosis and apoptosis: distinct modes of cell death with fundamentally different significance. Pathol Annu 17:229–259
4. Ucker DS (1991) Death by suicide: one way to go in mammalian cellular development? New Biol 3:103–109
5. Arends MJ, Wyllie AH (1991) Apoptosis: mechanisms and roles in pathology. Int Rev Exp Pathol 32:223–254
6. Brown KA, Page MT, Nguyen C et al (1996) Tumor necrosis factor alpha-induced apoptosis in cardiac myocytes. Involvement of the sphingolipid signaling cascade in cardiac cell death. J Clin Invest 98:2854–2865
7. Xu YJ, Saini HK, Zhang M et al (2006) MAPK activation and apoptotic alterations in hearts subjected to calcium paradox are attenuated by taurine. Cardiovasc Res 72:163–174
8. Nakamura T, Ueda Y, Juan Y et al (2000) Fas-mediated apoptosis in adriamycin-induced cardiomyopathy in rats: in vivo study. Circulation 102:572–578
9. Azhar G, Liu L, Zhang X, Wei JY (1999) Influence of age on hypoxia/reoxygenation-induced DNA fragmentation and bcl-2, bcl-xl, bax and fas in the rat heart and brain. Mech Ageing Dev 112:5–25
10. Chao W, Shen Y, Li L et al (2002) Importance of FADD signaling in serum-deprivation- and hypoxia-induced cardiomyocyte apoptosis. J Biol Chem 277:31639–31645
11. Ing DJ, Zang J, Dzau VJ et al (1999) Modulation of cytokine-induced cardiac myocyte apoptosis by nitric oxide. Bak and Bcl-x. Circ Res 84:21–33
12. Bialik S, Cryns VL, Drincic A et al (1999) The mitochondrial apoptotic pathway is activated by serum and glucose deprivation in cardiac myocytes. Circ Res 85:403–414
13. de Moissac D, Gurevich RM, Zheng H et al (2000) Caspase activation and mitochondrial cytochrome c release during hypoxia-mediated apoptosis of adult ventricular myocytes. J Mol Cell Cardiol 32:53–63
14. Adams JW, Pagel AL, Means CK et al (2000) Cardiomyocyte apoptosis induced by Gαq signaling is mediated by permeability transition pore formation and activation of the mitochondrial death pathway. Circ Res 87:1180–1187

15. Itoh G, Tamura J, Suzuki M et al (1995) DNA fragmentation of human infarcted myocardial cells demonstrated by the nick end labeling method and DNA agarose gel electrophoresis. Am J Pathol 146:1325–1331
16. Bishopric NH, Andreka P, Slepak T, Webster KA (2001) Molecular mechanisms of apoptosis in the cardiac myocyte. Curr Opin Pharmacol 1:141–150
17. Kitsis RN, Narula J (2008) Introduction – cell death in heart failure. Heart Fail Rev 13:107–109
18. Hengartner MO (2000) The biochemistry of apoptosis. Nature 407:770–776
19. Antonsson B, Martinou JC (2000) The Bcl-2 protein family. Exp Cell Res 256:50–57
20. Adams JM, Cory S (1998) The Bcl-2 protein family: arbiters of cell survival. Science 281:1322–1326
21. Condorelli G, Morisco C, Stassi G et al (1999) Increased cardiomyocyte apoptosis and changes in proapoptotic and antiapoptotic genes bax and bcl-2 during left ventricular adaptations to chronic pressure overload in the rat. Circulation 99:3071–3078
22. Regula KM, Kirshenbaum LA (2005) Apoptosis of ventricular myocytes: a means to an end. J Mol Cell Cardiol 38:3–13
23. Tamarappoo BK, John BT, Reinier K et al (2012) Vulnerable myocardial interstitium in patients with isolated left ventricular hypertrophy and sudden cardiac death: a post-mortem histological evaluation. J Am Heart Assoc 1:1–9
24. Dent MR, Das S, Dhalla NS (2007) Alterations in both death and survival signals for apoptosis in heart failure due to volume overload. J Mol Cell Cardiol 43:726–732
25. Dent MR, Tappia PS, Dhalla NS. Gender differences in apoptotic signaling in heart failure due to volume overload. Apoptosis 15: 499–510, 2010a. Erratum in: Apoptosis 16, 757–758, 2011
26. Das S, Babick AP, Xu YJ et al (2010) TNF-alpha-mediated signal transduction pathway is a major determinant of apoptosis in dilated cardiomyopathy. J Cell Mol Med 14:1988–1997
27. Hassan AF, Kamal MM (2013) Effect of exercise training and anabolic androgenic steroids on hemodynamics, glycogen content, angiogenesis and apoptosis of cardiac muscle in adult male rats. Int J Health Sci. 7:47–60
28. Kajstura J, Cheng W, Reiss K et al (1996) Apoptotic and necrotic myocyte cell deaths are independent contributing variables to infarct size in rats. Lab Invest 74:86–107
29. Baldi A, Abbate A, Bussani R et al (2002) Apoptosis and post-infarction left ventricular remodelling. J Mol Cell Cardiol 34:165–174
30. Kurrelmeyer K, Michael L, Baumgarten G et al (2000) Endogenous myocardial tumor necrosis factor protects the adult cardiac myocyte against ischemic-induced apoptosis in a murine model of acute myocardial infarction. Proc Natl Acad Sci USA 290:5456–5461
31. Cheng W, Kajstura J, Nitahara JA et al (1996) Programmed cell death affects the viable myocardium after infarction in rats. Exp Cell Res 226:316–327
32. Narula J, Haider N, Virmani R et al (1996) Apoptosis in myocytes in end-stage heart failure. N Engl J Med 335:1182–1189
33. Olivetti G, Abbi R, Quaini F et al (1997) Apoptosis in the failing human. N Engl J Med 336:1131–1141
34. Mihailidou AS, Loan Le TY, Mardini M, Funder JW (2009) Glucocorticoids activate cardiac mineralocorticoid receptors during experimental myocardial infarction. Hypertension 54:1306–1312
35. Gottlieb R (2005) ICE-ing the heart. Circ Res 96:1036–1038
36. Yaoita H, Ogawa K, Maehara K et al (1998) Attenuation of ischemia/reperfusion injury in rats by a caspase inhibitor. Circulation 97:276–281
37. Meldrum DR, Wang M, Tsai BM et al (2005) Intracelllular signaling mechanisms of sex hormones in acute myocardial inflammation and injury. Front Biosci 10:1835–1867
38. Booij HG, Yu H, De Boer RA et al (2016) Overexpression of A kinase interacting protein 1 attenuates myocardial ischaemia/reperfusion injury but does not influence heart failure development. Cardiovasc Res 111:217–226
39. Bing OH (1994) Hypothesis: apoptosis may be a mechanism for the transition to heart failure with chronic pressure overload. J Mol Cell Cardiol 26:943–948

40. Fu HY, Okada K, Liao Y et al (2010) Ablation of C/EBP homologous protein attenuates endoplasmic reticulum-mediated apoptosis and cardiac dysfunction induced by pressure overload. Circulation 122:361–369
41. O'Connell TD, Swigart PM, Rodrigo MC et al (2006) Alpha1-adrenergic receptors prevent a maladaptive cardiac response to pressure overload. J Clin Invest 116:1005–1015
42. Zankov DP, Sato A, Shimizu A, Ogita H (2017) Differential effects of myocardial afadin on pressure overload-induced compensated cardiac hypertrophy. Circ J 81:1862–1870
43. Chen C, Zou LX, Lin QY et al (2019) Resveratrol as a new inhibitor of immunoproteasome prevents PTEN degradation and attenuates cardiac hypertrophy after pressure overload. Redox Biol 20:390–401
44. Gogiraju R, Hubert A, Fahrer J et al (2019) Endothelial leptin receptor deletion promotes cardiac autophagy and angiogenesis following pressure overload by suppressing Akt/mTOR signaling. Circ Heart Fail 12:1–16
45. Chen YW, Pat B, Gladden JD et al (2011) Dynamic molecular and histopathological changes in the extracellular matrix and inflammation in the transition to heart failure in isolated volume overload. Am J Physiol Heart Circ Physiol 300:H2251–H2260
46. Moorjani N, Westaby S, Narula J et al (2009) Effects of left ventricular volume overload on mitochondrial and death-receptor-mediated apoptotic pathways in the transition to heart failure. Am J Cardiol 103:1261–1268
47. Treskatsch S, Shakibaei M, Feldheiser A et al (2015) Ultrastructural changes associated with myocardial apoptosis, in failing rat hearts induced by volume overload. Int J Cardiol 197:327–332
48. Kolpakov MA, Seqqat R, Rafiq K et al (2009) Pleiotropic effects of neutrophils on myocyte apoptosis and left ventricular remodeling during early volume overload. J Mol Cell Cardiol 47:634–645
49. Mohamed BA, Schnelle M, Khadjeh S et al (2016) Molecular and structural transition mechanisms in long-term volume overload. Eur J Heart Fail 18:362–371
50. Reddy S, Zhao M, Hu DQ et al (2013) Physiologic and molecular characterization of a murine model of right ventricular volume overload. Am J Physiol Heart Circ Physiol 304:H1314–H1327
51. Shaqura M, Mohamed DM, Aboryag NB et al (2017) Pathological alterations in liver injury following congestive heart failure induced by volume overload in rats. PLoS ONE 12:1–19
52. Aboryag NB, Mohamed DM, Dehe L et al (2017) histopathological changes in the kidney following congestive heart failure by volume overload in rats. Oxid Med Cell Longev 1–10:2017
53. Wang X, Ren B, Liu SY et al (2003) Characterization of cardiac hypertrophy and heart failure due to volume overload in the rat. J Appl Physiol 94:752–763
54. Dent MR, Dhalla NS, Tappia PS (2004) Phospholipase C gene expression, protein content, and activities in cardiac hypertrophy and heart failure due to volume overload. Am J Physiol Heart Circ Physiol 287:H719–H727
55. Guerra S, Leri A, Wang X et al (1999) Myocyte death in the failing human heart is gender dependent. Circ Res 85:856–866
56. Chen SN, Lombardi R, Karmouch J et al (2019) dna damage response/tp53 pathway is activated and contributes to the pathogenesis of dilated cardiomyopathy associated with LMNA (Lamin A/C) mutations. Circ Res 124:856–873
57. Takagi C, Urasawa K, Yoshida I et al (1999) Enhanced GRK5 expression in the hearts of cardiomyopathic hamsters, J2N-k. Biochem Biophys Res Commun 262:206–210
58. Cesselli D, Jakoniuk I, Barlucchi L et al (2001) Oxidative stress-mediated cardiac cell death is a major determinant of ventricular dysfunction and failure in dog dilated cardiomyopathy. Circ Res 89:279–286
59. Takeda N (2003) Cardiomyopathy: molecular and immunological aspects. Int J Mol Med 11:13–26
60. Bennett, Martin R (2002) Apoptosis in the cardiovascular system. Heart 87:480–487

61. Rodrigues PG, Miranda-Silva D, Costa SM et al (2019) Early myocardial changes induced by doxorubicin in the nonfailing dilated ventricle. Am J Physiol Heart Circ Physiol 316:H459–H475
62. Xia Y, Chen Z, Chen A et al (2017) LCZ696 improves cardiac function via alleviating Drp1-mediated mitochondrial dysfunction in mice with doxorubicin-induced dilated cardiomyopathy. J Mol Cell Cardiol 108:138–148
63. Xie D, Liao Y, Wu B et al (2018) Cardiac nestin + cells derived from early stage of dilated cardiomyopathy enhanced the survival of the doxorubicin-injured cardiac muscle hl-1 cells. Int Heart J 59:180–189
64. Major JL, Salih M, Tuana BS (2017) E2F6 protein levels modulate drug induced apoptosis in cardiomyocytes. Cell Signal 40:230–238
65. Wu J, Guo W, Lin SZ et al (2016) Gp130-mediated STAT3 activation by S-propargyl-cysteine, an endogenous hydrogen sulfide initiator, prevents doxorubicin-induced cardiotoxicity. Cell Death Dis 7:1–13
66. Wu L, Chen Y, Chen Y et al (2019) Effect of HIF-1α/miR-10b-5p/PTEN on hypoxia-induced cardiomyocyte apoptosis. J Am Heart Assoc 8:1–15
67. De Angelis A, Piegari E, Cappetta D et al (2010) Anthracycline cardiomyopathy is mediated by depletion of the cardiac stem cell pool and is rescued by restoration of progenitor cell function. Circulation 121:276–292
68. Li Z, Bing OH, Long X et al (1997) Increased cardiomyocyte apoptosis during the transition to heart failure in the spontaneously hypertensive rat. Am J Physiol 272:H2313–H2319
69. Mitsuhashi S, Saito N, Watano K et al (2003) Defect of delta-sarcoglycan gene is responsible for development of dilated cardiomyopathy of a novel hamster strain, J2N-k: calcineurin/PP2B activity in the heart of J2N-k hamster. J Biochem 134:269–276
70. Sakamoto A, Ono K, Abe M et al (1997) Both hypertrophic and dilated cardiomyopathies are caused by mutation of the same gene, delta-sarcoglycan, in hamster: an animal model of disrupted dystrophin-associated glycoprotein complex. Proc Natl Acad Sci USA 94:13873–13878
71. Ren J (2007) Influence of gender on oxidative stress, lipid peroxidation, protein damage and apoptosis in hearts and brains from spontaneously hypertensive rats. Clin Exp Pharmacol Physiol 34:432–438
72. Liu JJ, Peng L, Bradley CJ et al (2000) Increased apoptosis in the heart of genetic hypertension, associated with increased fibroblasts. Cardiovasc Res 45:729–735
73. Bennett MR, Evan GI, Schwartz SM (1995) Apoptosis of human vascular smooth muscle cells derived from normal vessels and coronary atherosclerotic plaques. J Clin Invest 95:2266–2274
74. Moe GW, Marin-Garcia J (2016) Role of cell death in the progression of heart failure. Heart Fail Rev 21:157–167
75. Perlman H, Maillard L, Krasinski K et al (1997) Evidence for the rapid onset of apoptosis in medial smooth muscle cells after balloon injury. Circulation 95:981–987
76. Pollman MJ, Hall JL, Mann MJ et al (1998) Inhibition of neointimal cell Bcl-x expression induces apoptosis and regression of vascular disease. Nature Med 4:222–227
77. Shekhar A, Heeger P, Reutelingsperger C et al (2018) Targeted imaging for cell death in cardiovascular disorders. JACC Cardiovasc Imaging 11:476–493
78. Brill A, Torchinsky A, Carp H, Toder V (1999) The role of apoptosis in normal and abnormal embryonic development. J Assist Reprod Genet 16:512–519
79. Reed, Douglas R (2011) Apoptosis: physiology and pathology. John C. Green (ed) Cambridge University Press, London. 978-0-521-88656-7
80. Singh R, Letai A, Sarosiek K (2019) Regulation of apoptosis in health and disease: the balancing act of BCL-2 family proteins. Nat Rev Mol Cell Biol 20:175–193
81. Cotter TG (2009) Apoptosis and cancer: the genesis of a research field. Nat Rev Cancer 9:501–507
82. Mattson MP (2000) Apoptosis in neurodegenerative disorders. Nat Rev Mol Cell Bio 1:120–129

83. Mughal W, Dhingra R, Kirshenbaum LA (2012) Striking a balance: autophagy, apoptosis, and necrosis in a normal and failing heart. Curr Hypertens Rep 14:540–547
84. Teringova E, Tousek P (2017) Apoptosis in ischemic heart disease. J Transl Med 87
85. Favaloro B, Allocati N, Graziano V et al (2012) Role of apoptosis in disease. Aging 4:330–349
86. Whelan RS, Kaplinskiy V, Kitsis RN (2010) Cell death in the pathogenesis of heart disease: mechanisms and significance. Annu Rev Physiol 72:19–44
87. Kessler EL, Rivaud MR, Vos MA, van Veen TAB (2019) Sex-specific influence on cardiac structural remodeling and therapy in cardiovascular disease. Biol Sex Differ 10:1–11
88. Meyer S, van der Meer P, van Tintelen JP, van den Berg MP (2014) Sex differences in cardiomyopathies. Eur J Heart Fail 16:238–247
89. Winham SJ, de Andrade M, Miller VM (2014) Genetics of cardiovascular disease: importance of sex and ethnicity. Atherosclerosis 241:219–228
90. Pedram A, Razandi M, Lubahn D et al (2008) Estrogen inhibits cardiac hypertrophy: role of estrogen receptor-β to inhibit calcineurin. Endocrinology 149:3361–3369
91. Voloshenyuk TG, Gardner JD (2010) Estrogen improves TIMP-MMP balance and collagen distribution in volume-overloaded hearts of ovariectomized females. Am J Physiol Integr Comp Physiol 299:683–693
92. Kararigas G, Fliegner D, Gustafsson J-A, Regitz-Zagrosek V (2011) Role of the estrogen/estrogen-receptor-beta axis in the genomic response to pressure overload-induced hypertrophy. Physiol Genomics 43:438–446
93. Dunlay SM, Roger VL (2012) Gender differences in the pathophysiology, clinical presentation, and outcomes of ischemic heart failure. Curr Heart Fail Rep 9:267–276
94. Chen C, Hu LX, Dong T et al (2013) Apoptosis and autophagy contribute to gender difference in cardiac ischemia-reperfusion induced injury in rats. Life Sci 93:265–270
95. Tower J (2015) Programmed cell death in aging. Ageing Res Rev 23:90–100
96. Mallat Z, Fornes P, Costagliola R et al (2001) Age and gender effects on cardiomyocyte apoptosis in the normal human heart. J Gerontol A Biol Sci Med Sci 56:M719–M723
97. Olivetti G, Giordano G, Corradi D et al (1995) Gender differences and aging: effects on the human heart. J Am Coll Cardiol 26:1068–1079
98. Zhang XP, Vatner SF, Shen YT et al (2007) Increased apoptosis and myocyte enlargement with decreased cardiac mass; distinctive features of the aging male, but not female, monkey heart. J Mol Cell Cardiol 43:487–491
99. Kang PM, Izumo S (2000) Apoptosis and heart failure. A critical review of the literature. Circ Res 86:1107–1113
100. Biondi-Zaccai GGL, Abbate A, Bussani R et al (2005) Reduced post-infarction myocardial apoptosis in women: a clue to their different clinical course? Heart 91:99–101
101. Shehata ML, Lossnitzer D, Skrok J et al (2011) Myocardial delayed enhancement in pulmonary hypertension: pulmonary hemodynamics, right ventricular function, and remodeling. Am J Roentgenol 196:87–94
102. Fairweather DL, Cooper LT, Blauwet LA (2013) Sex and gender differences in myocarditis and dilated cardiomyopathy. Curr Probl Cardiol 38:7–46
103. Akdis D, Saguner AM, Shah K et al (2017) Sex hormones affect outcome in arrhythmogenic right ventricular cardiomyopathy/dysplasia: from a stem cell derived cardiomyocyte-based model to clinical biomarkers of disease outcome. Eur Heart J 38:1498–1508
104. Arimura T, Onoue K, Takahashi-Tanaka Y et al (2013) Nuclear accumulation of androgen receptor in gender difference of dilated cardiomyopathy due to Lamin A/C mutations. Cardiovasc Res 99:382–394
105. Calkins H (2015) Arrhythmogenic right ventricular dysplasia/cardiomyopathy – three decades of Progress. Circ J 79:901–913
106. Bauce B, Frigo G, Marcus FI et al (2008) Comparison of clinical features of arrhythmogenic right ventricular cardiomyopathy in men versus women. Am J Cardiol 102:1252–1257
107. Benjamin EJ, Blaha MJ, Chiuve SE et al (2017) Heart disease and stroke statistics-2017 update: a report from the american heart association. Circulation 135:1–775

108. Bozkurt B, Shaden K (2017) Heart failure in women. Methodist Debakey Cardiovasc J 13:216–223
109. Cohn JN, Tognoni G (2001) A randomized trial of the angiotensin-receptor blocker valsartan in chronic heart failure. N Engl J Med 345:1667–1675
110. Pfeffer MA, Swedberg K, Granger CB et al (2003) Effects of candesartan on mortality and morbidity in patients with chronic heart failure: the CHARM-Overall programme. Lancet 362:759–766
111. Pitt B, Poole-Wilson PA, Segal R et al (2000) Effect of losartan compared with captopril on mortality in patients with symptomatic heart failure: randomised trial – the losartan heart failure survival study ELITE II. Lancet 355:1582–1587
112. Blenck CL, Harvey PA, Reckelhoff JF, Leinwand LA (2016) the importance of biological sex and estrogen in rodent models of cardiovascular health and disease. Circ Res 118:1294–1312
113. Wang F, He Q, Sun Y et al (2010) Female adult mouse cardiomyocytes are protected against oxidative stress. Hypertension 55:1172–1178
114. Cavasin MA, Tao Z, Menon S et al (2004) Gender differences in cardiac function during early remodeling after acute myocardial infarction in mice. Life Sci 75:2181–2192
115. Shaw LJ, Bairey Merz CN, Pepine CJ et al (2006) Insights from the NHLBI-sponsored women's ischemia syndrome evaluation (WISE) study: Part I: gender differences in traditional and novel risk factors, symptom evaluation, and gender-optimized diagnostic strategies. J Am Coll Cardiol 47:S4–S20
116. Kanaya AM, Grady D, Barrett-Connor E (2002) Explaining the sex difference in coronary heart disease mortality among patients with type 2 diabetes mellitus: a meta-analysis. Arch Intern Med 62:1737–1745
117. Gregg EW, Gu Q, Cheng YJ et al (2007) Mortality trends in men and women with diabetes, 1971 to 2000. Ann Intern Med 147:149–155
118. Taylor AL (2015) Heart failure in women. Curr Heart Failure Rep 12:187–195
119. Schulman-Marcus J, Hartaigh BO, Gransar H et al (2016) Sexspecific associations between coronary artery plaque extent and risk of major adverse cardiovascular events the CONFIRM long-term registry. JACC Cardiovasc Imaging 9:364–372
120. Wang M, Baker L, Tsai BM et al (2005) Sex differences in the myocardial inflammatory response to ischemia-reperfusion injury. Am J Physiol Endocrinol Metab 288:E321–E326
121. Wang M, Crisostomo PR, Markel TA et al (2008) Mechanisms of sex differences in TNFR2-mediated cardioprotection. Circulation 118:S38–S45
122. Wang M, Tsai BM, Kher A et al (2005) Role of endogenous testosterone in myocardial proinflammatory and proapoptotic signaling after acute ischemia-reperfusion. Am J Physiol Heart Circ Physiol 288:H221–H226
123. Alves MG, Machado NG, Sardao VA et al (2011) Anti-apoptotic protection afforded by cardioplegic celsior and histidine buffer solutions to hearts subjected to ischemia and ischemia/reperfusion. J Cell Biochem 112:3872–3881
124. Eefting F, Rensing B, Wigman J et al (2004) Role of apoptosis in reperfusion injury. Cardiovasc Res 61:414–426
125. Gill C, Mestril R, Samali A (2002) Losing heart: the role of apoptosis in heart disease-a novel therapeutic target? FASEB J 16:135–146
126. Liu H, Pedram A, Kim JK (2011) Oestrogen prevents cardiomyocyte apoptosis by suppressing p38 alpha-mediated activation of p53 and by down-regulating p53 inhibition on p38 beta. Cardiovasc Res 89:119–128
127. Regitz-Zagrosek V, Kararigas G (2017) Mechanistic pathways of sex differences in cardiovascular disease. Physiol Rev 97:1–37
128. Hartman RJG, Huisman SE, den Ruijter HM (2018) Sex differences in cardiovascular epigenetics-a systematic review. Biol Sex Differ 9:1–8
129. Chella Krishnan K, Mehrabian M, Lusis AJ (2018) Sex differences in metabolism and cardiometabolic disorders. Curr Opin Lipidol 29:404–410
130. Appelman Y, van Rijn BB, ten Haaf ME et al (2014) Sex differences in cardiovascular risk factors and disease prevention. Atherosclerosis 241:211–218

131. Salton CJ, Chuang ML, O'Donnell CJ, et al (2002) Gender differences and normal left ventricular anatomy in an adult population free of hypertension. A cardiovascular magnetic resonance study of the Framingham heart study offspring cohort. J Am Coll Cardiol 39:1055–1060
132. Sandstede J, Lipke C, Beer M et al (2000) Age- and gender-specific differences in left and right ventricular cardiac function and mass determined by cine magnetic resonance imaging. Eur Radiol 10:438–442
133. Narula J, Kharbanda S, Khaw BA (1997) Apoptosis and the heart. Chest 112:1358–1362
134. Narula J, Hajjar RJ, Dec GW (1998) Apoptosis in the failing heart. Cardiol Clin 16:691–710
135. Mehilli J, Ndrepepa G, Kastrati A et al (2005) Gender and myocardial salvage after reperfusion treatment in acute myocardial infarction. J Am Coll Cardiol 45:828–831
136. Cleland JG, Swedberg K, Follath F, et al (2003) The EuroHeart failure survey programme–a survey on the quality of care among patients with heart failure in Europe. Part 1: patient characteristics and diagnosis. Eur Heart J 24:442–463
137. Carroll JD, Carroll EP, Feldman T et al (1992) Sex-associated differences in left ventricular function in aortic stenosis of the elderly. Circulation 86:1099–1107
138. Rohini A, Agrawal N, Koyani CN, Singh R (2010) Molecular targets and regulators of cardiac hypertrophy. Pharmacol Res 61:269–280
139. Petrov G, Regitz-Zagrosek V, Lehmkuhl E et al (2010) Regression of myocardial hypertrophy after aortic valve replacement: faster in women? Circulation 122:S23–S28
140. Dent MR, Tappia PS, Dhalla NS (2011) Gender differences in beta-adrenoceptor system in cardiac hypertrophy due to arteriovenous fistula. J Cell Physiol 226:181–186
141. Crabbe DL, Dipla K, Ambati S et al (2003) Gender differences in post-infarction hypertrophy in end-stage failing hearts. J Am Coll Cardiol 41:300–306
142. Regitz-Zagrosek V, Seeland U (2011) Sex and gender differences in myocardial hypertrophy and heart failure. Wien Med Wochenschr 161:109–116
143. Schipke J, Grimm C, Arnstein G et al (2016) Cardiomyocyte loss is not required for the progression of left ventricular hypertrophy induced by pressure overload in female mice. J Anat 229:75–81
144. Fliegner D, Schubert C, Penkalla A et al (2010) Female sex and estrogen receptor-beta attenuate cardiac remodeling and apoptosis in pressure overload. Am J Physiol Regul Integr Comp Physiol 298:R1597–R1606
145. Aurigenvna GP, Gaasch WH (1995) Gender differences in older patients with pressure-overload hypertrophy of the left ventricle. Cardiol 86:310–317
146. Villari B, Campbell SE, Schneider J et al (1995) Sex-dependent differences in left ventricular function and structure in chronic pressure overload. Eur Heart J 16:1410–1419
147. Weinberg EO, Thienelt CD, Katz SE et al (1999) Gender differences in molecular remodeling in pressure overload hypertrophy. J Am Coll Cardiol 34:264–273
148. Previlon M, Pezet M, Vinet L et al (2014) Gender-specific potential inhibitory role of Ca^{2+}/calmodulin dependent protein kinase phosphatase (CaMKP) in pressure-overloaded mouse heart. PLoS ONE 9:1–12
149. Douglas PS, Katz SE, Weinberg EO et al (1998) Hypertrophic remodeling: gender differences in the early response to left ventricular pressure overload. J Am Coll of Cardiol 32:1118–1125
150. Gardner JD, Brower GL, Janicki JS (2002) Gender differences in cardiac remodeling secondary to chronic volume overload. J Card Fail 8:101–107
151. Beaumont C, Walsh-Wilkinson E, Drolet MC et al (2017) Female rats with severe left ventricle volume overload exhibit more cardiac hypertrophy but fewer myocardial transcriptional changes than males. Sci Rep 7:729
152. Dent MR, Tappia PS, Dhalla NS (2010) Gender differences in cardiac dysfunction and remodeling due to volume overload. J Card Fail 16:439–449
153. Van Eickels M, Grohe C, Cleutjens JP et al (2001) 17 β-estradiol attenuates the development of pressure-overload hypertrophy. Circulation 104:1419–1423
154. Skavdahl M, Steenbergen C, Clark J et al (2005) Estrogen receptor-β mediates male-female differences in the development of pressure overload hypertrophy. Am J Physiol Circ Physiol 288:H469–H476

155. Patten RD, Pourati I, Aronovitz MJ et al (2008) 17-Estradiol differentially affects left ventricular and cardiomyocyte hypertrophy following myocardial infarction and pressure overload. J Card Fail 14:245–253
156. Gardner JD, Murray DB, Voloshenyuk TG et al (2010) Estrogen attenuates chronic volume overload induced structural and functional remodeling in male rat hearts. Am J Physiol Heart Circ Physiol 298:H497–H504
157. Bouma W, Noma M, Kanemoto S et al (2010) Sex-related resistance to myocardial ischemia-reperfusion injury is associated with high constitutive ARC expression. Am J Physiol 298:H1510–H1517
158. Huang C, Gu H, Zhang W et al (2010) Testosterone-down-regulated Akt pathway during cardiac ischemia/reperfusion: a mechanism involving BAD, Bcl-2 and FOXO3a. J Surg Res 64:1–11
159. Appiah D, Schreiner PJ, Demerath EW et al (2016) Association of age at menopause with incident heart failure: a prospective cohort study and meta-analysis. J Am Heart Assoc 5:1–10
160. Schonfelder G (2005) The biological impact of estrogens on gender differences in congestive heart failure. Cardiovasc Res 67:573–574
161. Rossouw JE (2002) Hormones, genetic factors, and gender differences in cardiovascular disease. Cardiovasc Res 53:550–557
162. Wang M, Wang Y, Weil B et al (2009) Estrogen receptor beta mediates increased activation of PI3K/Akt signaling and improved myocardial function in female hearts following acute ischemia. Am J Physiol Regul Integr Comp Physiol 296:R972–R978
163. Richards RG, Di Augustine RP, Petrusz P et al (1996) Estradiol stimulates tyrosine phosphorylation of the insulin-like growth factor-1 receptor and insulin receptor substrate-1 in uterus. Proc Natl Acad Sci USA 93:12002–12007
164. Mendelsohn ME, Karas RH (1999) The protective effects of estrogen on the cardiovascular system. N Engl J Med 340:1801–1811
165. Mahmoodzadeh S, Fliegner D, Dworatzek E (2012) Sex differences in animal models for cardiovascular diseases and the role of estrogen. Handb Exp Pharmacol 214:23–48
166. Regitz-Zagrosek V, Oertelt-Prigione S, Seeland U, Hetzer R (2010) Sex and gender differences in myocardial hypertrophy and heart failure. Circ J 74:1265–1273
167. Ponikowski P, Voors AA, Anker SD et al (2016) 2016 ESC guidelines for the diagnosis and treatment of acute and chronic heart failure. Eur Heart J 37:2129–2200
168. Olivetti G, Melissari M, Capasso JM, Anversa P (1991) Cardiomyopathy of the aging human heart. Myocyte loss and reactive cellular hypertrophy. Circ Res 68:1560–1568
169. Mahmoodzadeh S, Eder S, Nordmeyer J et al (2006) Estrogen receptor alpha up-regulation and redistribution in human heart failure. FASEB J 20:926–934
170. Mendelsohn ME (2002) Genomic and nongenomic effects of estrogen in the vasculature. Am J Cardiol 90:3F–6F
171. Ventetuolo CE, Ouyang P, Bluemke DA et al (2011) Sex hormones are associated with right ventricular structure and function: the MESA-right ventricle study. Am J Respir Crit Care Med 183:659–667
172. Kuroski De Bold ML (1999) Estrogen, natriuretic peptides and the reninangiotensin system. Cardiovasc Res 41:524–531

Chapter 8
Sex-Specific Differences in β-Adrenoceptor Signal Transduction in Heart Failure Due to Volume-Overload

Naranjan S. Dhalla, Vijayan Elimban, Anureet K. Shah, and Mohamad Nusier

Abstract Volume-overload upon inducing arteriovenous shunt was found to produce sex-dependent changes in cardiac hypertrophy and heart failure. These alterations are accompanied by sex-dependent changes in β-adrenoceptor (AR)—guanine protein—adenylyl cyclase complex in the heart. Unlike females, prolonged exposure of animals to volume-overload resulted in heart failure in males. Protein content of β_1-AR and β_2-AR as well as their mRNA levels in the heart were lower in males and higher in females upon inducing volume-overload. Hearts from females also showed higher adenylyl cyclase protein content as well as epinephrine stimulated adenylyl cyclase activity in comparison to male hearts subjected to volume-overload. Marked increase in β_1-AR, β_2-AR and adenylyl cyclase content and depression in epinephrine-stimulated adenylyl cyclase activity as well as plasma catecholamines, were seen due to ovariectomy. All these changes except those in plasma catecholamines were fully or partially prevented upon treatment of ovariectomized animals with estrogen. These sex-specific alterations in β-AR signal transduction system may explain the reduced risk of heart failure in females, unlike males, under stressful conditions.

Keywords Volume-overload · Heart failure · β-Adrenoceptors · Adenylyl cyclase activity · β-Adrenoceptor mRNA · Adenylyl cyclase mRNA

N. S. Dhalla (✉) · V. Elimban
Institute of Cardiovascular Sciences, St. Boniface Hospital Albrechtsen Research Centre, Max Rady College of Medicine, University of Manitoba, Winnipeg, Manitoba R2H 2A6, Canada
e-mail: nsdhalla@sbrc.ca

A. K. Shah
Department of Kinesiology, Nutrition and Food Science, California State University, Los Angeles, USA

M. Nusier
School of Medicine, Jordan University of Science and Technology, Irbid, Jordan

B. Ostadal and N. S. Dhalla (eds.), *Sex Differences in Heart Disease*, Advances in Biochemistry in Health and Disease 21,
https://doi.org/10.1007/978-3-030-58677-5_8

Introduction

β-adrenoceptors (β-AR) signal transduction system is now well known to be intimately involved in the regulation of heart function by the sympathetic nervous system [1–10]. This pathway has been shown to be mainly composed of different components such as β_1- and β_2- AR receptors, guanine nucleotide stimulatory and inhibitory proteins (Gs- and Gi proteins) and adenylyl cyclase, which are present in the sarcolemmal membrane. In addition to G-proteins, β-AR kinase (GRK proteins) and β-arrestins are also known to play a regulatory role for the activation of β-AR—adenylyl cyclase complex. Both norepinephrine (NE) (released from the sympathetic nerve endings) and epinephrine (EPI) (released from the adrenal medulla) activate the β-AR—adenylyl cyclase complex to form cyclic AMP upon stimulation of the sympathetic nervous system. This signal transducing molecule activates protein kinase A (PKA), phosphorylates various Ca^{2+}—handling proteins, promotes Ca^{2+}—transport in cardiomyocytes and increases the extent of cardiac contraction and relaxation. The activation of β-AR signal transduction system has also been implicated in the development of several biochemical processes such as cardiac hypertrophy, cardiac apoptosis and myocardial metabolism.

Although the activated β-AR signal transduction system is considered to provide critical support for the maintenance of cardiac function at initial stages of heart disease, varying degrees of defects in different components of this pathways have been reported to occur during later stages of heart disease [11–22]. On the basis of research work carried out in various experimental models of heart failure such as cardiomyopathic hamsters [23–25], myocardial infarction [26–29] and volume-overload [30, 31], it has been suggested that derangements in the β-AR—G-proteins—adenylyl cyclase system may serve as one of the important mechanisms for the development of cardiac dysfunction in heart failure [9, 10]. In fact, changes in β-AR signal transduction were also seen to depend upon the type and stage of cardiac hypertrophy because catecholamine-induced increases in cardiac function, activation of adenylyl cyclase and intracellular Ca^{2+} concentration as well as β_1-AR density were unaltered by pressure- overload and augmented by volume-overload at early stages; however, changes in these parameters were attenuated by both pressure- and volume-overloads at late stages [32]. The present article is intended to focus on discussion whether alterations in the β-AR signal transduction pathway in cardiac hypertrophy and heart failure are gender dependent.

Gender/Sex Dependent Differences in β-AR Signal Transduction

Although general characteristics of healthy hearts from both males and females are not different from each other, unaltered, depressed or augmented responses to β-AR stimulation have been reported in males in comparison to females [33–37]. The enhanced response of male cardiomyocytes to β-AR stimulation was shown to be due to greater formation of cyclic AMP and greater influx of Ca^{2+} [38]. Gender differences in β-AR responsiveness has also been observed in the diabetic rat heart [39]. The cardiac contractile responses to β-AR stimulation were preserved in old females, unlike old males [40]. Furthermore, marked differences between males and females for cardiac remodeling and cardiac dysfunction have been reported to occur as a consequence of β_1-AR and β_2-AR overexpressions [41–45]. In an experimental model of catecholamine-induced cardiac remodeling, the extent of increase in cardiac wall thickness, myocardial cross-sectional area and collagen deposition in females was less than that of males [46]. Thus, it is evident that gender differences with respect to β-AR signal transduction become evident only after exposure of the animal to stressful situations.

Several studies have been carried out to reveal the mechanisms of gender/sex differences for the occurrence of cardiac dysfunction during the development of heart disease [47, 48]. These investigations during cardiac remodeling in heart failure have identified two major mechanisms namely (i) the role of gender/sex-dependent subcellular defects and Ca^{2+} cycling; (ii) the role of gender/sex-dependent alterations in β-AR signal transduction pathways. It has also been demonstrated that females have greater α-AR sensitivity than males, whereas β_1-AR sensitivity is similar between genders [49]. A marked reduction in β-AR responsiveness in heart failure has been reported in males compared to females [50]. It should be recognized that there occurs a differential regulation of β-AR density and cyclic AMP production with respect to age and sex [51]. Strenuous exercise has also been shown to depress ventricular function to a greater degree in males in comparison to females as a consequence of corresponding changes in the β-AR responsiveness [52]. The susceptibility of animals to AR stimulation-induced arrhythmias is also dependent upon sex [53].

Gender/Sex-Dependent Alterations of β-AR Signal Transduction in Heart Failure

Since heart failure is invariably preceded by cardiac hypertrophy, this section will be devoted to discussion for β-AR signal transduction changes in both cardiac hypertrophy and heart failure induced by volume-overload as a consequence of arteriovenous fistula or AV shunt in male and female rats [54, 55]. At early stages of inducing volume-overload, the degree of cardiac hypertrophy was greater, the increase in cardiac output was larger and the elevation of plasma epinephrine and norepinephrine was higher in males than in females [54]. On the other hand, increases in both the density of β_1-AR and epinephrine-stimulated adenylyl cyclase activities in hypertrophied hearts are higher in females than those in males [54]. A wide variety of alterations in mRNA levels for different regulatory functions for β-AR—adenylyl cyclase complex were also observed in both male and female hypertrophied hearts. For example, unlike female hearts, mRNA levels of β_1-AR proteins were decreased and those for Gi-proteins were increased in male hearts due to cardiac hypertrophy. Furthermore, mRNA levels for GRK2 and β-arrestins 2 proteins were increased to equal extent in both male and female hypertrophied hearts, whereas mRNA for β_2-AR, GRK3 and β-arrestins 1 were increased only in female hypertrophied heart [54]. These results have been suggested to indicate sex-dependent differences in β-AR signal transduction pathway in males and females upon the induction of cardiac hypertrophy.

Alterations in β-AR signal transduction pathway were also examined in volume-overload heart failure due to induction of AV shunt for 16 weeks in both male and female rats [55]. The results in Fig. 8.1 show that protein content for β_1-AR and β_2-AR were depressed in male failing hearts but those in female hearts were increased. It may also be noted that protein content for β_1-AR in female sham control hearts, unlike β_2-AR, was higher than those in control males (Fig. 8.1). Likewise, the pattern of changes in β_1-and β_2- mRNA levels was similar to that seen with respect to the protein content in the failing hearts, although no differences in the sham control values for males and females were seen (Fig. 8.2). The values for both basal and epinephrine-stimulated adenylyl cyclase activities in female failing hearts were found to be higher than those in males (Fig. 8.3). Furthermore, adenylyl cyclase protein content in female experimental hearts, unlike male hearts, were higher (Fig. 8.3). It may be noted from

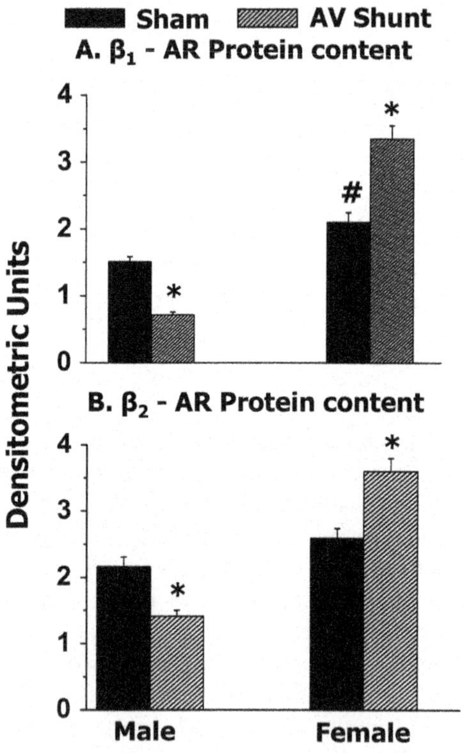

Fig. 8.1 Sex-specific changes in β_1- and β_2-adrenoceptor protein content in sham and failing (16 weeks AV shunt) hearts due to volume-overload. Data are taken from our paper—Dent MR, Tappia PS and Dhalla NS. J. Cell. Physiol. 227: 3080–3087, 2012. *_ $P < 0.05$ versus sham; #_ $P < 0.05$ versus corresponding value for male

Fig. 8.2 Sex-specific changes in β_1- and β_2-adrenoceptor mRNA levels in sham and failing (16 weeks AV shunt) hearts due to volume-overload. Data are taken from our paper—Dent MR, Tappia PS and Dhalla NS. J. Cell. Physiol. 227: 3080–3087, 2012. *_ P < 0.05 versus sham

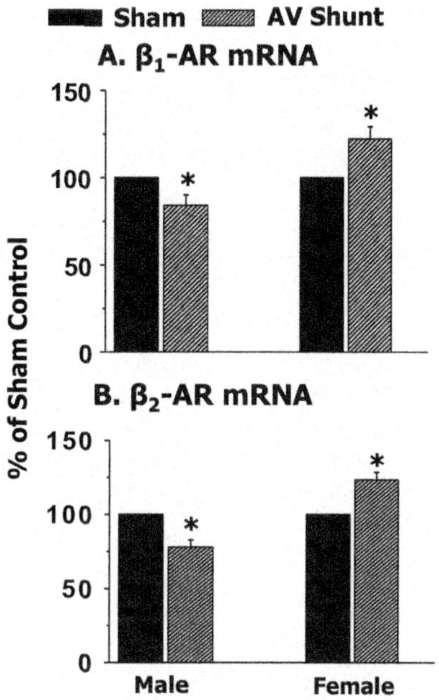

Fig. 8.3 Sex-specific changes in adenylyl cyclase activities and protein content in sham and failing (16 weeks AV shunt) hearts due to volume-overload. Data are taken from our paper—Dent MR, Tappia PS and Dhalla NS. J. Cell. Physiol. 227: 3080–3087, 2012. *_ P < 0.05 versus sham; #_ P < 0.05 versus corresponding value for male. Basal and epinephrine (EPI)—stimulated adenylyl cyclase activities were determined in the absence and presence of 1 μM epinephrine

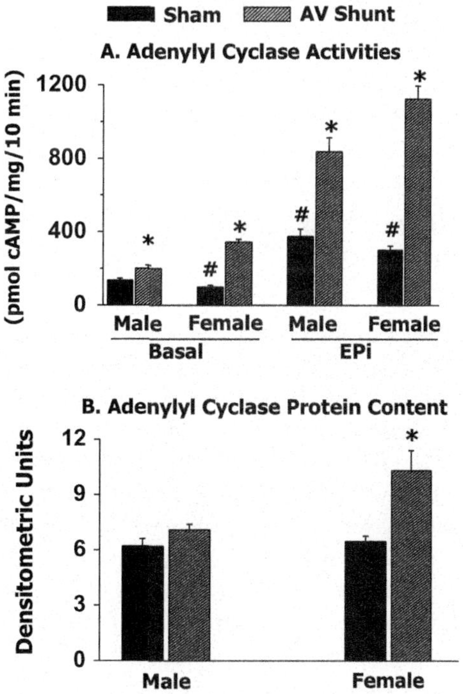

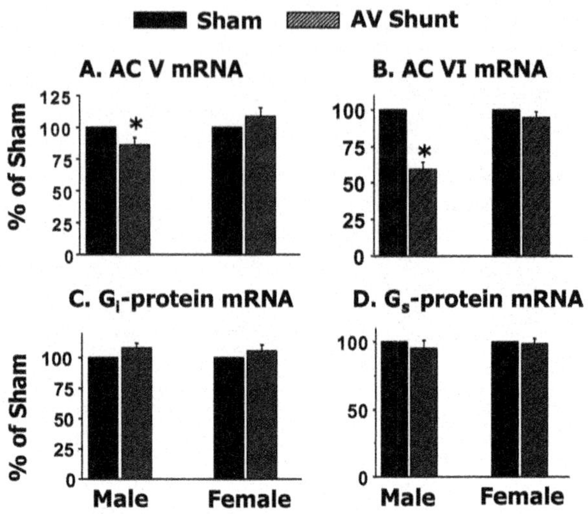

Fig. 8.4 Sex-specific changes in mRNA levels for adenylyl cyclase (V and V1) as well as Gi- Gs- proteins in sham and failing (16 weeks AV shunt) hearts due to volume-overload. Data are taken from our paper—Dent MR, Tappia PS and Dhalla NS. J. Cell. Physiol. 227: 3080–3087, 2012. *_ P < 0.05 versus sham

Fig. 8.4 that heart failure due to volume-overload depressed the mRNA levels for both adenylyl cyclase V and VI isoforms in males only. Furthermore, no change in mRNA levels for Gi- and Gs- proteins was seen in both males and females [4]. These results are consistent with the view that β-AR—adenylyl cyclase system is induced by heart failure and the sex- specific changes in these observations can also explain the role of β-AR—adenylyl cyclase complex in maintaining heart function in females and the loss of adrenergic support in heart failure in males.

Role of Estrogen in β-AR Signal Transduction System in Females

Since estrogen has been shown to serve as a cardioprotective agent responsible for reducing the incidence and severity of cardiovascular disease risk in women [56–60], the effect of estrogen on β-AR signal transduction was tested upon inducing volume-overload in ovariectomized female rats according to the procedure described earlier [55]. Ovariectomy was observed to increase protein content of β_1-AR, β_2-AR and adenylyl cyclase (Figs. 8.5 and 8.6). The epinephrine stimulated adenylyl cyclase activity was markedly reduced by ovariectomy without changes in the basal activity (Fig. 8.6). All these changes in ovariectomized animals were partially prevented by treatment with estrogen (Figs. 8.5 and 8.6). It is pointed out that ovariectomy produced a marked depression in the adenylyl cyclase protein content in the experimental (AV shunt) animal and this may explain the attenuated responses of adenylyl cyclase upon treatment of ovariectomized animals with estrogen (Fig. 8.6).

It may be noted from data in Fig. 8.7 that plasma levels of both norepinephrine and epinephrine were depressed by ovariectomy in the sham and experimental animals

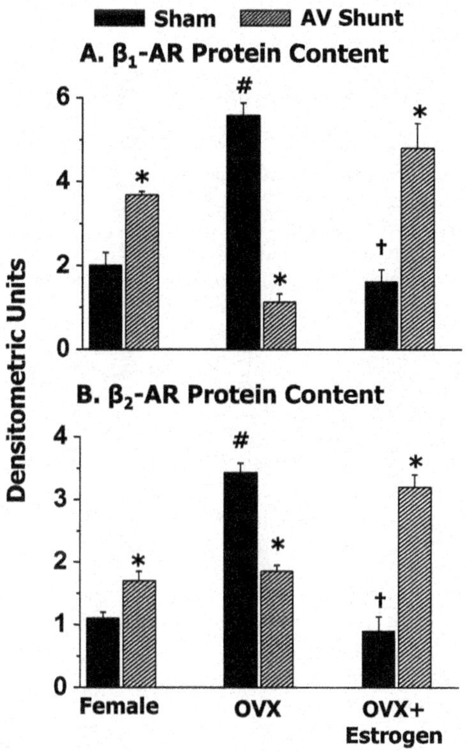

Fig. 8.5 Protein content of β_1- and β_2- adrenoceptors in sham and experimental (16 weeks AV shunt) hearts in control female, ovariectomized (OVX) and OVX + estrogen treated rats. OVX animals were implanted with or without a slowly releasable pellet (1.5 mg estrogen) twice during 16 weeks period. Data are taken from our paper—Dent MR, Tappia PS and Dhalla NS. J. Cell. Physiol. 227: 3080–3087, 2012. *_ $P < 0.05$ versus sham; #_ $P < 0.05$ versus control (Female); †_ $P < 0.05$ versus OVX

and these changes were not affected significantly upon treatment with estrogen. These observations indicate that other hormones in addition to estrogen may also be involved in offering cardioprotection for reducing the risk of heart failure in females. Nevertheless, it is pointed out that the plasma levels of both norepinephrine and epinephrine in control and experimental females were markedly lower than those in the males. These changes may support the view that prolonged activation of the sympathetic nervous system along with associated changes in the β-AR signal transduction pathway may play a critical role in determining sex-dependent differences in the development of heart failure.

Conclusions

In comparison to males, the activation of β-AR—G-protein—adenylyl cyclase complex by stimulation of the sympathetic nervous system and subsequent release of catecholamines has been demonstrated to produce a greater increase in cardiac function, larger degree of cardiac hypertrophy and higher extent of cardiac output in

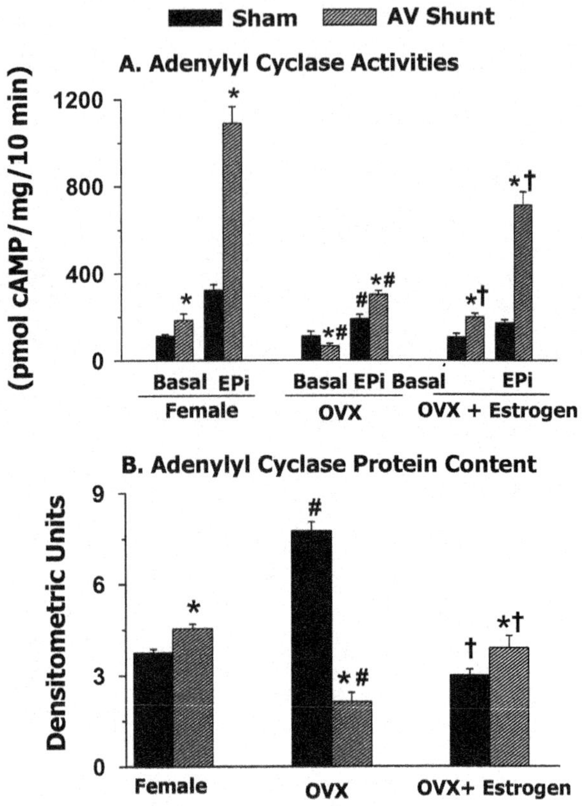

Fig. 8.6 Adenylyl cyclase activities and protein contents in sham and experimental (16 weeks AV shunt) hearts in control (intact female), ovariectomized (OVX) and estrogen treated ovariectomized rats. Adenylyl cyclase activities were determined in the absence (Basal) and presence of 1 μM epinephrine (EPI). OVX animals were implanted with or without a slowly releasable pellet (1.5 mg estrogen) twice during 16 weeks period. Data are taken from our paper—Dent MR, Tappia PS and Dhalla NS. J. Cell. Physiol. 227: 3080–3087, 2012. *_ P < 0.05 versus sham; #_ P < 0.05 versus corresponding value for female; †_ P < 0.05 versus corresponding value for OVX

females. This β-AR signal pathways provides support to maintaining cardiac function during the development of heart disease in females whereas varying degrees of defects in different components of β-AR—G-protein—adenylyl cyclase complex upon prolonged activation contributes to the occurrence of cardiac dysfunction in males. Several events associated with the development of cardiac hypertrophy and heart failure due to volume-overload have been discussed to emphasize gender- or sex-specific differences with respect to β-AR signal transduction pathways. Attenuations of β-AR and adenylyl cyclase by ovariectomy in females subjected to volume-overload, and partial recovery of these alterations upon treatment with estrogen support the view for the important role of this hormone in reducing the risk of heart disease in female population.

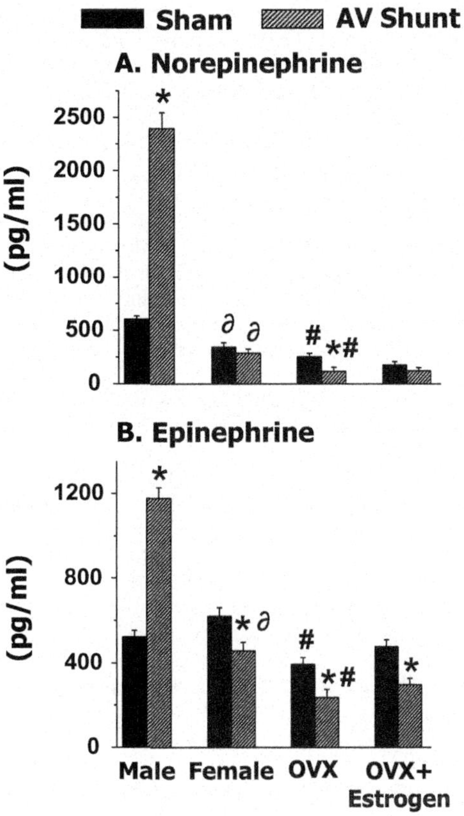

Fig. 8.7 Plasma levels of norepinephrine and epinephrine in sham and experimental (16 weeks AV shunt) male, female, ovariectomized (OVX), and OVX + estrogen treated rats. Data are taken from our paper—Dent MR, Tappia PS and Dhalla NS. J. Cell. Physiol. 227: 3080–3087, 2012. OVX animals were implanted with or without slowly releasable pellet (1.5 mg estrogen) twice during 16 weeks period. *_ P < 0.05 versus sham; ∂_ P < 0.05 versus corresponding value for male; #_ P < 0.05 versus P < 0.05 versus corresponding value for female

Acknowledgments The infrastructure during the preparation of this article was provided by the St. Boniface Hospital Research Foundation, Winnipeg, Canada.

References

1. Stiles GL, Caron MG, Lefkowitz RJ (1984) Beta-adrenergic receptors: biochemical mechanisms of physiological regulation. Physiol Rev 64:661–743
2. Bristow MR, Hershberger RE, Port JD et al (1990) Beta-adrenergic pathways in nonfailing and failing human ventricular myocardium. Circulation 82:I12–I25
3. Brodde OE (1991) Beta 1- and beta 2-adrenoceptors in the human heart: properties, function, and alterations in chronic heart failure. Pharmacol Rev 43:203–242
4. Bristow MR, Feldman AM (1992) Changes in the receptor-G protein-adenylyl cyclase system in heart failure from various types of heart muscle disease. Basic Res Cardiol 87:15–35
5. Bohm M (1995) Alterations of beta-adrenoceptor-G-protein-regulated adenylyl cyclase in heart failure. Mol Cell Biochem 147:147–160

6. Homcy CJ, Vatner SF, Vatner DE (1991) Beta-adrenergic receptor regulation in the heart in pathophysiologic states: abnormal adrenergic responsiveness in cardiac disease. Annu Rev Physiol 53:137–159
7. Post SR, Hammond HK, Insel PA (1999) Beta-adrenergic receptors and receptor signaling in heart failure. Annu Rev Pharmacol Toxicol 39:343–601
8. Vatner DE, Asai K, Iwase M et al (1999) Beta-adrenergic receptor-G protein-adenylyl cyclase signal transduction in the failing heart. Am J Cardiol 83:80H–85H
9. Dhalla NS, Wang X, Sethi R, Das PK (1997) β-adrenergic linked signal transduction mechanisms in failing hearts. Heart Fail Rev 2:55–65
10. Wang X, Dhalla NS (2000) Modification of β-adrenoceptor signal transduction pathway by genetic manipulation and heart failure. Mol Cell Biochem 214:131–155
11. Bristow MR, Ginsburg R, Minobe W et al (1982) Decreased catecholamine sensitivity and b-adrenergicreceptor density in failing human heart. New Engl J Med 307:205–211
12. Brodde OE (1991) Pathophysiology of the β-adrenoceptor system in chronic heart failure: Consequences for treatment with agonists, partial agonists or antagonists. Eur Heart J 12:54–62
13. Steinfath M, Geertz B, Schmitz W et al (1991) Distinct down-regulation of cardiac β_1- and β_2-adrenoceptors in different human heart diseases. Naunyn-Schmiedeberg's Arch Pharmacol 343:217–220
14. Bohm M, Beuckelmamn D, Brown L et al (1988) Reduction of β-adrenoceptor density and evaluation of positive inotropic responses in isolated, diseased human myocardium. Eur Heart J 9:844–852
15. Calderone A, Bouvier M, Li K et al (1991) Dysfunction of the β- and α-adrenergic systems in a model of congestive heart failure: the pacing-overdrive dog. Circ Res 69:332–343
16. Bristow MR, Kantrowitz NE, Ginsburg R, Fowler MB (1985) β-adrenergic function in heart muscle disease and heart failure. J Mol Cell Cardiol 17:41–52
17. Bristow MR, Sandoval AB, Gilbert EM et al (1988) Myocardial α- and β-adrenergic receptors in heart failure: Is cardiac derived norepinephrine the regulatory signal? Eur Heart J 9:35–40
18. Bristow MR, Ginsburg R, Ymans V et al (1986) β_1-and β_2-adrenergic-receptor subpopulations in nonfailing and failing human ventricular myocardium. Circ Res 69:297–309
19. Bristow MR, Hershberger RE, Port JD et al (1989) β_1-and β_2-adrenergic receptor-mediated adenylate cyclase stimulation in nonfailing and failing human ventricular myocardium. Mol Pharmacol 35:295–303
20. Bristow MR, Minobe W, Rasmussen R et al (1992) β-Adrenergic neuroeffector abnormalities in failing human heart are produced by local rather than systemic mechanisms. J Clin Invest 89:803–815
21. Altschuld RA, Starling RC, Hamlin RL et al (1995) Responses of failing canine and human heart cells to β_2-adrenergic stimulation. Circulation 92:1512–1618
22. Bohm M, Gierschik P, Knorr A et al (1992) Role of altered G-protein expression in the regulation of myocardial adenylyl cyclase activity and force of contraction in spontaneous hypertensive cardiomyopathy in rats. J Hypertens 10:1115–1128
23. Kaura D, Takeda N, Sethi R et al (1996) Beta-adrenoceptor mediated signal transduction in congestive heart failure in cardiomyopathic (UM-X7.1) hamsters. Mol Cell Biochem 157:191–196
24. Sethi R, Panagia V, Dhalla KS et al (1994) In: Nagano M, Takeda N, Dhalla NS (eds) The Cardiomyopathic Heart. Raven, New York, pp 73–86
25. Sethi R, Bector N, Takeda N, et al. Alterations in G-proteins in congestive heart failure in cardiomyopathic (UM-X7.1) hamsters. Mol Cell Biochem 140: 163–170, 1994
26. Sethi R, Elimban V, Chapman D et al (1998) Differential alterations in left and right ventricular G-proteins in congestive heart failure due to myocardial infarction. J Mol Cell Cardiol 30:2153–2163
27. Sethi R, Saini HK, Wang X et al (2006) Differential changes in β-adrenoceptor signal transduction in left and right ventricles of infarcted rats. Can J Physiol Pharmacol 84:747–754
28. Sethi R, Shao Q, Ren B et al (2004) Changes in beta-adrenoceptors in heart failure due to myocardial infarction are attenuated by blockade of renin-angiotensin system. Mol Cell Biochem 263:11–20

29. Sethi R, Shao Q, Takeda N, Dhalla NS (2003) Attenuation of changes in G_i-proteins and adenylyl cyclase in heart failure by an ACE inhibitor, imidapril. J Cell Mol Med 7:277–286
30. Wang X, Sentex E, Chapman D, Dhalla NS (2004) Alterations of adenylyl cyclase and G proteins in aortocaval shunt-induced heart failure. Am J Physiol Heart Circ Physiol 287:H118–H125
31. Wang X, Sentex E, Saini HK et al (2005) Upregulation of beta-adrenergic receptors in heart failure due to volume overload. Am J Physiol Heart Circ Physiol 289:H151–H159
32. Sethi R, Saini HK, Guo X et al (2007) Dependence of changes in β-adrenoceptor signal transduction upon type and stage of cardiac hypertrophy. J Appl Physiol 102:978–984
33. Monasky MM, Varian KD, Janssen PM (2008) Gender comparison of contractile performance and beta-adrenergic response in isolated rat cardiac trabeculae. J Comp Physiol B 178:307–313
34. Turner MJ, Mier CM, Spina RJ et al (1999) Effects of age and gender on the cardiovascular responses to isoproterenol. J Gerontol A Biol Sci Med Sci 54:B393–B400
35. Chu SH, Sutherland K, Beck J et al (2005) Sex differences in expression of calcium-handling proteins and beta-adrenergic receptors in rat heart ventricle. Life Sci 76:2735–2749
36. Capasso JM, Remily RM, Smith RH, Sonnenblick EH (1983) Sex differences in myocardial contractility in the rat. Basic Res Cardiol 78:156–171
37. Biondi-Zoccai GG, Baldi A, Biasucci LM, Abbate A (2004) Female gender, myocardial remodeling and cardiac failure: are women protected from increased myocardiocyte apoptosis? Ital Heart J 5:498–504
38. Vizgirda VM, Wahler GM, Sondgeroth KL et al (2002) Mechanisms of sex differences in rat cardiac myocyte response to beta-adrenergic stimulation. Am J Physiol Heart Circ Physiol 282:H256–H263
39. Bilginoglu A, Cicek FA, Ugur M et al (2007) The role of gender differences in beta-adrenergic receptor responsiveness of diabetic rat heart. Cell Biochem 305:63–69
40. Takagi G, Asai K, Vatner SF et al (2003) Gender differences on the effects of aging on cardiac and peripheral adrenergic stimulation in old conscious monkeys. Am J Physiol Heart Circ Physiol 285:H527–H534
41. Engelhardt S, Hein L, Wiesmann F, Lohse MJ (1999) Progressive hypertrophy and heart failure in 1-adrenergic receptor transgenic mice. Proc Natl Acad Sci USA 96:7059–7064
42. Bisognano JD, Weinberger HD, Bohlmeyer TJ et al (2000) Myocardial-directed overexpression of the human 1-adrenergic receptor in transgenic mice. J Mol Cell Cardiol 32:817–830
43. Liggett SB, Tepe NM, Lorenz JN et al (2000) Early and delayed consequences of 2-adrenergic receptor overexpression in mouse hearts: critical role for expression level. Circulation 101:1707–1714
44. Du XJ, Gao XM, Wang B et al (2000) Age dependent cardiomyopathy and heart failure phenotype in mice overexpressing 2-adrenergic receptors in the heart. Cardiovasc Res 48:448–454
45. Gao XM, Agrotis A, Autelitano DJ et al (2003) Sex hormones and cardiomyopathic phenotype induced by cardiac beta 2-adrenergic receptor overexpression. Endocrinology 144:4097–4105
46. Zhu B, Liu K, Yang C et al (2016) Gender-related differences in β-adrenergic receptor-mediated cardiac remodeling. Can J Physiol Pharmacol 94:1349–1355
47. Dhalla NS, Dent MR, Elimban V, Zieroth S (2012) Mechanisms of gender differences in cardiac dysfunction due to heart disease. In: Kimchi A (ed) Proceedings of the 16th World Congress on Heart Disease. Medimond International Proceedings, Bologna, pp 239–245
48. Dhalla NS, Malik A, Zieroth S, Tappia PS (2013) In: Jugdutt BI, Dhalla NS (eds) Cardiac Remodeling - Molecular Mechanisms. Springer Science + Business Media, LLC, New York, pp 47–64
49. Luzier AB, Nawarskas JJ, Anonuevo J et al (1998) The effects of gender on adrenergic receptor responsiveness. Clin Pharmacol 38:618–624
50. Wei SK, McCurley JM, Hanlon SU, Haigney MC (2007) Gender differences in Na/Ca exchanger current and beta-adrenergic responsiveness in heart failure in pig myocytes. Ann N Y Acad Sci 1099:183–189

51. Hoffmann S, Bohme J, Kube C et al (2016) Differential regulation of the β-adrenoceptor density and cyclic AMP level with age and sex in turkey cardiac chambers. Eur J Pharmacol 777:88–95
52. Scott JM, Esch BT, Haykowsky MJ et al (2007) Sex differences in left ventricular function and beta-receptor responsiveness following prolonged strenuous exercise. J Appl Physiol 102:681–687
53. Teplitz L, Igic R, Berbaum ML, Schwertz DW (2005) Sex differences in susceptibility to epinephrine-induced arrythmias. J Cardiovasc Pharmacol 46:548–555
54. Dent MR, Tappia PS, Dhalla NS (2011) Gender differences in β-adrenoceptor system in cardiac hypertrophy due to arteriovenous fistula. J Cell Physiol 226:181–186
55. Dent MR, Tappia PS, Dhalla NS (2012) Gender related alterations of β-adrenoceptor mechanisms in heart failure due to arteriovenous fistula. J Cell Physiol 227:3080–3087
56. Schonfelder G (2005) The biological impact of estrogens on gender differences in congestive heart failure. Cardiovasc Res 67:573–574
57. Rossouw JE (2002) Hormones, genetic factors, and gender differences in cardiovascular disease. Cardiovasc Res 53:550–557
58. Wang M, Wang Y, Weil B et al (2009) Estrogen receptor beta mediates increased activation of PI3K/Akt signaling and improved myocardial function in female hearts following acute ischemia. Am J Physiol Regul Integr Comp Physiol 296:R972–R978
59. Van Eickels M, Grohe C, Cleutjens JP et al (2001) 17β-estradiol attenuates the development of pressure-overload hypertrophy. Circulation 104:1419–1423
60. Richards RG, Diaugustine RP, Petrusz P et al (1996) Estradiol stimulates tyrosine phosphorylation of the insulin-like growth factor-1 receptor and insulin receptor substrate-1 in uterus. Proc Natl Acad Sci USA 93:12002–12007

Chapter 9
Sex and Heart Transplantation

Ana Ayesta and Manuel Martínez-Sellés

Abstract Heart transplantation is the treatment of choice in selected patients with end-stage heart disease. When allocating a graft several factors must be considered. Among these donor and recipient sex are relevant and might influence the prognosis. The influence of donor and recipient sex on prognosis after heart transplantation has been the focus of several studies. In most of them, sex mismatch between donor and recipient was identified as a determinant of worst prognosis, particularly in the combination female donor to male recipient. A recently published metanalysis confirmed this relationship. Different physiological reasons could explain these findings, including rejection, cardiovascular allograft vasculopathy, primary graft failure, and important confounding variables as age, urgent transplantation, and size mismatch. Also, pulmonary hypertension of the recipient seems to have a key role in this association. When allocating a graft, sex mismatch should be considered as a variable that will potentially influence on prognosis, especially in male recipients with previous pulmonary hypertension.

Keywords Sex · Mismatch · Heart · Transplantation · Female donor · Male recipient · Pulmonary hypertension

Introduction

Heart failure is the common final pathway of most heart diseases. It affects 1–2% of adult population and this rate is higher than 10% in those >70 years. The development and worsening of symptoms leads to a decrease in functional class, quality of life, hospitalizations and premature death. Although some treatments have decreased hospitalization rates and mortality, heart failure still has a poor prognosis [1]. Heart

A. Ayesta
Heart Area, Hospital Universitario Central de Asturias, Oviedo, Asturias, Spain

M. Martínez-Sellés (✉)
Cardiology Department, Hospital Universitario Gregorio Marañón, CIBERCV, Universidad Europea, Universidad Complutense, C/Dr. Esquerdo 46, 28007 Madrid, Spain
e-mail: mmselles@ecardiologia.es

B. Ostadal and N. S. Dhalla (eds.), *Sex Differences in Heart Disease*, Advances in Biochemistry in Health and Disease 21,
https://doi.org/10.1007/978-3-030-58677-5_9

transplantation is the treatment of choice in selected patients with end-stage heart disease [1]. However, patient survival is still suboptimal [2, 3] due to immunosuppressive treatment (infections and tumors), acute rejection and cardiac allograft vasculopathy. However, since late 90s and early 21st century [3] there has been a decrease in the number of heart transplantations. Shortage of optimal donors and improvement in heart diseases prognosis are probably the main causes. This has also led to a worsening of recipient clinical profile, with higher comorbidity rates and a progressive extension of the donation criteria (elderly donors, deaths from stroke, and greater ischemia times) [2, 3].

When allocating a graft, several serological and immunological factors must be considered. Among these, donor and recipient sex have been studied for years. Single-center and small studies were the first ones but we now have data from multi-center studies with large number of patients from international registries. These studies have analyzed the influence that both donor and recipient sex have on different events and, more recently, the interaction between donor and recipient sex. In this chapter we will address the issue of donor/recipient sex influence on heart transplantation results and how this could influence clinical decisions.

Donor/Recipient Sex Influence on Heart Transplantation Prognosis

Several small and single-center studies relating donor/recipient sex with mortality were first published with heterogeneous results. Most of them reported worst survival of recipients from female donor (FD) [4–6], while others related female recipient (FR) with higher rates of rejection [4, 7–10]. In 1988 Crandall et al. [7] analyzed 140 transplants and found no differences in survival between male recipients (MR) and FR. Similar results were found in 1991 [4], although recipients from male donors (MD) had greater survival. In 1994 the results of heart transplantation in the USA reported a greater survival of those recipients from MD [11]. The first study that specifically analyzed survival based on donor sex was published in 1996 [12]. After studying 283 transplants, the authors showed that recipients from FD doubled mortality and had higher rejection rates compared to those of MD. No differences in survival were noted according to recipient sex. Other studies also demonstrated a lower survival of recipients from FD. Mc Carthy et al. [5] identified 5 significant variables as risk factors, including FD. Likewise, a multicenter study in 2002 [6] identified female sex as a risk factor. In this same study, FD was related to intra-hospital mortality. On the contrary, Solomon et al. [13] identified FD as a risk factor for low cardiac output and need for intra-aortic balloon pump but not for mortality. Subsequent studies failed to show the relationship between donor and/or recipient sex with mortality. In 2005 Radovancevic et al. [14] published xxx the their single-center results and did not identify donor/recipient sex or sex mismatch as a predictor of survival. In a more recent study, female sex was not identified as a risk factor for worse long-term

Table 9.1 Single-center studies that analyze the influence of donor and/or recipient sex on survival after heart transplantation

Female donor is related with mortality
– Fabbri et al. [4] – Mc Carthy et al. [5] – Tsai et al. [6]
Donor sex does not influence on mortality
– Fiorelli et al. [15] – Brock et al. [16]
Female recipient is related with better survival
– Fabbri et al. [4] – Mastrobouni et al. [20]
Recipient sex does not influence on mortality
– Del Rizzo et al. [17]
Donor and recipient sex do not influence on survival
– Solomon et al. [13] (only analyses female sex) – Randovancevic et al. [14] – Chou et al. [18]

Table 9.2 Multicentric studies that analyze the influence of donor and/or recipient sex on survival after heart transplantation

Female donor increases mortality
– Keck et al. [11] – Bryan et al. [12]
Recipient sex does not influence on survival
– Crandall et al. [7] – Bocchi et al. [19] – Bryan et al. [12]

survival [15]. Two other studies analyzing early mortality failed to show influence of donor [16] or recipient [17] x) sex. Chou et al. [18] did not find, in an analysis of 214 patients from a single center, an independent influence of donor and recipient sex on survival. A multicenter study summarizing the Brazilian experience published in 2001 also found no differences in survival based on recipient sex [19]. The results of these studies are summarized in Tables 9.1 and 9.2.

Sex Mismatch Influence on Heart Transplantation Prognosis

After the initial results that failed to show an association between donor/recipient sex and survival [5, 6, 12–19], the influence of donor/recipient sex mismatch on prognosis was studied. Single-center or small sample studies were followed by large multi-center studies and the analysis of large international registries. The first studies that reported the influence of sex mismatch on survival were published in 1998. Kirsh et al. [21] showed an influence on early mortality while Prendregast et al. [22] found worst annual and censored survival. This was due to worse rates in the female to male (F/M) group and attributed to size mismatch. Higher rates of acute rejection were also found in recipients from sex-mismatched heart. The worst survival in F/M was later demonstrated in several studies [23–30]. Kittleson et al. [29] showed better survival in patients receiving a same sex heart. However, they failed to demonstrate statistically significant worse survival of F/M group although when analyzing survival curves an increase in early mortality in F/M is suggested. Eiffert et al. [28] reported that F/M had lower survival first year after heart transplantation, with a better survival of the female to female (F/F) group. Likewise, a smaller analysis [30] of male recipients showed worst survival in F/M group, mainly due to worst survival in the initial post-transplant stages. Despite these results, other studies proved otherwise. Some of them related sex mismatch with mortality (regardless of the recipient sex) [21, 31–33], while others failed to show this relation [34–38]. Thus, a study [39] that analyzed 200 male recipients between November 2003 and December 2013 in a Portuguese center failed to show higher mortality in the sex mismatch group, although they observed a trend. The results of the analysis of large registries also reported heterogeneous results [23–26, 40, 41]. Most of them showed a worse survival in F/M [23, 25, 26], although some of them found similar results for M/F [40, 41]. The results of the analysis of the Collaborative Transplant Study were reported by Zeier et al. in 2002 [26]. They analyzed 25,432 transplanted patients between 1985 and 2000 and reported worse survival of F/M, while there was no difference in FR survival. The results of the analysis of United Network for Organ Sharing (UNOS) database [25, 41] showed that patients receiving a sex mismatched heart were 14% more likely to have died at 5 years than those with a same sex heart. A more thorough analysis of these data showed that this was not valid for FR and the lowest survival of 4 groups (created based on sex mismatch) showed lower survival at 5 years in the F/M and greater survival in the male to male (M/M) group [25]. The database of the International Society for Heart and Lung Transplantation (ISHLT) is the biggest heart transplant database. It has been analyzed several times [23, 40, 42]. In 2012 Khush et al. [40] published an analysis of this database. They created two groups of recipients (men and women) and analyzed the differences in survival of both the recipient and the graft in mismatch/no mismatch different groups. In a cohort of 61,938 transplanted patients between January 1990 and December 2008 (60,584 patients after ruling out those with more than one transplant or with incomplete data regarding baseline characteristics), they reported an increase in mortality in F/M, compared to M/M. However, they suggested that this was also valid for male to female (M/F). In survival curves we can realize that difference in survival MR is influenced by early survival. Subsequently, Kackmarek et al. [23] analyzed 67,855 transplanted patients between January 1980 and June 2009 and divided them into 4 groups according to donor and recipient sex. The worst annual survival rates were found in F/M. In the most recent analysis, which included 52,455 patients between 1994 and 2013, Bergenfeldt et al. [42] found that sex mismatch increased mortality independently of weight match. In 2014 we analyzed 4625 patients from the Spanish Heart Transplantation Registry and found that sex mismatch influence on early mortality only in male recipients, particularly in patients with pulmonary gradient >13 mmHg [43]. On the other hand, the results of the University of Alabama (CTRS) base, previously published, failed to demonstrate this association [24]. Interaction between sex, weight mismatch and survival was found, especially in F/M. However, these differences were not observed when the weight mismatch was minimum. The analysis of the results of the UNOS database [41] of 42,765 transplanted patients between October 1989 and June 2011 showed that survival differences associated with sex mismatch of a male recipient were modified and could be attributed to differences in cardiac predicted mass by a mathematical model. The authors demonstrated that M/F increased mortality, although this was not shown in the case of F/M. The authors suggested that the assessment of cardiac size may not be adequate. Recently, we have published a meta-analysis analyzing sex mismatch influence on one-year survival [44]. A total of 76,175 patients were included in our quantitative meta-analysis and we found a poor prognosis in F/M compared with other groups. Our results were strongly influenced by the data of the largest international registry [23]. The results of these studies are summarized in Tables 9.3 and 9.4.

Table 9.3 Single-center studies that analyze the influence of sex mismatch on survival after heart transplantation

Sex mismatch influence on worst survival
– Kirsch et al. [21] – Schelechta et al. [33]
Sex mismatch influence on worst survival only in male recipients
– Prendergast et al. [22] – Al- Khaldi et al. [27] – Welp et al. [30] – Kittleson et al. [29] – Eiffert et al. [28]
Sex mismatch does not influence on survival
– Tsai et al. [6] – Del Rizzo et al. [17] – Radovancevic et al. [14] – Keogh et al. [45] – De Santo et al. [34] – Yamani et al. [37] – Izquierdo et al. [46] – Mastrobouni et al. [20] – Tsao et al. [36] – Aliabadi et al. [38] – Correia et al. [39]ara>

Tabla 4 Multicentric studies that analyze the influence of sex mismatch on survival after heart transplantation

Sex mismatch influence on worst survival
– Khush et al. [40] – Maltais et al. [32] – Bello et al. [31]
Sex mismatch influence on worst survival only in male recipients
– Zeier et al. [26] – Weiss et al. [25] – Stehlik et al. [24] – Kaczmarek et al. [23] – Martínez- Sellés et al. [43] – Ayesta et al. [44]
Sex mismatch influence on worst survival only in female recipients
– Reed et al. [41]
Sex mismatch does not influence on survival
– Jalowiec et al. [35]

Possible Physiopatological Mechanisms

The influence of sex mismatch on prognosis after heart transplantation could be due to different reasons. Anatomical, genetic, immunological, and hormonal differences between men and women could influence on survival or complications after heart transplantation. Moreover, the relationship with confounding factors and the heterogeneity among studies should be considered. Anatomic, hormonal, genetic differences, differences in ventricular arrhythmias and greater protection of female heart against necrosis and signs of cell death could be involved in better survival in women with heart failure [47, 48], specially in those with previous pregnancies [48]. This situation could be comparable to better survival of M/F. Differences in the structure of both ventricles and, mainly in the right ventricle, lead to a better adaptation to the hemodynamics of left ventricular dysfunction [49–52]. Also, women have a special ability to adapt to volume overload due to pregnancy, especially in those with several previous pregnancies [53, 54]. Anatomical responses to heart disease are also different. While women develop concentric hypertrophy and later clinical presentation, men develop eccentric hypertrophy and earlier dilatation [55–58]. In addition, women have a lower rate of ventricular arrhythmias than men (in the same state of heart failure) probably due to lower dispersion of the conductive tissue through a non-homogeneous myocardial tissue and dominance of vagal system [59]. Moreover, left ventricular systolic dysfunction is a predictor of mortality in men but not in women [60]. Hormonal differences may be also involved. Estrogen increases endothelial cell growth and greater resistance to endothelial damage [61], and both estrogen and progesterone produce angiogenic effects and vasodilation [62]. In addition, sex hormones could affect the release of calcium to contractile proteins and, consequently, contractile function [63]. There is also a possible influence of sex hormones on gene expression [64]. Differences in endocrine and immune system between men and women [65] could also lead to a different adaption to sex mismatch. Women have a greater immune response [7, 66–68] with higher incidence of autoimmune diseases, higher levels of circulating immunoglobulins, and lower tissue sensitivity to androgens than men [68]. In women with previous pregnancies this is more striking due to the development of antibodies during pregnancy [68]. These mechanisms could influence on higher rejection rates [12], although this do not always entail higher mortality [4, 7, 8]. On the contrary, the presence of male cells in the explanted heart of transplanted women with previous male pregnancies [69] could explain a better survival of M/F. Published results in this regard are also heterogeneous. This could be due to the lack of a systematic method to diagnose rejection and the lack of agreement on the level of abnormality of the biopsy, especially in those published before the ISHLT consensus [70]. As shown in Table 9.5, acute rejection has also been related to female donor [71], sex mismatch between donor and recipient in male [12] and female recipients [35, 45, 72] and, in some cases, has not been related to donor and/or recipient sex [27, 34, 73] or sex mismatch [31].

Table 9.5 Influence of donor/recipient sex on rejection after heart transplantation

Female recipient influence on rejection
– Crandall et al. [7] – Esmore et al. [8] – Kobashigawa et al. [10] – Kirklin et al. [9] – Sharples et al. [71]
Female donor influence on rejection
– Sharples et al. [71]
Sex mismatch influence on rejection
– Prendergast et al. [22] – Aliabadi et al. [38]
Sex mismatch influence on rejection only in female recipients
– Jalowiec et al. [35] – Keogh et al. [45] – Patel et al. [72]
Sex mismatch influence on rejection only in male recipients
– Bryan et al. [12]
Sex mismatch does not influence on rejection
– Bello et al. [31]
Donor/Recipient sex do not influence on rejection
– Al-Khaldi et al. [27] – De Santo et al. [34] – George et al. [73]

Cardiac Allograft Vasculopathy

Cardiac allograft vasculopathy (CAV) could also influence survival in patients receiving mismatched hearts. Immunological or size mismatch could be the reasons under this association [71, 74]. However, the lack of uniformity in its description until the consensus definition [75] led to heterogeneous results (Table 9.6). Sharples et al. [71] showed in 323 patients, 19 graft losses with higher risk of F/M. In a subsequent study Mehra et al. [74] confirmed these results by intravascular ultrasound (IVUS), although it was a study with a small sample and some technical limitations. Years later, a study by Enric et al. [76] with volumetric intravascular ultrasound showed that the highest rate of graft disease was seen in female donors, with no differences between male and female recipients. Constanzo et al. [77] analyzed 2,609 patients from 39 centers who had survived one year after heart transplantation. They tried to identify preoperative donor/recipient factors that could influence on the development of this disease and on death and retransplantation. The early development of CAV was related with MD and MR. In other study [78], 505 transplanted patients were analyzed between November 1985 and February 2001 and MD was related to the appearance of CAV. There were other studies that linked this disease with sex mismatch [38] and others that failed to find a relationship between CAV and donor/recipient sex [28].

Table 9.6 Influence of donor/recipient sex on cardiac allograft vasculopathy after heart transplantation

Female donor/Male recipient influence on the development of CAV
– Sharples et al. [71] – Mehra et al. [74]
Female donor influence on the development of CAV
– Enric et al. [76]
Male sex (donor and/or recipient) influence on the development of CAV
– Constanzo et al. [77]
Male donor influence on the development of CAV
– Caforio et al. [78]
Female donor/Female recipient influence on the development of CAV
– Yamani et al. [37]
Sex mismatch influence on the development of CAV
– Aliabadi et al. [38]
Donor/Recipient sex do not influence on the development of CAV
– Eiffert et al. [28]

Primary Graft Failure

Primary graft failure (PGF) is an impairment of the transplanted heart that occurs in the first 24 h after transplantation; it has idiopathic etiology. It is classified as left heart failure (mild, moderate or severe), right heart failure or both. Secondary graft failure is an organ dysfunction due to a known cause as hyperacute rejection, pulmonary hypertension or surgical complications [79]. It is known that death during early postoperative period causes a decrease in survival of around 10–15% [80]. The main cause of death in the first 30 days is PGF with up to 22% mortality in the series published by the Spanish Registry of Cardiac Transplantation [81]. Subsequent data from this registry showed an increase in mortality in F/M only in the first 30 days [43] but PGF was related with FD, as previously noted [82], but not with sex mismatch. On the contrary, some studies showed an association between sex mismatch and PGF. An analysis of 16,716 patients from the UNOS database [83] showed that sex mismatch was a risk factor for the development of PGF in MR. However, this study used an extensive definition of this disease that could include cases of humoral rejection. Other previous studies showed similar results. A multicenter study of 7,259 patients [84] showed a higher risk of early graft failure in F/M and this was particularly important when the size exceeded 30%. Again the definition used included cases of secondary graft failure [79]. Other confounding factors or heterogeneity among the studies should be seriously considered. Among these, D age, R age D/R size mismatch and urgent transplant are the most important.

Donor and Recipient Age

Lately, increasing donor age has led to a progressive decrease in survival, especially with 60 years and older donors [85]. Advanced donor age has been related to mortality, mainly during the first year after HT [86]. Analyses from different studies showed that FD were older than MD [23, 27, 33, 34, 37, 39, 43]. Al-Khaldi et al. (n = 849) [27], found worse survival F/M in older donors (>45 years) and older recipients (>45 years), regardless of donor age. In multivariate analysis, FD was only a risk factor when in MR > 45 years. Correia et al. [39] analyzed 200 MR and found that FD were older than MD but survival was similar in both groups. De Santo et al. (n = 99) [34] did not identified sex mismatch or donor sex as a risk factor in spite of donor age difference between the groups (older FD). Survival was also similar among different groups despite older FD in the published results by Yamani et al. (n = 361) [37]. We have found (n = 4,625) [43] that F/M had a worst survival than M/M in univariate analysis but this association lost statistical significance when adjusting by donor age (being FD older). However, there was a trend towards worst survival in F/M. Kaczmarek et al. [23] also found older FD and reported worse one and ten-year survival in F/M, F/F than in M/F, M/M. Multivariate analysis still confirmed the worst survival of F/M group, so age was discarded as a confounding factor. The age of FR could also act as a confounding factor. Kaczmarek et al. [23] showed that FR were younger than MR. However, M/M with the oldest recipients had better one-year survival than M/F. Jalowiec et al. [35] also reported younger recipients in M/F group but they did not found differences on survival among the groups. Other studies did not found younger FR [33, 34] or these data were not reported [36]. In our previous analysis [43] from the Spanish Registry, FR were younger than MR, but recipient age did not change multivariate analysis. Al-Khaldi et al. [38] found an interaction between age and donor/recipient sex. FR (younger) had no impact on multivariate analysis and the M/M group was the one with the best one-year survival. This confirmed previously published data from the UNOS registry that recipient < 55 years and donor <30 years had the best long-term survival [87]. In other analysis of this same registry, an older age of the donor was associated with a higher risk of developing CAV throughout his life [88]. However, other previously published analyzes had observed similar survival rates among older marginal donors and the standard donor [89] or just a slight increase in mortality [90].

Undersizing and Oversizing

The "under-sizing" effect could also interact with sex mismatch influence on survival, especially in F/M. Female heart would not be able to keep the perfusion required by a man who previously had a larger heart. This could result in pulmonary hypertension and immediate right ventricular failure [91]. In a situation of previous pulmonary hypertension with high transpulmonary gradient this effect could be more important resulting in higher mortality [43]. The analysis of the ISHLT database performed in 2013 [23] was focused on the influence of donor and recipient body mass index (BMI). The existence of an undersizing effect is suggested due to F/M worst results after correction of weight and height. On the contrary, better short-term results were reported in the combination of M/F, suggesting a beneficial effect of oversizing, especially when the recipient had high pulmonary pressures. In a previous analysis of this same database, Khush et al. [40] adjusted the results based on weight mismatch, using three different parameters: donor and recipient weight, donor and recipient weight difference and weight ratio of recipient regarding donor weight. The authors found worse survival in F/M, but they did not find an interaction of the difference in weight in this survival. However, they also suggested a worse survival of M/F, which, eventually, could influence on not finding interaction between these variables. The most recent analysis of ISHLT database [42] using data from from 1994 until 2013 found that sex mismatch increased mortality independently of weight match. UNOS data published in 2009 [25] studied BMI ratio and body surface area (BSA) ratio between donors and recipients, finding a quite precise adjustment, probably due to a deliberate move to allocate the graft adjusting by cardiac size. Other studies were consistent with this adjustment and showed no difference among the four groups in donor/recipient BSA ratio [27, 33, 34]. On the contrary, Correia et al. [39] and Jalowiec et al. [35] found a difference in this ratio. However, this was due to oversized hearts in some recipients (not to undersized hearts in MR) and could not explain a higher early mortality in F/M. Indeed, Correia et al. [39] described a deliberate aim to avoid the use of FD (smaller) in recipients with some degree of pulmonary hypertension when allocating a graft. Previously, a single-center study [22] had shown a worse annual survival in F/M, being this group the one with a minor BSA ratio between donor and recipient, suggesting that female hearts were not able to support male circulation. However, authors of this same group had not previously found deleterious effects of size mismatch [91, 92], suggesting a different adaptation of that smaller heart with greater left ventricular hypertrophy [92]. Other authors related size mismatch greater than 20% with a significant reduction in graft survival, a generally accepted criterion for donor selection in heart transplantation [93, 94]. In 2014 Reed et al. [41] suggested that the measures used to assess the relationship between donor and recipient cardiac size were not adequate, having a poor correlation between weight and heart size. They studied a new way of assessing this relationship by combining mathematical formula. They conducted a retrospective study of 31,634 patients in the UNOS registry and used predictive models, identifying undersizing pairs with increased risk. The formula calculated the predicted cardiac mass combining the predicted left ventricular and right ventricular cardiac mass. They found that a difference of 10–15% (undersized heart) resulted in higher risk of mortality. In the adjusted analysis the risk attributed to sex mismatch in F/M disappeared and higher mortality was observed in M/F. Looking at the pediatric population could help assess the role of size-mismatch as a confounding factor in sex mismatch, as recently done by Tosi et al. [95]. In their study of 3,630 heart transplant recipients <18 years-old donor/recipient BSA ratio was >1.2 in all sex groups (meaning paediatric subjects tend to receive adequate or oversized hearts). Sex matching was not associated with advanced survival and M/F had an increased mortality compared to F/M. This could be partially explained by the fact that females have an increased risk of mortality after paediatric cardiac surgery. Later on, similar results were reported in a 5,795 recipients [96]. These results would agree with the theory that cardiac size mismatch is interacting with worst survival in F/M, which would mean that traditional measures do not accurately address this issue. Also, they might be an interaction between cardiac size mismatch, sex and age. The cardiac transplant research database (CTRD) analysis included 7,321 patients transplanted between 1990 and 2007 [24]. The authors found that 56% of patients weighed less than donors, while 44% weighed more. Oversizing was not related to risk or benefit, even in the group with pulmonary hypertension, while undersizing was related with mortality. They also found an interaction between weight difference, age, and recipient sex, with a higher one-year mortality in F/M with an older organ (more than forty years) and a 30% weight difference. As we can see, the results about the possible influence of the size mismatch on survival are heterogeneous, and its interaction with sex mismatch is not clear. In our previous analysis [43] we showed that sex mismatch increased mortality only in men with pulmonary hypertension the first month after heart transplantation. Thus, it is possible that MR with high transpulmonary gradient that receives a smaller heart may not be able to maintain pulmonary artery pressure, resulting in right ventricular impairment and mortality. However, there were not significant differences in weight relationship between donor and recipient in M/M vs. F/M, so those results cannot be attributed to a difference in cardiac size. Previous studies had shown higher mortality with increasing pulmonary vascular resistance [94, 97] and an increased risk of graft failure in those with pulmonary vascular resistance greater than 4 UW and with a donor/recipient weight ratio of less than 0.8 [32, 98, 100]. The analysis of CTRD database explored a possible interaction between the percentage of weight difference and high pulmonary resistance [24]. In the univariate analysis, mortality appeared to be higher in patients with high pulmonary resistance and a percentage of weight difference above 20%. However, this interaction was not confirmed as significant in the multivariate model. However, clinical practice has adapted graft allocation to these circumstances [21, 22, 39, 94, 99]. Thus, a single-center Portuguese study [39] showed same survival in those patients with sex mismatch due to a good selection of grafts based on cardiac size and in those patients with high transpulmonary gradient. However, it is a single-center and small sample study so their results cannot be considered superior to those observed in large international bases.

Urgent Transplant

The UNOS registry data published in 2009 [25] showed higher mortality in F/M. The subgroup analysis showed that these data were only valid for those transplanted in maximum urgency. Previously, an analysis published in Spain [46] showed higher mortality in the male recipient of female heart, being attributed to the higher rates of urgent transplant.

Conclusion

Heart transplantation is the treatment of choice in selected patients with end-stage heart disease. When allocating a graft several factors must be considered. Among these donor and recipient sex are relevant and might influence the prognosis. Several studies have analyzed the influence of donor and recipient sex on prognosis. Sex mismatch, particularly FM in recipients with pulmonary hypertension seems to have a deleterious effect on prognosis. Further studies regarding the role of sex on heart transplantation would be welcome.

References

1. Ponikowski P, Voors AA, Anker SD et al (2016) ESC Guidelines for the diagnosis and treatment of acute and chronic heart failure: The Task Force for the diagnosis and treatment of acute and chronic heart failure of the European Society of Cardiology (ESC). Developed with the special contribution of the Heart Failure Association (HFA) of the ESC. Eur J Heart Fail 18:891
2. González-Vílchez F, Almenar-Bonet L, Crespo-Leiro MG et al (2018) Spanish heart transplantation registry. 29th official report of the spanish society of cardiology working group on heart failure. Rev Esp Cardiol 71:952–960
3. Khush KK, Cherikh WS, Chambers DC et al (2018) The international thoracic organ transplant registry of the international society for heart and lung transplantation: thirty-fifth adult heart transplantation report-2018; focus theme: multiorgan transplantation. J Heart Lung Transplant 37:1155–1168
4. Fabbri A, Bryan AJ, Sharples LD et al (1992) Influence of recipient and donor gender on outcome after heart transplantation. J Heart Lung Transplant 11:701–707
5. McCarthy JF, McCarthy PM, Massad MG et al (1998) Risk factors for death after heart transplantation: does a single-center experience correlate with multicenter registries? Ann Thorac Surg 65:1574–1578
6. Tsai FC, Marelli D, Bresson J et al (2002) Recent trends in early outcome of adult patients after heart transplantation: a single-institution review of 251 transplants using standard donor organs. Am J Transplant 2:539–545
7. Crandall BG, Renlund DG, O'Connell JB et al (1998) Increased cardiac allograft rejection in female heart transplant recipients. J Heart Transplant 7:419–423
8. Esmore D, Keogh A, Spratt P et al (1991) Heart transplantation in females. J Heart Lung Transplant 10:335–341
9. Kirklin JK, Naftel DC, Bourge RC et al (1992) Rejection after cardiac transplantation. A time-related risk factor analysis. Circulation 86:II236–II241
10. Kobashigawa JA, Kirklin JK, Naftel DC et al (1993) Pretransplantation risk factors for acute rejection after heart transplantation: a multiinstitutional study. The transplant cardiologists research database group. J Heart Lung Transplant 12:355–366
11. Keck BM, White R, Breen TJ et al (1994) Thoracic organ transplants in the United States: a report from the UNOS/ISHLT scientific registry for organ transplants. United network for organ sharing. International society for heart and lung transplantation. Clin Transpl 1994:37–46
12. Bryan CF, Mitchell SI, Borkon AM et al (1996) Influence of donor gender on patient mortality after heart transplantation. Transplant Proc 28:149–151
13. Solomon NA, McGiven JR, Alison PM et al (2004) Changing donor and recipient demographics in a heart transplantation program: influence on early outcome. Ann Thorac Surg 77:2096–2102
14. Radovancevic B, Konuralp C, Vrtovec B et al (2005) Factors predicting 10-year survival after heart transplantation. J Heart Lung Transplant 24:156–159
15. Fiorelli AI, Branco JN, Dinkhuysen JJ et al (2012) Risk factor analysis of late survival after heart transplantation according to donor profile: a multi-institutional retrospective study of 512 transplants. Transplant Proc 44:2469–2472
16. Brock MV, Salazar JD, Cameron DE et al (2001) The changing profile of the cardiac donor. J Heart Lung Transplant 20:1005–1009
17. Del Rizzo DF, Menkis AH, Pflugfelder PW et al (1999) The role of donor age and ischemic time on survival following orthotopic heart transplantation. J Heart Lung Transplant 18:310–319
18. Chou NK, Chi NH, Wu IH et al (2008) Medium-term outcomes in cardiac transplant recipients in a single cardiac transplant center in Taiwan. Transplant Proc 40:2626–2628
19. Bocchi EA, Fiorelli A (2001) The Brazilian experience with heart transplantation: a multicenter report. J Heart Lung Transplant 20:637–645

20. Mastrobuoni S, Dell'Aquila AM, Azcarate PM et al (2012) Long-term survival (> 20 years) following heart transplantation. J Cardiovasc Surg (Torino)53:677–684
21. Kirsch M, Baufreton C, Naftel DC et al (1998) Pretransplantation risk factors for death after heart transplantation: the Henri Mondor experience. J Heart Lung Transplant 17:268–277
22. Prendergast TW, Furukawa S, Beyer AJ III et al (1998) The role of gender in heart transplantation. Ann Thorac Surg 65:88–94
23. Kaczmarek I, Meiser B, Beiras-Fernandez A et al (2013) Gender does matter: gender-specific outcome analysis of 67,855 heart transplants. Thorac Cardiovasc Surg 61:29–36
24. Stehlik J, Feldman DS, Brown RN et al (2010) Interactions among donor characteristics influence post-transplant survival: a multi-institutional analysis. J Heart Lung Transplant 29:291–298
25. Weiss ES, Allen JG, Patel ND et al (2009) The impact of donor-recipient sex matching on survival after orthotopic heart transplantation: analysis of 18 000 transplants in the modern era. Circ Heart Fail 2:401–408
26. Zeier M, Dohler B, Opelz G, Ritz E (2002) The effect of donor gender on graft survival. J Am Soc Nephrol 13:2570–2576
27. Al-Khaldi A, Oyer PE, Robbins RC (2006) Outcome analysis of donor gender in heart transplantation. J Heart Lung Transplant 25:461–468
28. Eifert S, Kofler S, Nickel T et al (2012) Gender-based analysis of outcome after heart transplantation. Exp Clin Transplant 10:368–374
29. Kittleson MM, Shemin R, Patel JK et al (2011) Donor-recipient sex mismatch portends poor 10-year outcomes in a single-center experience. J Heart Lung Transplant 30:1018–1022
30. Welp H, Spieker T, Erren M et al (2009) Sex mismatch in heart transplantation is associated with increased number of severe rejection episodes and shorter long-term survival. Transplant Proc 41:2579–2584
31. Bello RA, D'Alessandro DA, Maybaum S, Goldstein DJ (2009) The impact of donor-recipient gender- matching on survival and rejection after cardiac transplantation. J Heart Lung Transplant 28:S185
32. Maltais S, Jaik NP, Feurer ID et al (2013) Mechanical circulatory support and heart transplantation: donor and recipient factors influencing graft survival. Ann Thorac Surg 96:1252–1258
33. Schlechta B, Kocher AA, Ofner P et al (1999) Impact of gender mismatch on the outcome of heart transplantation. Transplant Proc 31:3340–3342
34. De Santo LS, Marra C, De FM et al (2002) The impact of gender on heart transplantation outcomes: a single center experience. Ital Heart J 3:419–423
35. Jalowiec A, Grady KL, White-Williams C (2012) First-year clinical outcomes in gender-mismatched heart transplant recipients. J Cardiovasc Nurs 27:519–527
36. Tsao CI, Chen RJ, Chou NK et al (2008) The influence of gender on survival after heart transplantation. Transplant Proc 40:2634–2635
37. Yamani MH, Erinc SK, McNeill A et al (2005) The impact of donor gender on cardiac peri-transplantation ischemia injury. J Heart Lung Transplant 24:1741–1744
38. Aliabadi AZ, Dunkler D, Eskandary FA et al (2011) Do boys and girls match?- The effect of gender mismatch in cardiac transplantation. J Heart Lung Transplant 30:S194
39. Correia P, Prieto D, Batista M, Antunes MJ (2014) Gender mismatch between donor and recipient is a factor of morbidity but does not condition survival after cardiac transplantation. Transpl Int 27:1303–1310
40. Khush KK, Kubo JT, Desai M (2012) Influence of donor and recipient sex mismatch on heart transplant outcomes: analysis of the International Society for Heart and Lung Transplantation Registry. J Heart Lung Transplant 31:459–466
41. Reed RM, Netzer G, Hunsicker L et al (2014) Cardiac size and sex-matching in heart transplantation: size matters in matters of sex and the heart. JACC Heart Fail 2:73–83
42. Bergenfeldt H, Stehlik J, Höglund P et al (2017) Donor -recipient size matching and mortality in heart transplantation: influence of body mass index and gender. J Heart Lung Transplant 36:940–947

43. Martinez-Selles M, Almenar L, Paniagua-Martin MJ et al (2015) Donor/recipient sex mismatch and survival after heart transplantation: only an issue in male recipients? An analysis of the Spanish Heart Transplantation Registry. Transpl Int 28:305
44. Ayesta A, Urrútia G, Madrid E et al (2019) Sex-mismatch influence on survival after heart transplantation: A systematic review and meta-analysis of observational studies. Clin Transplant 20:e13737
45. Keogh AM, Valantine HA, Hunt SA et al (1991) Increased rejection in gender-mismatched grafts: amelioration by triple therapy. J Heart Lung Transplant 10:106–110
46. Izquierdo MT, Almenar L, Martinez-Dolz L et al (2007) Analysis of the impact of donor gender on early mortality. Transplant Proc 39:2375–2376
47. Martinez-Selles M, Doughty RN, Poppe K et al (2012) Gender and survival in patients with heart failure: interactions with diabetes and aetiology. Results from the MAGGIC individual patient meta-analysis. Eur J Heart Fail 14:473–479
48. Martínez- Sellés M, Díez- Villanueva P, Álvarez-García J et al (2018) Influence of sex and pregnancy on survival in patients admitted with heart failure. Data from prospective multicenter registry. Clin Cardiol. 41:924–930
49. Lorenz CH, Walker ES, Morgan VL et al (1999) Normal human right and left ventricular mass, systolic function, and gender differences by cine magnetic resonance imaging. J Cardiovasc Magn Reson 1:7–21
50. Marcus JT, DeWaal LK, Gotte MJ et al (1999) MRI-derived left ventricular function parameters and mass in healthy young adults: relation with gender and body size. Int J Card Imaging 15:411–419
51. Salton CJ, Chuang ML, O'Donnell CJ et al (2002) Gender differences and normal left ventricular anatomy in an adult population free of hypertension. A cardiovascular magnetic resonance study of the Framingham Heart Study Offspring cohort. J Am Coll Cardiol 39:1055–1060
52. Martinez-Selles M, Munoa MD, Martinez E et al (2006) The influence of sex on right ventricular dysfunction in patients with severely depressed left ventricular ejection fraction. Eur J Heart Fail 8:400–403
53. Clapp JF III, Capeless E (1997) Cardiovascular function before, during, and after the first and subsequent pregnancies. Am J Cardiol 80:1469–1473
54. Duvekot JJ, Peeters LL (1994) Maternal cardiovascular hemodynamic adaptation to pregnancy. Obstet Gynecol Surv 49:S1–S14
55. Hudsmith LE, Petersen SE, Francis JM et al (2005) Normal human left and right ventricular and left atrial dimensions using steady state free precession magnetic resonance imaging. J Cardiovasc Magn Reson 7:775–782
56. Guerra S, Leri A, Wang X et al (1999) Myocyte death in the failing human heart is gender dependent. Circ Res 85:856–866
57. Sandstede J, Lipke C, Beer M et al (2000) Age- and gender-specific differences in left and right ventricular cardiac function and mass determined by cine magnetic resonance imaging. Eur Radiol 10:438–442
58. Jessup M, Pina IL (2004) Is it important to examine gender differences in the epidemiology and outcome of severe heart failure? J Thorac Cardiovasc Surg 127:1247–1252
59. Crespo Leiro MG (2006) Paniagua Martin MJ. [Heart failure. Are women different?]. Rev Esp Cardiol 59:725–735
60. Martinez-Selles M, Garcia Robles JA, Prieto L et al (2003) Systolic dysfunction is a predictor of long term mortality in men but not in women with heart failure. Eur Heart J 24:2046–2053
61. McSweeney J, Pettey C, Lefler LL, Heo S (2012) Disparities in heart failure and other cardiovascular diseases among women. Womens Health (Lond Engl) 8:473–485
62. Bani D (2008) Relaxin as a natural agent for vascular health. Vasc Health Risk Manag 4:515–524
63. Martinez-Selles M (2007) What do women have in their hearts? Rev Esp Cardiol 60:1118–1121
64. Heidecker B, Lamirault G, Kasper EK et al (2010) The gene expression profile of patients with new-onset heart failure reveals important gender-specific differences. Eur Heart J 31:1188–1196

65. Bird MD, Karavitis J, Kovacs EJ (2008) Sex differences and estrogen modulation of the cellular immune response after injury. Cell Immunol 252:57–67
66. Katz MR, Barnhart GR, Szentpetery S et al (1987) Are steroids essential for successful maintenance of immunosuppression in heart transplantation? J Heart Transplant 6:293–297
67. Renlund DG, O'Connell JB, Gilbert EM et al (1987) Feasibility of discontinuation of corticosteroid maintenance therapy in heart transplantation. J Heart Transplant 6:71–78
68. Csete M (2008) Gender issues in transplantation. Anesth Analg 107:232–238
69. Bayes-Genis A, Bellosillo B, de la Calle O et al (2005) Identification of male cardiomyocytes of extracardiac origin in the hearts of women with male progeny: male fetal cell microchimerism of the heart. J Heart Lung Transplant 24:2179–2183
70. Stewart S, Winters GL, Fishbein MC et al (2005) Revision of the 1990 working formulation for the standardization of nomenclature in the diagnosis of heart rejection. J Heart Lung Transplant 24:1710–1720
71. Sharples LD, Caine N, Mullins P et al (1991) Risk factor analysis for the major hazards following heart transplantation–rejection, infection, and coronary occlusive disease. Transplantation 52:244–252
72. Patel J, Kittleson MM, Kawano M et al (2011) Does gender mismatch increase the risk of antibody-mediated rejection (AMR) after heart transplantation? J Heart Lung Transplant 30:S178
73. George JF, Pamboukian SV, Tallaj JA et al (2010) Balancing rejection and infection with respect to age, race, and gender: clues acquired from 17 years of cardiac transplantation data. J Heart Lung Transplant 29:966–972
74. Mehra MR, Ventura HO, Escobar A et al (1995) Does donor and recipient sex influence the development of cardiac allograft vasculopathy? Transplant Proc 27:1926–1929
75. Mehra MR, Crespo-Leiro MG, Dipchand A et al (2010) International Society for Heart and Lung Transplantation working formulation of a standardized nomenclature for cardiac allograft vasculopathy-2010. J Heart Lung Transplant 29:717–727
76. Erinc K, Yamani MH, Starling RC et al (2004) The influence of donor gender on allograft vasculopathy: evidence from intravascular ultrasound. Transplant Proc 36:3129–3131
77. Costanzo MR, Naftel DC, Pritzker MR et al (1998) Heart transplant coronary artery disease detected by coronary angiography: a multiinstitutional study of preoperative donor and recipient risk factors. Cardiac Transplant Research Database. J Heart Lung Transplant 17:744–753
78. Caforio AL, Tona F, Fortina AB et al (2004) Immune and nonimmune predictors of cardiac allograft vasculopathy onset and severity: multivariate risk factor analysis and role of immunosuppression. Am J Transplant 4:962–970
79. Kobashigawa J, Zuckermann A, Macdonald P et al (2014) Report from a consensus conference on primary graft dysfunction after cardiac transplantation. J Heart Lung Transplant 33:327–340
80. Stehlik J, Edwards LB, Kucheryavaya AY et al (2011) The Registry of the International Society for Heart and Lung Transplantation: Twenty-eighth Adult Heart Transplant Report–2011. J Heart Lung Transplant 30:1078–1094
81. Cosio Carmena MD, Gomez BM, Almenar L et al (2013) Primary graft failure after heart transplantation: characteristics in a contemporary cohort and performance of the RADIAL risk score. J Heart Lung Transplant 32:1187–1195
82. Hong KN, Iribarne A, Worku B et al (2011) Who is the high-risk recipient? Predicting mortality after heart transplant using pretransplant donor and recipient risk factors. Ann Thorac Surg 92:520–527
83. Russo MJ, Iribarne A, Hong KN et al (2010) Factors associated with primary graft failure after heart transplantation. Transplantation 90:444–450
84. Young JB, Hauptman PJ, Naftel DC et al (2001) Determinants of early graft failure following cardiac transplantation, a 10-year, multi-institutional, multivariable analysis. J Heart Lung Transplant 20:212

85. Lund LH, Edwards LB, Kucheryavaya AY et al (2013) The Registry of the International Society for Heart and Lung Transplantation: Thirtieth Official Adult Heart Transplant Report–2013; focus theme: age. J Heart Lung Transplant 32:951–964
86. Lund LH, Kush KK, Cherikh WS et al (2017) The registry of the International Society for Heart and Lung Transplantation: thirty-fourth official adult heart transplant report–2017; focus theme: allograft ischemic time. J Heart Lung Transplant 20:S1053
87. Whitson BA, Ravi Y, Emani S et al (2014) Heart transplant recipient and donor age mismatching: should the older recipient be paired with the older donor? J Heart Lung Transplant 33:S163
88. Nagji AS, Hranjec T, Swenson BR et al (2010) Donor age is associated with chronic allograft vasculopathy after adult heart transplantation: implications for donor allocation. Ann Thorac Surg 90:168–175
89. Prieto D, Correia P, Baptista M, Antunes MJ (2015) Outcome after heart transplantation from older donor age: expanding the donor pool. Eur J Cardiothorac Surg 47:672–678
90. Reiss N, Leprince P, Bonnet N et al (2007) Results after orthotopic heart transplantation accepting donor hearts >50 years: experience at La Pitie Salpetriere. Paris. Transplant Proc 39:549–553
91. Jeevanandam V, Furukawa S, Prendergast TW et al (1996) Standard criteria for an acceptable donor heart are restricting heart transplantation. Ann Thorac Surg 62:1268–1275
92. Mather PJ, Jeevanandam V, Eisen HJ et al (1995) Functional and morphologic adaptation of undersized donor hearts after heart transplantation. J Am Coll Cardiol 26:737–742
93. Chen JM, Weinberg AD, Rose EA et al (1996) Multivariate analysis of factors affecting waiting time to heart transplantation. Ann Thorac Surg 61:570–575
94. Murali S, Kormos RL, Uretsky BF et al (1993) Preoperative pulmonary hemodynamics and early mortality after orthotopic cardiac transplantation: the Pittsburgh experience. Am Heart 126:896–904
95. Tosi l, Federman M, Markovic D et al (2013) The effect of gender and gender match on mortality in pediatric heart transplantation. Am J Transplant 13:2996–3002
96. Kemma M, Albers E, Bradford MC et al (2016) Impact of donor-recipient sex match on long-term survival after heart transplantation in children: An analysis of 5797 pediatric heart transplants. Pediatr Transplant 20:249–255
97. Kirklin JK, Naftel DC, Kirklin JW et al (1998) Pulmonary vascular resistance and the risk of heart transplantation. J Heart Transplant 7:331–336
98. Russo LM, Webber SA (2005) Idiopathic restrictive cardiomyopathy in children. Heart 91:1199–1202
99. Patel ND, Weiss ES, Nwakanma LU et al (2008) Impact of donor-to-recipient weight ratio on survival after heart transplantation: analysis of the United Network for Organ Sharing Database. Circulation 118:S83–S88
100. Tallaj JA, Pamboukian SV, George JF et al (2014) Have risk factors for mortality after heart transplantation changed over time? Insights from 19 years of Cardiac Transplant Research Database study. J Heart Lung Transplant 33:1304–1311

Part III
Sex Differences in Risk Factors

Chapter 10
Sex, Age and Gene Interactions in Cardiometabolic Diseases

Pavel Hamet, Candan Hizel, and Johanne Tremblay

Abstract Crosstalk between environmental factors, genes and health outcomes is gaining more attention in the pathophysiology of cardiometabolic diseases. The genome and environment convergence is a major focus of personalized medicine that became subject of interest in preventive/predictive healthcare. The collective knowledge related to multifactorial and polygenic nature of diabetes, hypertension and cardiovascular diseases obtained from recent breakthroughs in "-omics" technologies and in bioinformatics provides compelling evidence on the functionality of the genome along with environmental and demographic factors, including sex, age and ethnicity. Accordingly, in the context of "genome–environment interactions", knowledge from "omics" and grasp of age/sex and gene interplay should contribute prospective analysis of complex diseases enabling risk stratification for more efficient and personalized treatment of individuals. To this end, this chapter aims to highlight the contribution of age/sex—genetic make-up interplay in the development of cardiovascular diseases and diabetes and hypertension complications.

Keywords Personalized medicine · Cardiometabolic diseases · Hypertension · Diabetes · Cardiovascular diseases · Gene-environment interaction, age and sex, risk assessment, genomics

P. Hamet (✉) · J. Tremblay
Department of Medicine, University of Montreal, CRCHUM, Montréal, QC, Canada
e-mail: pavel.hamet@umontreal.ca

P. Hamet · C. Hizel · J. Tremblay
OPTITHERA, Montreal, QC, Canada

B. Ostadal and N. S. Dhalla (eds.), *Sex Differences in Heart Disease*,
Advances in Biochemistry in Health and Disease 21,
https://doi.org/10.1007/978-3-030-58677-5_10

Introduction: Insights from Genetic Studies

Since the completion of the Human Genome Project (HGP) in 2003 [1, 2] and the international haplotype map (HapMap) project in 2005 [3], our understanding of health and diseases, being rare or common, monogenic or polygenic [4] underwent deep modifications in its conceptualization of predictive tests and tailored interventions during the pre-symptomatic phase of the disease [5].

As a natural extension of the HGP, genome-wide association studies (GWAS) [4] representing "hypothesis-free" approach, accelerated the identification of genomic variants, mostly in the form of single nucleotide polymorphisms (SNPs), associated with multifactorial disorders or with adverse drug reactions (ADR)s/toxicities or resistance to drug response [6, 7]. The publication of the first GWAS in 2007 and following GWAS studies [8–11] led to the identification of many genes unpredictably involved in the initiation and/or progression of common diseases. Many of these diseases such as type 2 diabetes (T2D) and hypertension are influenced by hundreds and even thousands of SNPs, dispersed throughout the genome, each one with small effect sizes [6]. Moreover, contrary to rare diseases in which the associated rare variants are usually causal and located within the coding region with protein modification, most SNPs relevant to complex diseases are located in non-coding regions, such as promoters, introns and in intergenic regions contributing indirectly to the disease [6].

Recent developments of polygenic risk scores (PRS) composed of thousands of common variants (mainly SNPs) identified by GWAS and that account for only modest effects on their own but when combined provide better prediction than traditional clinical risk factors, are becoming important genomic tools of personalized medicine (PM) [12, 13]. Henceforth, the utility of PRS for different cancers, including prostate [14] and breast cancer [15], for response to statin therapy [16], for coronary artery disease [17], and for microvascular and macrovascular complications of T2D [18] addresses a personalized healthcare approach that considers individual's unique clinical, genomic, environmental and demographic (sex, age, and ethnicity) information for risk assessments, risk stratified prevention and optimal drug treatments [19–21]. Accumulating bulk of evidence clearly acknowledge that complex diseases are the result of the combined effect of genes, environmental factors and their interactions generating an evermore intricate feature of human disease at the molecular level [22–26]. To this end, genomics of complex diseases, such as T2D, hypertension and their cardiovascular complications should be viewed through a lens of the interplay between genetic and environmental factors [23–26].

In this chapter, we are specifically discussing the critical impact of sex and age as internal environmental drivers of interactions with genetic backbone and resulting risk prediction with relevant clinical utility.

Genes and Age in Disease Penetrance and Outcomes

The evidence for a genetic basis of essential hypertension dates to studies of parent-child concordance of blood pressure in natural vs adopted children by Mongeau and Biron in Montreal where systolic blood pressure was estimated as sustained 61% by genes and 39% by shared household environment [27]. Our contribution related to the importance of genes, sex, age and environmental origin stems from studies of 120 multigenerational families, ascertained by the presence of hypertension and dyslipidemia, of French-Canadian individuals living in North-East of Quebec with founder effect [28]. Our studies demonstrated that the penetrance of hypertension, in families with genetic predisposition, appears at a younger age than in families of the general population living in the same geographical area. As illustrated in Fig. 10.1, at the youngest age studied (<20 years old) the ratio of hypertensive individuals within hypertensive families (and presumably sharing higher density of hypertensive alleles) exceeds 20-fold that in families of the general population, while at the age of >65 years

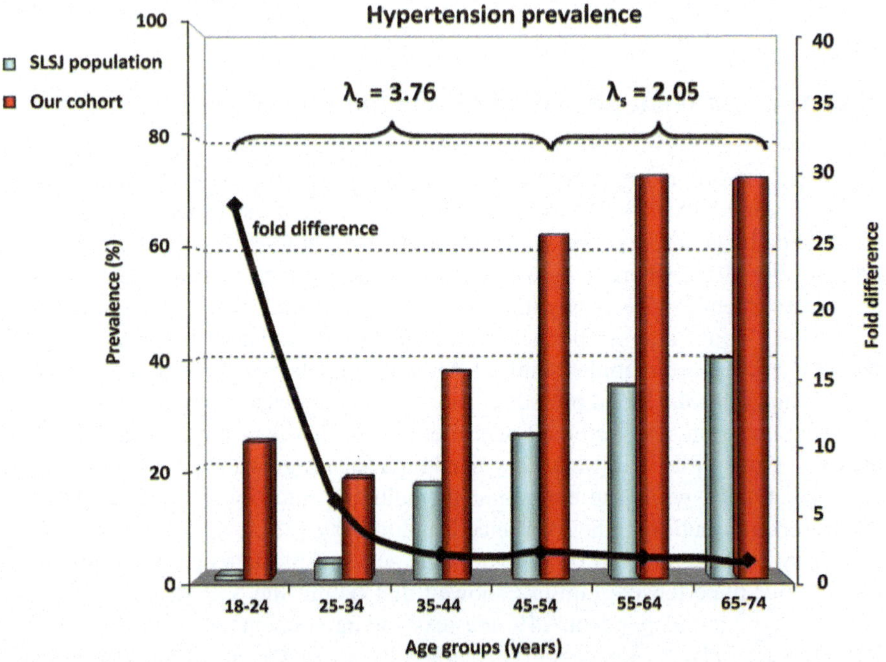

Fig. 10.1 The sibling risk ratio (λs) of hypertension per age group in French–Canadians. We observed a strong familial clustering in younger individuals, the highest fold difference in λs being for cases occurring between 18 and 24 years, ~27. Augmented λs for hypertension reflects increased genetic susceptibilities of our cohort, compared to the general population of Saguenay–Lac-St-Jean (SLSJ). However, the siblings may differ by environmental risk factors for hypertension, thus increasing the risk of hypertension independently of genetic effects (*Source* Simon et al. Am J Hypertens. 29:891–9, 2016 reproduced with permission) [24]

old, this ratio is less than 2-fold, underlying the increasing impact of environment and ageing [24]. These studies demonstrated that the genetic contribution (enriched in ascertained families) leads to an early penetrance of hypertension, while in the general population, environmental exposure requires a prolonged period for hypertension to appear.

Moreover, the impact of age (progression and ageing) may be competing with age of onset of the disease (penetrance). For instance, in the context of the ADVANCE clinical trial, where all participants had T2D, we observed a dominant effect of age on incidence of macrovascular events contrasting with microvascular events that were more dependent on early onset of T2D [29]. A similar situation was recently noted by Sattar et al. [30] as illustrated in Fig. 10.2. The study analyzed three hundred thousand sex- and age-matched controls to more than two hundred thousand individuals with T2D and showed that T2D is a major risk factor of cardiovascular and total mortality, AMI and stroke but that its impact is fading with age. These studies demonstrate that genetic factors have more impact at a younger age and that G × E interactions are not static but vary throughout the lifespan of an individual and should therefore be evaluated as a function of age [24, 28].

Sex, Age and Genes Relative Impact on Phenotypes

Our early studies using multigenerational families counted an intensive phenotyping totaling 539 traits. We estimated the proportion of the variance of these traits contributed by genes, sex, age and environment (residual). A few examples are illustrated in Fig. 10.3, where it shows that BMI is determined mainly by genes (60%) and environment (~35%) while total body fate (evaluated by impedance) is mainly determined by age and sex but genes' contribution dominates for specific fat accumulation such as subscapular skinfold. Most illuminating is the difference between systolic and diastolic blood pressures: other than genetics and environment, age has a major impact on systolic blood pressure, impact that is minor for diastolic blood pressure, compatible with well recognized biphasic progression of diastolic BP with age. And finally, while the environment (including nutrition) is relevant for LDL levels, its contribution to their variance is below 50% [28].

Our systematic analysis of genome-wide sex-specific linkage of cardiovascular traits in our hypertensive families showed that within the 539 studied traits, 47% are sex- and age-independent, 24% are sex- and age-dependent, 15% are only age- and 14% are only sex-dependent as defined by 2 LOD score differences between sexes [31]. Figure 10.4 illustrates a progressive gradient of sex- and age-specific and independent blood pressure and heart rate traits, contrasting with mainly age- and sex-independent humoral and biochemical traits.

In the same study, we identified a locus on chromosome 12, in significant linkage and association with systolic blood pressure, exclusively in men. A major lesson from this finding is that as seen in Fig. 10.5, sex-adjusted values completely hid this genetic locus, which resulted in the highest systolic BP in homozygous males

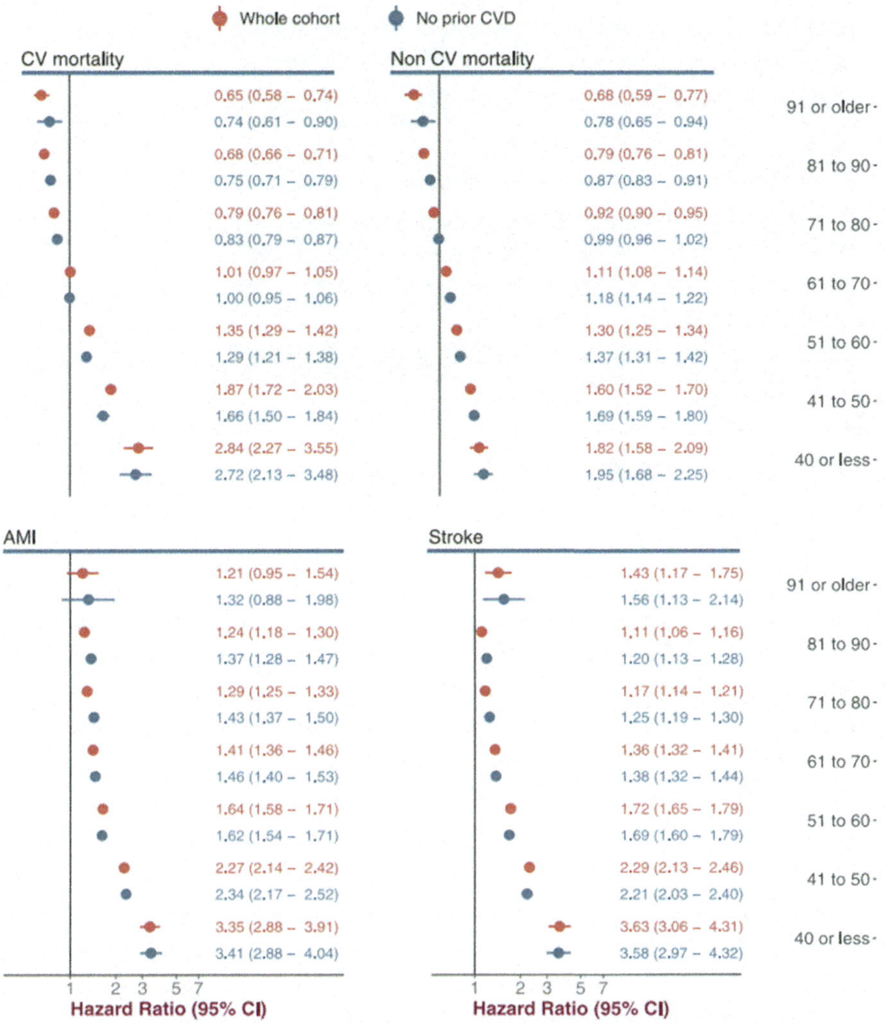

Fig. 10.2 Adjusted hazard ratios (95% CI) for patients with type 2 diabetes mellitus according to age at diagnosis, in comparison with matched controls in those without previous cardiovascular disease (blue) and in the whole cohort (red). The models used in the main analyses (blue) used age as the underlying time scale and include only sex, yearly time-updated duration (which is zero for the controls who are persons who have not yet been diagnosed with diabetes mellitus), and diabetes status, which is coded as either control or different categories of persons with diabetes according to their age at onset. The models used for the supporting analyses in the entire cohort (red) also include persons with prevalent cardiovascular disease at cohort entry, and, in addition, contain binary indicators for per index acute myocardial infarction (AMI), and stroke (*Sources* Sattar et al. Circulation 139:2228–2237, 2019—reproduced with permission) [30]

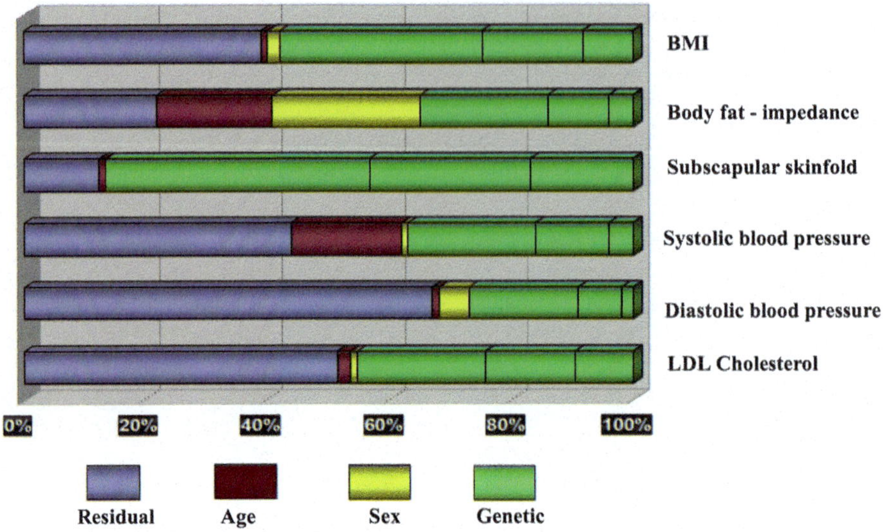

Fig. 10.3 Partitioning of Variance of a Set of Hemodynamic, Anthropometric, and Metabolic Traits. The oligogenic and multifactorial nature of these traits is, as demonstrated by the relatively large contribution of both genetic (from a few QTLs) and environmental (residual) factors to the variance. This analysis suggested the presence of QTLs with a large effect for several phenotypes. Age was found to contribute, to a large extent, to the variance of %body fat measured by impedance. On the other hand, the covariate sex explained 12–24% of the variance of waist circumference, LDL levels, and %body fat. These results thus confirmed that age and sex are important covariates for cardiovascular-related phenotypes and that a sizable sexual dimorphism has to be considered for several of these traits. (*Sources* Hamet et al. Am J Hum Genet. 76:815–32, 2005 with permission) [28]

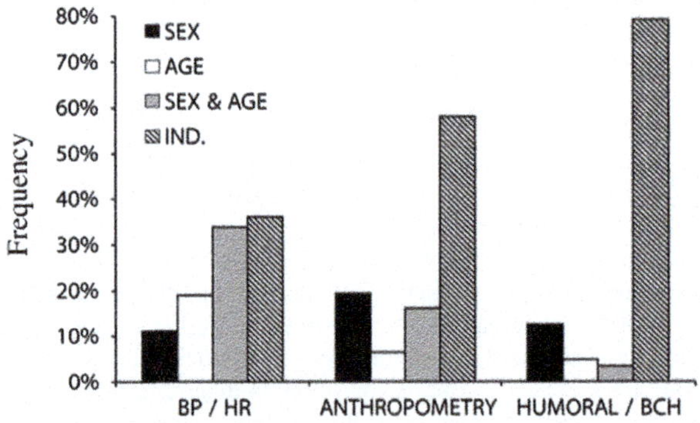

Fig. 10.4 Distribution of sex, age and sex, and age-related traits in major phenotype categories. Relative frequencies of phenotypes for which sex (■), age (□), and sex and age (■)are significant covariates in the linear regression model after correction for multiple testing ($P < 9.3 \times 10^{-5}$). The percentage of traits in each major phenotype category not passing the threshold is indicated as independent (IND.,). BP/HR indicate cardiac indices; HUMORAL/BCH, humoral and biochemical measures (*Source* Šeda et al. Hyepertension 51:1156–62, 2008 with permission) [31]

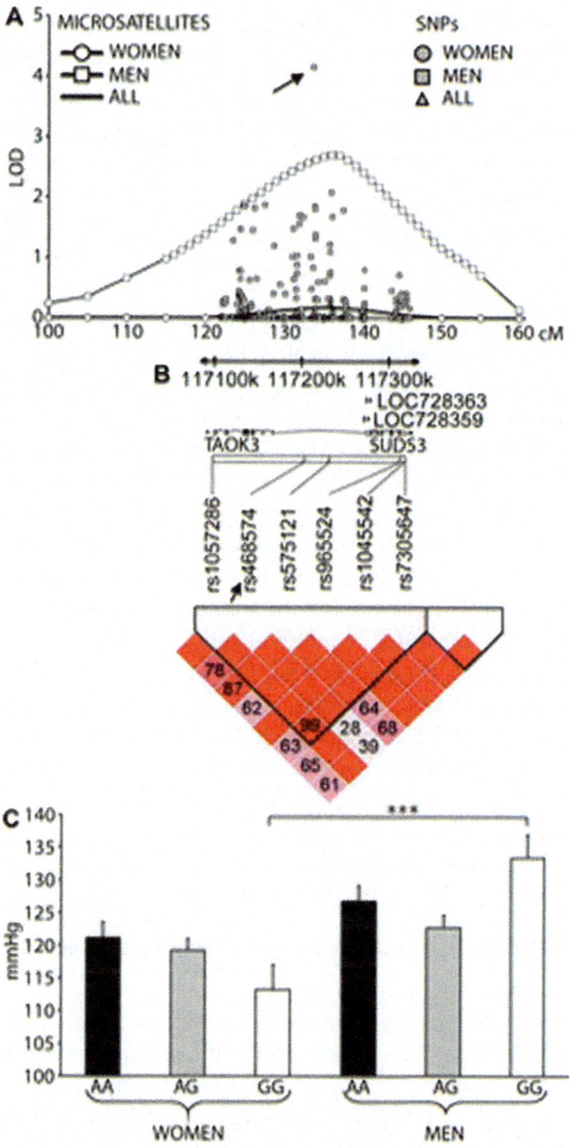

Fig. 10.5 A, Male-specific linkage of supine SBP on chromosome 12 by microsatellite-based multipoint linkage and 2-point linkage with SNPs. B, Detailed view of linkage disequilibrium at the peak of 2-point linkage is shown using Haploview 4.0-generated linkage disequilibrium blocks. The values of D′ are indicated in number and by color (for LOD ≥ 2: bright red: D′ = 1, shades of red/pink: D′ < 1; for LOD < 2: white), with indication of gene tracks in the region. C, SBP (age-adjusted mean ± SE) in the supine position according to sex and genotype of rs575121. Only significant results of Tukey's posthoc tests between women and men of identical genotypes are shown. ***$p < 0.001$. (*Source* Šeda et al. Hyepertension 51:1156–62, 2008 reproduced with permission) [31]

(135 mmHg) and lowest in homozygous females (115 mmHg) within the same families. It is furthermore intriguing, that the gene potentially implicated is FAOK3, a gene demonstrated to have an androgen sensitive expression [32].

Most recently we developed a PRS by selecting from publicly available GWAS data performed in 1.2 million individuals, 600 SNPs associated to risk factors and outcomes of T2D complications and together with principal component of geo-ethnicity (using 35 000 SNP), sex, and age at the onset of T2D, was used to stratify

Table 10.1 Predictive performance (AUC of the ROC) of PRS in males and females

Males			Females Difference		
Phenotype	AUC	95%CI	AUC	95%CI	P-Value
Microvascular events	0.66	0.65–0.70	0.71	0.66–0.77	0.15
Macrovascular events	0.65	0.62–0.68	0.72	0.68–0.76	0.01
All cause death	0.68	0.65–0.71	0.72	0.67–0.76	0.20
Cardiovascular death	0.71	0.67–0.75	0.77	0.72–0.82	0.09

T2D patients into low, medium and high risk categories for diabetic nephropathy, stroke, myocardial infarction, heart failure, and cardiovascular and total death [18]. The prediction performance of the PRS, expressed in c-statistics as AUCs is significantly better in women than men with T2D for macrovascular events and cardiovascular death (Table 10.1). We have also evaluated the clinically recognized impact of age at the onset of T2D on microvascular and macrovascular events, discussed above [29] in its genetic component. We observed a more than 2-fold gradient of both micro- and macrovascular events between individuals carrying low and high PRS in the youngest tertile of age at disease onset (<55 y/o) contrasting with overall increase in macrovascular event rates with age [18]. Finally, in agreement with Winkler et al. who demonstrated the influence of age and gene association for such traits as BMI and waist/hip ratio (WHR), illustrated in Fig. 10.6, we can conclude that assessment of genetic risk must include age of disease onset and that in most situations, polygenic risk scoring should be performed in men and women separately and not only in "adjusted for sex" condition [33]. This is further underlined in Fig. 10.7, depicting population distribution along sex and ethnicity (separating Celtic and Slavic participants in the ADVANCE trial) [34].

Lifestyle, Medication, Disease and Gene Interactions

Lifestyle and environmental factors have a massive impact on disease prevalence. One of most robust examples is the prevalence of diabetes in Pima Indians residing in Mexico *versus* USA. While both groups have very similar genetic profile, the prevalence of T2D is 7-fold higher in USA, accompanied by obesity and associated with sedentary lifestyle. In addition, the prevalence of diabetes in Pima Indians is similar in both sexes, while in non-Pima Mexicans it is much more prevalent in women [35].

Gene encoding fat mass and obesity-associated protein (FTO) is one the major gene associated with obesity and has been identified in over 20 GWAS for being significantly associated to BMI, obesity, T2D and waist circumference. Accordingly, two facts are relevant in the context of our discussion: None of the published GWAS detected its association with BP, until we reported it in the French-Canadian population [36]. One of the possible explanations could be that in the majority of GWAS

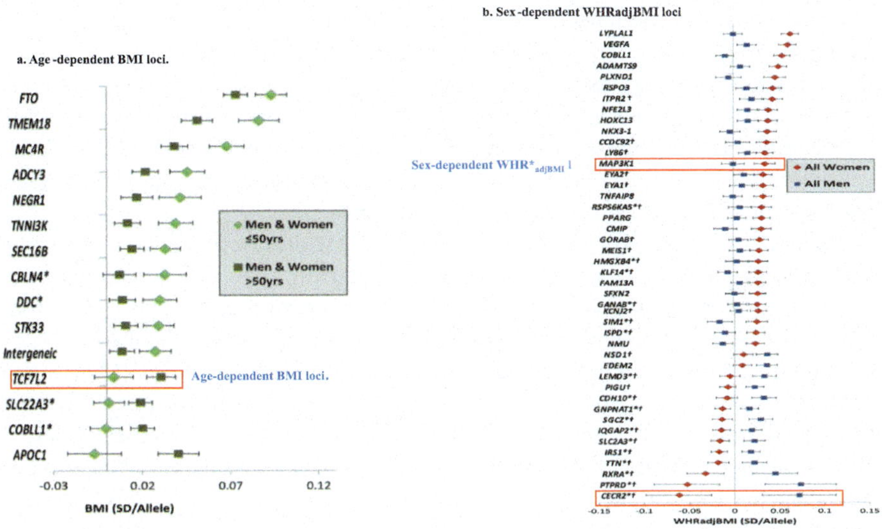

Fig. 10.6 **a** Age-dependent BMI loci. Effect estimates (beta ± 95CI) per standard deviation in BMI and risk allele for loci showing age-differences in men & women ≤ 50y compared to men & women > 50y. Loci are ordered by greater magnitude of effect in men & women ≤ 50y compared to men & women > 50y. (95%CI: 95% confidence interval; BMI: body mass index; SD: standard deviation, *Newly identified loci). (*Source* Winkler et al. PLoS Genet. 11: e1005378, 2015 Reproduced with permission) **b** Sex-dependent WHRadjBMI loci. Effect estimates (beta ± 95CI) per standard deviation in WHRadjBMI and risk allele for loci showing sex-differences in women compared to men. Loci are ordered by greater magnitude of effect in women compared to men. (95%CI: 95% confidence interval; SD: standard deviation. *Newly identified loci. † Newly identified sex-differences (*Source* Winkler et al. PLoS Genet. 11: e1005378, 2015 Reproduced with permission) [33]

Fig. 10.7 Distribution in deciles of PRS (wGRS, sex and PC1) of global risk in the two sexes and Celtic and Slavic ethnic groups of participants in ADVANCE trial—Global risk includes decline of renal function, myocardial infarction (MI), stroke and cardiovascular death. Data from [18]

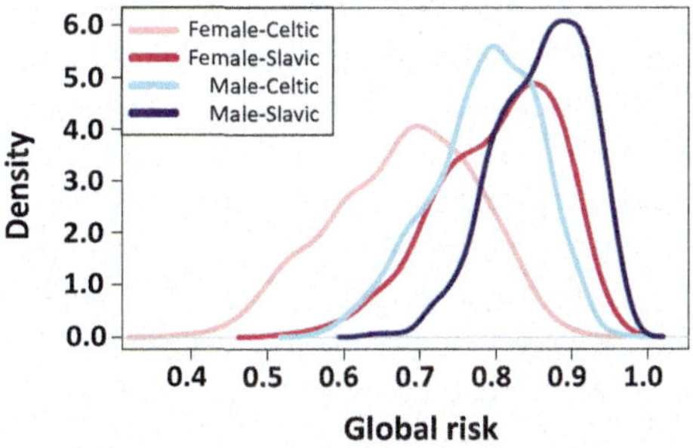

that used massive cohorts of subjects, blood pressure was determined in medicated subjects. Our observation revealed that genetic variants of FTO (specifically SNP rs7196791 on chr16) was significantly associated with systolic blood pressure ($p = 0.009$) in hypertensive subjects after 3 weeks of antihypertensive medication withdrawal, while in the same subjects under standard antihypertensive medication, there was no significant association ($p = 0.7$). Thus, drug–gene–outcome relationships

[37] that is strongly influenced by "environmental modulation" through medication history leading to phenotypic plasticity can obscure the genetic impact of such an important trait as hypertension [24]. Another fine example of gene-environment (G × E) interactions involving FTO gene is physical activity. Kilpeläinen et al. [38] demonstrated that the association between FTO gene (rs99339606) and BMI and waist circumference detected in inactive individuals is greatly attenuated by physical activity, particularly in North-American individuals.

Finally, we would like to mention the fact that disease itself can obscure or unveil the genetic contribution to a trait. GWAS from the CKDGen Consortium, that includes over 70 international groups and over one million subjects with chronic kidney diseases [39] showed that among the hundreds of genes associated to kidney function, a few were unique to diabetic nephropathy. One of them is RAB38 gene (rs649529) which was shown to be associated with albuminuria, but only in individuals with diabetes. Experimental evidence led by H. Jacob from Medical College of Wisconsin showed that in RAB38 knock-out rats, no abnormality could be observed until diabetes was induced by streptozotocin injections and massive albuminuria appeared exclusively in rats deprived of RAB38. It could be concluded from these studies that "diabetogenic genes" were not required but a "diabetic environment" resulted in the manifestation of genetic effects on albuminuria [40].

Conclusion

We are at dawn of application of first polygenic risk scores in clinical practice as a significant component of personalized medicine. In order to benefit fully from clinical utility of genomic information we must complete several steps: 1. clearly define sex-dependent differences and either determine the score in each sex separately (especially when in opposite direction) and adjust for sex only when appropriate. 2. Distinguish between impact of age, usually an accumulation of environmental exposures, and age at the onset of disease, reflecting rather age dependent susceptibility impacting on penetrance. 3. Realize that most frequent "environmental exposure" in disease is its treatment and that disease itself may represent an "exposure" to high blood pressure, hyperglycemia or hyperlipidemia. 4. Dynamics of phenotypes may modify profoundly the genomic determinant: genes controlling renin activity in the upright or recumbent positions are not the same genes, as we have fully analyzed previously [41]. 5. The diseases risk must by adjusted for ethnicity using principal component analysis as illustrated in Fig. 10.2 [18]. Other environmental factors must be analyzed for their additive effects or for their eventual interactions with genomic factors, such as physical activity, smoking, nutrition. We have discussed some of these aspects in this chapters, but they are all to be pondered in our path to clinical application of personalized medicine.

References

1. Venter JC, Adams MD, Myers EW et al (2001) The sequence of the human genome. Science 291:1304–1351
2. Moraes F, Góes A (2016) A decade of human genome project conclusion: Scientific diffusion about our genome knowledge. Biochem Mol Biol Educ 44:215–223
3. International HapMap Consortium (2005) A haplotype map of the human genome. Nature 437:1299–1320
4. Manolio TA, Collins FS (2009) The HapMap and genome-wide association studies in diagnosis and therapy. Annu Rev Med 60:443–456
5. Hizel C, Tremblay J, Bartlet G, Hamet P (2017) In: Verma M, Barh D (eds) Progress and Challenges in Precision Medicine, 1st edn. Elsevier Press, San Diego, pp 1–34
6. Visscher PM, Wray NR, Zhang Q et al (2017) 10 years of GWAS discovery: biology, function, and translation. Am J Hum Genet 101:5–22
7. Daly AK (2010) Genome-wide association studies in pharmacogenomics. Nat Rev Genet 11:241–246
8. Wellcome Trust Case Control Consortium (2007) Genome-wide association study of 14,000 cases of seven common diseases and 3,000 shared controls. Nature 447:661–678
9. Scott LJ, Mohlke KL, Bonnycastle LL et al (2007) A genome-wide association study of type 2 diabetes in Finns detects multiple susceptibility variants. Science 316:1341–1345
10. Diabetes Genetics Initiative of Broad Institute of Harvard and MIT, Lund University and Novartis Institutes of BioMedical Research, Saxena R, Voight BF, Lyssenko V, et al. Genome-wide association analysis identifies loci for type 2 diabetes and triglyceride levels. Science 316: 1331–1336, 2007
11. International Consortium for Blood Pressure Genome-Wide Association Studies, Ehret GB, Munroe PB, Rice KM, et al. Genetic variants in novel pathways influence blood pressure and cardiovascular disease risk. Nature 478: 103–109, 2011
12. Gibson G (2019) On the utilization of polygenic risk scores for therapeutic targeting. PLoS Genet 15:e1008060
13. Lambert SA, Abraham G, Inouye M (2019) Towards clinical utility of polygenic risk scores. Hum Mol Genet 28(R2):R133–R142
14. Gourd E (2018) New genetic discoveries in prostate cancer. Lancet Oncol 19:e383
15. Willoughby A, Andreassen PR, Toland AE (2019) Genetic testing to guide risk-stratified screens for breast cancer. J Pers Med 9:15
16. Mega JL, Stitziel NO, Smith JG et al (2015) Genetic risk, coronary heart disease events, and the clinical benefit of statin therapy: an analysis of primary and secondary prevention trials. Lancet 385:2264–2271
17. Damask A, Steg PG, Schwartz GG, et al (2019) Patients with high genome-wide polygenic risk scores for coronary artery disease may receive greater clinical benefit from Alirocumab treatment in the Odyssey outcomes trial. Circulation https://doi.org/10.1161/circulationaha.119.044434. [Epub ahead of print], 2019
18. Tremblay J, Haloui M, Harvey F, et al (2020) Polygenic risk scores predict diabetic complications and their response to therapy. 2020 medRxiv preprint. https://doi.org/10.1101/19010785
19. Hamburg MA, Collins FS (2010) The path to personalized medicine. N Engl J Med 363:301–304
20. Goetz LH, Schork NJ (2018) Personalized medicine: motivation, challenges, and progress. Fertil Steril 109:952–963
21. Tremblay J, Hamet P (2013) Role of genomics on the path to personalized medicine. Metabolism 62(Suppl 1):S2–5
22. Pausova Z, Tremblay J, Hamet P (1999) Gene-environment interactions in hypertension. Curr Hypertens Rep 1:42–50
23. Hamet P (2012) Future needs in exploration of gene-environment interactions. J Hypertens 30:1915–1916

24. Simon PH, Sylvestre MP, Tremblay J, Hamet P (2016) Key considerations and methods in the study of gene-environment interactions. Am J Hypertens 29:891–899
25. Kokubo Y, Padmanabhan S, Iwashima Y et al (2019) Gene and environmental interactions according to the components of lifestyle modifications in hypertension guidelines. Environ Health Prev Med 24:19
26. Tremblay J, Hamet P (2019) Environmental and genetic contributions to diabetes. Metabolism 100S:153952
27. Mongeau JG, Biron P, Sing CF (1986) The influence of genetics and household environment upon the variability of normal blood pressure: the Montreal adoption survey. Clin Exp Hypertens A 8:653–660
28. Hamet P, Merlo E, Seda O et al (2005) Quantitative founder-effect analysis of French Canadian families identifies specific loci contributing to metabolic phenotypes of hypertension. Am J Hum Genet 76:815–832
29. Zoungas S, Woodward M, Li Q et al (2014) Impact of age, age at diagnosis and duration of diabetes on the risk of macrovascular and microvascular complications and death in type 2 diabetes. Diabetologia 57:2465–2474
30. Sattar N, Rawshani A, Franzén S et al (2019) Age at diagnosis of type 2 diabetes mellitus and associations with cardiovascular and mortality risks. Circulation 139:2228–2237
31. Seda O, Tremblay J, Gaudet D et al (2008) Systematic, genome-wide, sex-specific linkage of cardiovascular traits in French Canadians. Hypertension 51:1156–1162
32. Romanuik TL, Wang G, Holt RA et al (2009) Identification of novel androgen-responsive genes by sequencing of LongSAGE libraries. BMC Genom 10:476
33. Winkler TW, Justice AE, Graff M et al (2015) The Influence of age and sex on genetic associations with adult body size and shape: A large-scale genome-wide interaction study. PLoS Genet 11:e1005378
34. Hamet P, Haloui M, Harvey F et al (2017) PROX1 gene CC genotype as a major determinant of early onset of type 2 diabetes in Slavic study participants from Action in Diabetes and Vascular Disease: Preterax and Diamicron MR Controlled Evaluation study. J Hypertens 35(Suppl 1):S24–S32
35. Schulz LO, Bennett PH, Ravussin E, et al. Effects of traditional and western environments on prevalence of type 2 diabetes in Pima Indians in Mexico and the U.S. Diabetes Care 29: 1866–1871, 2006
36. Pausova Z, Syme C, Abrahamowicz M et al (2009) A common variant of the FTO gene is associated with not only increased adiposity but also elevated blood pressure in French Canadians. Circ Cardiovasc Genet 2:260–269
37. Baye TM, Abebe T, Wilke RA (2011) Genotype-environment interactions and their translational implications. Per Med 8:59–70
38. Kilpeläinen TO, Qi L, Brage S et al (2011) Physical activity attenuates the influence of FTO variants on obesity risk: a meta-analysis of 218,166 adults and 19,268 children. PLoS Med 8:e1001116
39. Wuttke M, Li Y, Li M et al (2019) A catalog of genetic loci associated with kidney function from analyses of a million individuals. Nat Genet 51:957–972
40. Teumer A, Tin A, Sorice R et al (2016) Genome-wide association studies identify genetic loci associated with albuminuria in diabetes. Diabetes 65:803–817
41. Arenas IA, Tremblay J, Deslauriers B et al (2013) Dynamic genetic linkage of intermediate blood pressure phenotypes during postural adaptations in a founder population. Physiol Genomics 45:138–150

Chapter 11
Sex Differences in Response to Fatty Acids in Cardiovascular Health and Disease

Paramjit S. Tappia and Anureet K. Shah

Abstract It has now been established that differences in the risk of cardiovascular diseases, including atherosclerosis, exist between men and women. The field of nutritional genomics focuses on the interaction between diet and gene and consequent risk of disease. It is conceivable that advances in this area of investigation could result in specific dietary recommendations and guidelines to individual's nutritional pattern as well as to populations at risk of cardiovascular abnormalities. Accordingly, this chapter will briefly describe and provide an update on sex dependent regulation of gene expression by different fatty acids in the diet in relation to heart disease risk. Such information could lead to novel sex-specific nutritional strategies for the management and prevention of cardiovascular disease.

Keywords Sex differences · Fatty acids · Gene expression · Metabolic responses · Cardiovascular disease

Introduction

A wealth of evidence has established that there are differences in cardiovascular disease (CVD) risk between men and women [1]. Most notable is that CVD develops much later in women as compared to men, and that heart disease is the leading cause of death in women over the age of 65 years, and in fact, heart disease risk in women may even be underestimated [2]. The European Heart Survey has found that women are more likely to be misdiagnosed for cardiovascular abnormalities [3]. The National Health and Nutrition Examination Surveys revealed that while an increase in myocardial infarction (MI) in women between the ages of 35 to 54 years occurred, a decrease was reported in male counterparts [4]. Hypertension in women

P. S. Tappia (✉)
Asper Clinical Research Institute, St. Boniface Hospital, Winnipeg, Canada
e-mail: ptappia@sbrc.ca

A. K. Shah
Department of Kinesiology, Nutrition and Food Sciences, California State University, Los Angeles, USA

B. Ostadal and N. S. Dhalla (eds.), *Sex Differences in Heart Disease*,
Advances in Biochemistry in Health and Disease 21,
https://doi.org/10.1007/978-3-030-58677-5_11

over the age of 65 years is higher than in men; in addition, the incidence of obesity and diabetes, which are major risk factors for CVD, are also increasing in women [5, 6].

The sex of an individual is an important factor in the determination of nutritional well-being, metabolic status, and risk to diet-induced diseases. Indeed, the field of nutrigenomics, or molecular nutrition in relation to human health and disease has grown exponentially over the last few years. Since nutrients can modulate the expression of a number of different genes [8], this chapter will describe how certain fatty acids in the human diet can affect gene expression and influence CVD risk in men and women (Fig. 11.1). Although some advances have been made, the relationship between the sex-dependent response to fatty acids and risk of CVD still remains to be completely understood.

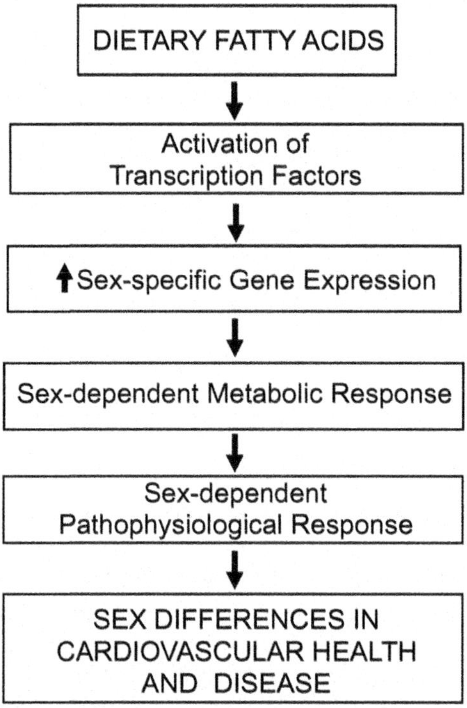

Fig. 11.1 Putative role of dietary fatty acids in sex differences in the regulation of gene transcription and cardiovascular health and disease

Influence of Sex in the Response to Fatty Acids and Risk of Cardiovascular Disease

The n-3 polyunsaturated fatty acids (PUFAs) such as eicosapentaenoic acid (EPA), docosahexaenoic acid (DHA), linolenic acid (LNA) and arachidonic acid (AA) and oleic acid (OA), a monounsaturated fatty acid, exhibit cardiovascular health properties [7]. A high intake of marine n-3 PUFA has been associated with lower risk of coronary heart disease (CHD) [8–14]. In a study by Joensen et al. [15], it was observed that men with > 0.39 g of n-3 PUFA in the diet/day exhibited lower occurrence of acute coronary syndrome.

According to the Japanese Health, Labour and Welfare Ministry, the Tochigi region of Japan has a high mortality rate due to CAD and low consumption of seafood. In fact, lower intake of EPA has been reported in women, particularly menopausal women, in this region [16]. In a Danish study with 8-11 yr old children, although high EPA levels were associated with low TG concentrations, they were, however; associated with higher cholesterol levels [17]. α-linolenic acid (ALA) is a plant-derived n-3 fatty acid that exerts anti-atherosclerotic [18, 19], antithrombotic [20], and anti-inflammatory [21, 22] actions. Although some observational studies have indicated beneficial effects of dietary ALA in CHD [23–27], others have reported inconsistencies [28–33], while no correlation between dietary ALA and risk of MI has been reported [34].

Although experimental studies have demonstrated that conjugated linoleic acid (CLA) can exert several beneficial cardiovascular health actions [35–40], human studies are inconsistent with respect to any beneficial effect of CLA in atherosclerosis [38, 39, 41–45]. On the other hand, high CLA content of adipose tissue has been correlated to low risk of MI [46], while Wannamethee et al. [47] reported that high circulating CLA levels were negatively linked to HF risk in older men.

Our earlier experimental studies have demonstrated that specific CLA isomers elicit different functional effects on heart that are related to sex [48]. We have further reported that supplementation of the diet with different CLA isomers can influence cardiac gene expression as well as sarcoplasmic reticulum (SR) function [49]. It was accordingly suggested that the sex-dependent cardiac contractile response to different CLA isomers may be related to not only modulation of gene expression of the heart, but also to altered SR function. An increase in mRNA expression and activity of sodium-calcium exchange (NCX) [50] and reduced SERCA2a expression [51, 52] has been observed in HF. In this regard, we have also reported an increase in SERCA2a gene expression in the heart of male rats, and a reduced NCX expression in the female heart due to t10,c12 CLA isomer. These data were taken to suggest that specific dietary isomers of CLA may alter Ca^{2+}-cycling proteins, and that CLA isomers exert beneficial cardiac effects according to sex. It should be mentioned that the dietary intake of trans fatty acids (TFAs) is correlated to adipose tissue content [53]. Although an adverse link between TFAs and CHD has been reported [54, 55], the relationship is still not conclusive. In a study by Jakobsen et al. [56], a higher

MI rate was observed in women with higher adipose tissue content of specific 18:1 TFAs as compared to men.

While dietary fish intake in women is linked to protective effects against CHD, a similar correlation was not reported in men [57]. Some lines of evidence have suggested that females respond differently to dietary fat with respect to CHD progression [58]. On the other hand, a greater lipemic response to a standardized fat meal has been reported to occur in men as compared to women [59]. Furthermore, in this study, men also presented with increased blood pressure as well as carotid intima-media thickness; interestingly, a slight correlation between carotid intima-media thickness and levels of ox-LDL was observed, but only in men [59].

It has been suggested that the cardioprotective effects of estrogen are mediated through cyclooxygenase activation and synthesis of prostacyclin (PGI2) [60–62]. Booth et al. [63] have demonstrated that myocardial protection against I-R injury provided by estrogen is due to an increase in the production of PGI2 via COX2 activity. Similarly, estrogen receptor β has been shown to mediate sex differences in the response to I-R [64]. In this study, baseline experiments revealed that less functional recovery was observed in the hearts from wild-type (WT) male mice as compared to the hearts from WT female counterparts subsequent to I-R. It was then further demonstrated that while no differences in the extent of functional recovery subsequent to I-R were observed between α estrogen receptor knock out female hearts and WT females, a lower functional recovery in β estrogen receptor knock out mice vs. WT females, which was also comparable to the recovery of heart function in WT males was seen. These observations provided the evidence that the cardioprotective effects of estrogen are mediated through the β estrogen receptor [64]. Since reperfusion of the ischemic heart results in an increased production of fatty acids [65], it is possible that these fatty acids could play an interactive role in the cardioprotective effects of estrogen.

Regulation of Gene Transcription, Apolipoproteins and Atherogenicity by Fatty Acids

Accumulating evidence has indicated that dietary practices can impact on the genetic determinants of CVD risk. The risk for ACS is increased in carriers of the chromosome 9p21 variant rs4977574 and low dietary n-3 fatty acids [66]. Consumption of a high fat diet can increase lipopolysaccharide (LPS) levels and induce subclinical inflammation [67]. In this study, consumption of a high-fat high-saturated meal by healthy women, resulted in endotoxemia and increased TNF-α and VCAM-1 expression levels. In addition, these diets were found to modulate the circulating miRNAs levels involved in the regulation of inflammatory and lipid metabolic proteins postprandially [67].

Atherothrombotic disease has been linked to the activation of the 5-lipoxygenase pathway; in addition MI risk has been associated with a repeat polymorphism in the arachidonate lipoxygenase-5 (*ALOX-5*) gene [68, 69]. Interestingly, consumption of AA and EPA in the diet have been reported to exert differential effects on 5-lipoxygenase expression [70, 71], suggesting that high AA in the diet increases the risk of atherosclerosis and MI, whereas EPA may lower the risk. On the other hand, Gammaelmark et al. [72] reported that the interaction of EPA and AA and *ALOX-5* polymorphism is not important enough to influence the risk of MI in men and women.

It is now well documented that PPARs are important transcription factors that regulate lipid metabolism and thus are considered as important therapeutic targets for atherosclerosis. It has been shown that the PPAR reporter homogene expression level in female mice is much lower than that in male mice [73]. Such a difference in the level of PPAR reporter homogene expression could help to explain the disparity in PPAR transcriptional activity in male and female mice as well as the ineffectiveness of hormone supplementation or dietary interventions in females to increase liver PPAR. These researchers also suggested a sex-specific risk for atherosclerosis. In a study by Morise et al. [74] examining the protective effects of n-3 PUFA and estrogens against atherosclerosis in male and female hamsters fed fat diets enriched in ALA or saturated fatty acids (SFAs), it was observed that the SFAs caused marked elevations in plasma triglycerides and VLDL concentrations in male hamsters as compared to females. No sex differences were reported in PPAR or sterol regulatory element binding protein in response to these diets [75].

Dietary flaxseed (high ALA content) has been shown to reduce hepatic and circulating levels of cholesterol in hypercholesterolemic male and female mice by attenuation of cholesterol absorption and/or bile acid resorption, but no sex variances in this response were observed [76]. While experimental models of atherosclerosis have demonstrated down regulation of leptin mRNA expression and protein levels [77], dietary supplementation with flaxseed results in an increase in leptin expression that is inversely associated with atherosclerosis. The activation of the nuclear factor-κB (NFκB) system of transcription factors in response to inflammatory and atherogenic stimuli can be attenuated by n-3 fatty acids [78, 79]. Some common oils have been reported to exert an anti-atherogenic effect in the development of atherosclerosis in apo-E knockout mice [80–82]. However, sex differences in the response to n-3 fatty acids and the development of atherosclerosis warrants further investigation.

Linoleic acid (LA) can activate specific endothelial transcription factors that initiate the formation of the atherosclerotic lesion, it is considered as an atherogenic fatty acid [83]. However, size reduction of the atherosclerotic lesions has been observed in male mice fed a sunflower oil diet, rich in LA, whereas lesion reduction in female mice has been observed with olive oil, rich in OA, as well as with palm oil, rich in both saturated fatty acid, palmitic acid, and OA [84]. These responses to different fatty acids were attributed to reduced triglyceride levels in male mice, while the atherosclerotic lesion reduction in female mice was linked to an increase in circulating ApoA-1 levels [84]. Although the authors suggested that different fatty acids can influence the development of atherosclerotic lesion, the impact of fatty acids on apolipoproteins needs to be fully explored, as depressed Apo-E gene has also been associated with increased atherosclerosis risk [85]. Furthermore, an elevated risk for the development of atherosclerosis was reported in female heterozygous Apo-E deficient mice on atherogenic diets as compared to male counterparts with a 2-fold increase in circulating levels of cholesterol in the female mice as compared to male mice [85]. It should be noted that sex-dependent expression of different subtypes of Apo-E could also be major confounding factors in determining CVD, a possibility that warrants further investigation.

It should be mentioned that the expression of fatty acid β-oxidation enzymes is decreased at the gene transcriptional level in animal models of LV hypertrophy and HF [86]. In humans with idiopathic dilated cardiomyopathy, myocardial fatty acid oxidation is decreased. In hypertensive heart disease, independent predictors of increased LV mass were male sex, decreased myocardial fatty acid oxidation, decreased ejection fraction and increase in systolic blood pressure. Although sex differences have not been determined, a decrease in fatty acid-binding protein 3 gene expression has been reported in the atrium of patients with post-operative atrial fibrillation [87], suggesting the potential association between altered fatty acid transport and increased atrial fibrillation onset after cardiac surgery.

Sex-Linked Protective Mechanisms of Fatty Acids Against Cardiovascular Disease

There are several sites where different fatty acids could interact to modify CVD disease progression and sex can be seen to influence the response (Fig. 11.2). The anti-atherogenic effects of n-3 PUFAs may be mediated through regulation of nitric oxide (NO) production and release [88, 89]. Nitric oxide can attenuate the atherosclerotic processes including the inhibition of platelet aggregation and vascular smooth muscle

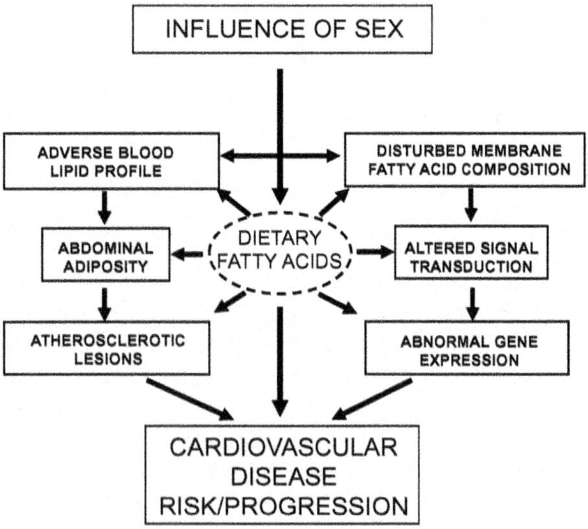

Fig. 11.2 Potential mechanistic sites where sex may be an important determinant of fatty acid action and CVD risk/progression

cell proliferation [96, 97]. In fact, n-3 PUFAs have been suggested to increase NO levels [90, 91]. Since estrogen was also reported to increase NO production, it is evident that the anti-atherogenic effects of both n-3 PUTAs and estrogen may be due to increases in NO levels [92].

An increase in the saturated fatty acid and reduced PUFA content (with the exception of LA) in the phospholipid fractions of human coronary arteries has been reported in sudden cardiac death, which has been suggested to be due to a decrease in delta6-desaturase activity in the coronary artery wall resulting in impairment of LA metabolism [93]. An alteration in the fatty acid composition of serum cholesterol esters, with a low arachidonic to dihomogammalinolenic ratio, attributed to depressed delta5 desaturase activity, was observed in middle-aged men who later develop a MI. Thus, it is conceivable the quality and type of fat in the diet, or altered fatty acid metabolism, may occur in the initial stages of CHD [94].

PGI_2 is an inhibitor of platelet aggregation and a vasodilator. Experimental studies have revealed that PGI_2 levels are inversely related to plaque size, and treatment of arteries of ovariectomized female monkeys with estrogen results in an increased production of PGI_2 [95]. Since the precursor for PGI_2 is AA, it was suggested that the levels of this fatty acid may play an important role in the risk of atherosclerosis in male and females.

While different fatty acids are known to exert a cardioprotective role, there is a lack of information in the literature regarding sex differences in cardiac gene expression in response to different fatty acids in the diet [96]. Microarray analysis of adult rat cardiomyocytes have demonstrated that fatty acids can differentially regulate the expression of genes [97]. Essentially, diets that are rich in OA are cardioprotective, whereas diets rich in saturated fatty acids such as palmitic acid are ineffective. While such a study to identify the influence of sex has not been conducted, it is conceivable that the response to fatty acids particularly in pre and postmenopausal women may exist. An increase in the bioavailability of epoxyeicosatrienoic acids (EETs), metabolites of AA with cardioprotective properties that are catabolised by soluble epoxide hydrolase (sEH) [98, 99], has been suggested to attenuate coronary myogenic constriction in sEH knock out female mice [100].

Conclusions

This chapter has provided some discussion on sex differences in the response to fatty acids that are an important determinant of the risk of CVD. From the evidence provided it is evident that dietary fatty acids can regulate gene expression and influence metabolic and cell function resulting in the development of atherosclerosis and cardiac dysfunction. The beneficial and protective role of estrogen against cardiovascular disorders has also been highlighted, which clearly demonstrates the important role of sex in adverse cardiovascular health outcomes. Although some advances on the sex differences in the responsiveness to dietary fatty acids have been made over recent years, this area of investigation still remains largely unexplored. For example, sex differences in the cellular and molecular mechanisms involved in the regulation of cell function still remain to be completely understood. The sex differences and the impact of dietary fatty acids on downstream signal transduction processes in relation to cardiovascular health remain to be examined. The interaction of specific fatty acids on epigenetic changes that may modulate gene expression and either increase or reduce the risk of CVD need to be determined. Another important factor to consider would be the influence of ethnicity on the responsiveness to fatty acids and CVD risk. These aspects would be of great value to advancements in personalized medicine, but are yet to be fully addressed. Accordingly, it is envisioned that new approaches for prevention and management of CVD constituting of dietary interventions will ultimately encompass sex into consideration.

Acknowledgements Infrastructural support was provided by the St. Boniface Hospital Research Foundation.

References

1. Ordovas JM (2007) Gender, a significant factor in the cross talk between genes, environment, and health. Gend Med 4(Suppl B):S111–S122
2. Maas AH, Appelman YE (2010) Gender differences in coronary heart disease. Neth Heart J 18:598–602
3. Daly CA, Clemens F, Sendon JL et al (2005) The clinical characteristics and investigations planned in patients with stable angina presenting to cardiologists in Europe: from the Euro Heart Survey of Stable Angina. Eur Heart J 26:996–1010
4. Towfighi A, Zheng L, Ovbiagele B (2009) Sex-specific trends in midlife coronary heart disease risk and prevalence. Arch Intern Med 169:1762–1766
5. Roger VL, Go AS, Lloyd-Jones DM et al (2011) Heart disease and stroke statistics-2011 update: a report from the American Heart Association. Circulation 123:e18–e209
6. Mosca L, Benjamin EJ, Berra K et al (2011) Effectiveness based guidelines for the prevention of cardiovascular disease in women-2011 update: A Guideline from the American Heart Association. Circulation 123:1243–1262
7. Ander BP, Dupasquier CMC, Prociuk MA, Pierce GN (2003) Polyunsaturated fatty acids and their effects on cardiovascular disease. Exp Clin Cardiol 8:164–172
8. Albert CM, Hennekens CH, O'Donnell CJ et al (1998) Fish consumption and risk of sudden cardiac death. JAMA 279:23–28
9. Marchioli R, Barzi F, Bomba E et al (2002) Early protection against sudden death by *n*-3 polyunsaturated fatty acids after myocardial infarction: time-course analysis of the results of the Gruppo Italiano per lo Studio della Sopravvivenza nell'Infarto Miocardico (GISSI)-Prevenzione. Circulation 105:1897–1903
10. Siscovick DS, Raghunathan TE, King I et al (1995) Dietary intake and cell membrane levels of long-chain *n*-3 polyunsaturated fatty acids and the risk of primary cardiac arrest. JAMA 274:1363–1367
11. Christensen JH, Schmidt EB (2001) *n*-3 Fatty acids and the risk of sudden cardiac death. Lipids 36(Suppl):S115–S118
12. Daviglus ML, Stamler J, Orencia AJ et al (1997) Fish consumption and the 30-year risk of fatal myocardial infarction. N Engl J Med 336:1046–1053
13. Yuan JM, Ross RK, Gao YT et al (2001) Fish and shellfish consumption in relation to death from myocardial infarction among men in Shanghai. China. Am J Epidemiol 154:809–816

14. Kromhout D, Bosschieter EB, de Lezenne CC (1985) The inverse relation between fish consumption and 20-year mortality from coronary heart disease. N Engl J Med 312:1205–1209
15. Joensen AM, Schmidt EB, Dethlefsen C et al (2010) Dietary intake of total marine n-3 polyunsaturated fatty acids, eicosapentaenoic acid, docosahexaenoic acid and docosapentaenoic acid and the risk of acute coronary syndrome - A cohort study. Br J Nutr 103:602–607
16. Kitagawa Y, Abe S, Toyoda S et al (2014) Gender differences in the ratio of eicosapentaenoic acid to arachidonic acid in an inland prefecture, Tochigi: Tochigi Ryomo EPA/AA Trial in coronary artery disease (TREAT-CAD). Intern Med 53:177–182
17. Damsgaard CT, Eidner MB, Stark KD et al (2014) Eicosapentaenoic Acid and Docosahexaenoic Acid in Whole Blood Are Differentially and Sex-Specifically Associated With Cardiometabolic Risk Markers in 8-11-year-old Danish Children. PLoS ONE 9:e109368
18. Sala-Vila A, Cofán M, Pérez-Heras A et al (2010) Fatty acids in serum phospholipids and carotid intima-media thickness in Spanish subjects with primary dyslipidemia. Am J Clin Nutr 92:186–193
19. Sala-Vila A, Cofán M, Nu´ñez I, et al. Carotid and femoral plaque burden is inversely associated with the a-linolenic acid proportion of serum phospholipids in Spanish subjects with primary dyslipidemia. Atherosclerosis 214: 209–214, 2011
20. Owren PA, Hellem A, Odegaard A (1964) Linolenic acid and for the prevention of thrombosis and myocardial infarction. Lancet 2:975–978
21. Rallidis LS, Paschos G, Liakos GK et al (2003) Dietary α-linolenic acid decreases C-reactive protein, serum amyloid A and interleukin-6 in dyslipidaemic patients. Atherosclerosis 67:237–242
22. Caughey GE, Montzioris E, Gibson RA et al (1996) The effect on human tumor necrosis factor a and interleukin 1-beta production of diets enriched in n-3 acids from vegetable oil or fish oil. Am J Clin Nutr 63:116–122
23. Ascherio A, Rimm EB, Giovannucci EL et al (1996) Dietary fat and risk of coronary heart disease in men: cohort follow up study in the United States. BMJ 313:84–90
24. Mozaffarian D, Ascherio A, Hu FB et al (2005) Interplay between different polyunsaturated fatty acids and risk of coronary heart disease in men. Circulation 111:157–164
25. Hu FB, Stampfer MJ, Manson JE et al (1999) Dietary intake of a-linolenic acid and risk of fatal ischemic heart disease among women. Am J Clin Nutr 69:890–897
26. Dolecek TA (1992) Epidemiological evidence of relationships between dietary polyunsaturated fatty acids and mortality in the multiple risk factor intervention trial. Proc Soc Exp Biol Med 200:177–182
27. Koh AS, Pan A, Wang R et al (2015) The association between dietary omega-3 fatty acids and cardiovascular death: the Singapore Chinese Health Study. Eur J Prev Cardiol 22:364–372
28. Pietinen P, Ascherio A, Korhonen P, et al. Intake of fatty acids and risk of coronary heart disease in a cohort of Finnish men. The α-Tocopherol, β-Carotene Cancer Prevention Study. Am J Epidemiol 145: 876–887, 1997
29. Oomen CM, Ocké MC, Feskens EJ et al (2001) α-Linolenic acid intake is not beneficially associated with 10-y risk of coronary artery disease incidence: the Zutphen Elderly Study. Am J Clin Nutr 74:457–463
30. Albert CM, Oh K, Whang W et al (2005) Dietary α-linolenic acid intake and risk of sudden cardiac death and coronary heart disease. Circulation 112:3232–3238
31. de Goede J, Verschuren WMM, Boer JM et al (2011) α-linolenic acid intake and 10-year incidence of coronary heart disease and stroke in 20,000 middle-aged men and women in the Netherlands. PLoS ONE 6:e17967
32. Vedtofte MS, Jakobsen MU, Lauritzen L, Heitmann BL (2011) Dietary a-linolenic acid, linoleic acid, and n–3 long-chain PUFA and risk of ischemic heart disease. Am J Clin Nutr 94:1097–1103
33. Fretts AM, Mozaffarian D, Siscovick DS et al (2014) Plasma phospholipid and dietary a-linolenic acid, mortality, CHD and stroke: the Cardiovascular Health Study. Br J Nutr 112:1206–1213

34. Bork CS, Jakobsen MU, Lundbye-Christensen S et al (2016) Dietary intake and adipose tissue content of α-linolenic acid and risk of myocardial infarction: A Danish cohort study. Am J Clin Nutr 104:41–48
35. Dilzer A, Park Y (2012) Implication of conjugated linoleic acid (CLA) in human health. Crit Rev Food Sci Nutr 52:488–513
36. Shokryzadan P, Rajion MA, Meng GY et al (2017) Conjugated linoleic acid: a potent fatty acid linked to animal and human health. Crit Rev Food Sci Nutr 13:2737–2748
37. Bhattacharya A, Banu J, Rahman M et al (2006) Biological effects of conjugated linoleic acids in health and disease. J Nutr Biochem 17:789–810
38. Mensink RP (2005) Metabolic and health effects of isomeric fatty acids. Curr Opin Lipidol 16:27–30
39. Tricon S, Yaqoob P (2006) Conjugated linoleic acid and human health: a critical evaluation of the evidence. Curr Opin Clin Nutr Metab Care 9:105–110
40. Alibin CP, Kopilas MA, Anderson DI (2008) Suppression of cardiac myocyte hypertrophy by conjugated linoleic acid. J Biol Chem 283:10707–10715
41. Wanders AJ, Brouwer IA, Siebelink E, Katan MB (2010) Effect of a high intake of conjugated linoleic acid on lipoprotein levels in healthy human subjects. PLoS ONE 5:e9000
42. Yang J, Wang HP, Zhou LM et al (2015) Effect of conjugated linoleic acid on blood pressure: a meta-analysis of randomized, double-blind placebo-controlled trials. Lipids Health Dis 14:11
43. Smit LA, Katan MB, Wanders AJ et al (2011) A high intake of trans fatty acids has little effect on markers of inflammation and oxidative stress in humans. J Nutr 141:1673–1678
44. Smedman A, Basu S, Jovinge S et al (2005) Conjugated linoleic acid increased C-reactive protein in human subjects. Br J Nutr 94:791–795
45. Sofi F, Buccioni A, Cesari F et al (2010) Effects of a dairy product (pecorino cheese) naturally rich in cis-9, trans-11 conjugated linoleic acid on lipid, inflammatory and haemorheological variables: a dietary intervention study. Nutr Metab Cardiovasc Dis 20:117–124
46. Smit LA, Baylin A, Campos H (2010) Conjugated linoleic acid in adipose tissue and risk of myocardial infarction. Am J Clin Nutr 92:34–40
47. Wannamethee SG, Jeffries BJ, Lennon L, et al. Serum conjugated linoleic acid and risk of incident heart failure in older men: The British regional heart study. J Am Heart Assoc 7: pii: e006653, 2018
48. Tappia PS, Mangat R, Gabriel C et al (2006) Gender differences in the cardiac response to dietary conjugated linoleic acid isomers. Can J Physiol Pharmacol 84:257–264
49. Tappia PS, Dent MR, Aroutiounova N et al (2007) Gender differences in the modulation of cardiac gene expression by dietary conjugated linoleic acid isomers. Can J Physiol Pharmacol 85:465–475
50. Wei SK, McCurley JM, Hanlon SU, Haigney MC (2007) Gender differences in Na/Ca exchanger current and beta-adrenergic responsiveness in heart failure in pig myocytes. Ann N Y Acad Sci 1099:183–189
51. Hovnanian A (2007) SERCA pumps and human diseases. Subcell Biochem 45:337–363
52. Butter C1, Rastogi S, Minden HH, et al. Cardiac contractility modulation electrical signals improve myocardial gene expression in patients with heart failure. J Am Coll Cardiol 51:1784–1799, 2008
53. Hodson L, Skeaff CM, Fielding BA (2008) Fatty acid composition of adipose tissue and blood in humans and its use as a biomarker of dietary intake. Prog Lipid Res 47:348–380
54. Brouwer IA, Wanders AJ, Katan MB (2013) Trans fatty acids and cardiovascular health: research completed? Eur J Clin Nutr 67:541–547
55. Souza RJ, Mente A, Maroleanu A et al (2015) Intake of saturated and trans unsaturated fatty acids and risk of all cause mortality, cardiovascular disease, and type 2 diabetes: systematic review and meta-analysis of observational studies. BMJ 351:h3978
56. Jakobsen MU, Gorst-Rasmussen A, Eriksen HH et al (2018) Trans fatty acids in adipose tissue and risk of myocardial infarction: A case-cohort study. PLoS ONE 13:e0202363
57. Jarvinen R, Knekt P, Rissanen H, Reunanen A (2006) Intake of fish and long-chain n-3 fatty acids and the risk of coronary heart mortality in men and women. Br J Nutr 95:824–829

58. Lapointe A, Balk EM, Lichtenstein AH (2006) Gender differences in plasma lipid response to dietary fat. Nutr Rev 64:234–249
59. Tentor J, Harada LM, Nakamura RT et al (2006) Sex- dependent variables in the modulation of postalimentary lipemia. Nutrition 22:9–15
60. Geary GG, Krause DN, Duckles SP (2000) Estrogen reduces mouse cerebral artery tone through endothelial NOS- and cyclooxygenase-dependent mechanisms. Am J Physiol Heart Circ Physiol 279:H511–H519
61. Huang A, Wu Y, Sun D et al (2001) Effect of estrogen on flow-induced dilation in NO deficiency: role of prostaglandins and EDHF. J Appl Physiol 91:2561–2566
62. Ospina JA, Duckles SP, Krause DN.17β-estradiol decreases vascular tone in cerebral arteries by shifting COX-dependent vasoconstriction to vasodilation. Am J Physiol Heart Circ Physiol 285: H241-H250, 2003
63. Booth EA, Flint RR, Lucas KL et al (2008) Estrogen protects the heart from ischemia-reperfusion injury via COX-2-derived PGI2. J Cardiovasc Pharmacol 52:228–235
64. Gabel SA, Walker VR, London RE et al (2005) Estrogen receptor β mediates gender differences in ischemia/reperfusion injury. J Mol Cell Cardiol 38:289–297
65. Lopaschuk GD, Collins-Nakai R, Olley PM et al (1994) Plasma fatty acid levels in infants and adults after myocardial ischemia. Am Heart J 128:61–67
66. Leung Yinko SS, Thanassoulis G, Stark KD et al (2014) Omega-3 fatty acids and the genetic risk of early onset acute coronary syndrome. Nutr Metab Cardiovasc Dis 24:1234–1239
67. Quintanilha BJ, Pinto Ferreira LR, Ferreira FM, et al. Circulating plasma microRNAs dysregulation and metabolic endotoxemia induced by a high-fat high-saturated diet. Clin Nutr pii: So261–5614(19)30092-2, 2019
68. Poeckel D, Funk CD (2010) The 5-lipoxygenase/leukotriene pathway in preclinical models of cardiovascular disease. Cardiovasc Res 86:243–253
69. Riccioni G, Back M (2012) Leukotrienes as modifiers of preclinical atherosclerosis? Sci World J 2012:490968
70. Dwyer JH, Allayee H, Dwyer KM et al (2004) Arachidonate 5-lipoxygenase promoter genotype, dietary arachidonic acid, and atherosclerosis. N Engl J Med 350:29–37
71. Allayee H, Baylin A, Hartiala J et al (2008) Nutrigenetic association of the 5-lipoxygenase gene with myocardial infarction. Am J Clin Nutr 88:934–940
72. Gammelmark A, Lundbye-Christensen S, Tjønneland A et al (2017) Interactions between 5-lipoxygenase polymorphisms and aipose tissue contents of arachidonic and eicosapentaenoic acids do not affect risk of myocardial infarction in middle-aged men and women in a Danish case-cohort study. J Nutr 147:1340–1347
73. Ciana P, Biserni A, Tatangelo L et al (2007) A novel peroxisome proliferator-activated receptor responsive element-luciferase reporter mouse reveals gender specificity of peroxisome proliferator-activated receptor activity in liver. Mol Endocrinol 21:338–400
74. Morise A, Serougne C, Gripois D et al (2004) Effects of dietary alpha linolenic acid on cholesterol metabolism in male and female hamsters of the LPN strain. J Nutr Biochem 15:51–61
75. Morise A, Mourot J, Boue C, et al Gender-related response of lipid metabolism to dietary fatty acids in the hamster. Br J Nutr 95: 709–720, 2006
76. Pellizzon MA, Billheimer JT, Bloedon LT et al (2007) Flaxseed reduces plasma cholesterol levels in hypercholesterolemic mouse models. J Am Coll Nutr 26:66–75
77. McCullough RS, Edel AL, Bassett CMC et al (2011) The α linolenic acid content of flaxseed is associated with an induction of adipose leptin expression. Lipids 46:1043–1052
78. De Caterina R, Liao JK, Libby P (2000) Fatty acid modulation of endothelial activation. Am J Clin Nutr 71:213S–223S
79. De Caterina R, Massaro M (2005) Omega-3 fatty acids and the regulation of expression of endothelial proatherogenic and pro-inflammatory genes. J Membr Biol 206:103–116
80. Acin S, Navarro MA, Perona JS et al (2007) Olive oil preparation determines the atherosclerotic protection in apolipoprotein E knockout mice. J Nutr Biochem 18:418–424

81. Cho YY, Kwon EY, Kim HJ et al (2011) Differential effect of corn oil-based low trans structural fat on the plasma and hepatic lipid profile in an atherogenic mouse model: comparison to hydrogenated trans fat. Lipids Health Dis 10:15
82. Wang HH, Hung TM, Wei J, Chiang AN (2004) Fish oil increases antioxidant enzyme activities in macrophages and reduces atherosclerotic lesions in apoE-knockout mice. Cardiovasc Res 61:169–176
83. Toborek M, Hennig B (1998) The role of linoleic acid in endothelial cell gene expression. Relationship to atherosclerosis. Subcell Biochem 30:415–436
84. Calleja L, Paris MA, Paul A et al (1999) Low-cholesterol and high-fat diets reduce atherosclerotic lesion development in ApoE-knockout mice. Arterioscler Throm Vasc Biol 19:2368–2375
85. van Ree JH, van den Broek WJ, Dahlmans VE (1994) Diet-induced hypercholesterolemia and atherosclerosis in heterozygous apolipoprotein E-deficient mice. Atherosclerosis 111:25–37
86. de las Fuentes L, Herrero P, Petereson LR, et al. Myocardial fatty acid metabolism: independent predictor of left ventricular mass in hypertensive heart disease. Hypertension 41: 83–87, 2004
87. Shingu Y, Yokota T, Takada S et al (2018) Decreased gene expression of fatty acid binding protein 3 in the atrium of patients with new onset of atrial fibrillation in cardiac perioperative phase. J Cardiol 71:65–70
88. Abeywarden MY, Head RJ (2001) Long chain n-3 polyunsaturated fatty acids and blood vessel function. Cardiovasc Res 52:361–371
89. Bhatnagar D, Durrington PN (2003) Omega-3 fatty acids: Their role in the prevention and treatment of atherosclerosis related risk factors and complications. Int J Clin Pract 57:305–314
90. Omura M, Kobayashi S, Mizukami Y et al (2001) Eicosapentaenoic acid (EPA) induces Ca^{2+}-independent activation and translocation of endothelial nitric oxide synthase and endothelium-dependent vasorelaxation. FEBS Lett 487:361–366
91. Hirafuji M, Machida T, Tsunoda M et al (2002) Docosahexaenoic acid potentiates interleikin-1β induction of nitric oxide synthase through mechanisms involving p44/42 MAPK activation in rat vascular smooth muscle cells. Br J Pharmacol 136:613–619
92. Das UN (2002) Estrogen, statins and polyunsaturated fatty acids: similarities in their actions and benefits-is there a common link. Nutrition 18:178–188
93. Luostarinen R, Boberg M, Saldeen T (1993) Fatty acid composition in total phospholipids of human coronary arteries in sudden cardiac death. Atherosclerosis 99:187–193
94. Ohrvall M, Berglund L, Salminen I et al (1996) The serum cholesterol ester fatty acid composition but not the serum concentration of α tocopherol predicts the development of myocardial infraction in 50-year-old men: 19 years follow up. Atherosclerosis 127:65–71
95. Egan KM, Lawson JA, Fries S et al (2004) COX-2-derived prostacyclin confers atheroprotection on female mice. Science 306:1954–1957
96. van Bilsen M, de Vries JE, Van der Vusse GJ (1997) Long-term effects of fatty acids on cell viability and gene expression of neonatal cardiac myocytes. Prostaglandins Leukot Essent Fatty Acids 57:39–45
97. Lockridge JB, Sailors ML, Durgan DJ et al (2008) Bioinformatic profiling of the transcriptional response of adult rat cardiomyocytes to distinct fatty acids. J Lipid Res 49:1395–1408
98. Imig JD, Hammock BD (2009) Soluble epoxide hydrolase as a therapeutic target for cardiovascular diseases. Nat Rev Drug Discov 8:794–805
99. Sinal CJ, Miyata M, Tohkin M et al (2000) Targeted disruption of soluble epoxide hydrolase reveals a role in blood pressure regulation. J Biol Chem 275:40504–40510
100. Froogh G, Qin J, Kandhi S et al (2016) Female-favorable attenuation of coronary myogenic constriction via reciprocal activations of epoxyeicosatrienoic acids and nitric oxide. Am J Physiol Heart Circ Physiol 310:H1448–H1454

Chapter 12
Nutritional Aspect of Sex-Dependent Difference in Heart Disease

Anureet K. Shah, Anabela Kim, and Carrie Adase

Abstract According to a 2019 report by the American Heart Association, cardiovascular disease is the leading cause of death worldwide, accounting for more than 17.6 million deaths per year in 2016; this number is expected to grow to more than 23.6 million by 2030. Body composition, which is influenced by nutritional status, is considered to have a great significance in determining the incidence of chronic diseases including heart disease. The difference in body composition is not only affected by age, race, and heredity, but also to a greater extent by sex. Biological males and females show variable body fat at different locations as males tend to accumulate more fat in the abdominal area and females in the thighs and buttocks. This sex difference in fat distribution is considered to determine the difference in the susceptibility of males and females to various pathological stimuli for the development of heart disease. Because males have higher muscle mass as compared to females of the same age, difference in the metabolic status of the body has also been shown to determine sex difference for the occurrence of heart disease. Furthermore, in view of the role of different sex hormones in determining the incidence of heart disease, the interaction of sex hormones with some nutritional factors and metabolic pathways has been explored to explain gender difference for the development of cardiovascular abnormalities.

Keywords Sex differences · Cardiovascular disease · Calorie restriction · Sex hormones · Cholesterol metabolic pathway

Introduction

Over the past decade, there has been increasing awareness about the leading cause of cardiovascular disease, and a considerable effort has been made to better understand

---

A. K. Shah (✉) · A. Kim · C. Adase
Department of Kinesiology, Nutrition and Food Science, California State University, Los Angeles, CA, USA
e-mail: akaur23@calstatela.edu

B. Ostadal and N. S. Dhalla (eds.), *Sex Differences in Heart Disease*, Advances in Biochemistry in Health and Disease 21,
https://doi.org/10.1007/978-3-030-58677-5_12

its implication with respect to sex differences. It is now well known that cardiovascular disease is the underlying cause of 1 in every 3 deaths in the United States (Table 12.1). Prior to the enactment of the National Institutes of Health (NIH) Revitalization Act of 1993, subjects of randomized controlled trials predominantly consisted of males, and the outcomes of early clinical studies were often founded on limited female representation [1]. However, it has now become evident that there are marked differences between males and females with respect to the incidences of heart disease and thus issues for both sexes are now handled separately. Furthermore, special attention is being paid to determine the role of nutrition in sex difference for the development of heart disease. It may be noted that some of the benefit of diet lie in the activation of various metabolic pathways including reduced inflammation, quick repair, less occurrence of cardiovascular problems and ultimately increased longevity [2]. Of these various metabolic pathways involved in longevity and sexual dimorphism are the nutrient—responsive mTOR [3], growth hormones [4, 5], Sirtuins [6], IGF-1 [7] and cholesterol metabolic pathway. Accordingly, this article will be focused on discussions of these aspects with respect to sex differences for the occurrence of heart disease. In addition, it is intended to discuss the sex differences in response to nutritional manipulation by calorie restriction, sex hormones, and cholesterol biosynthetic pathway. It should be noted that cholesterol is an integral part of the cell membrane of all organs and is considered to play a critical role in maintaining their functional status.

Table 12.1 Cardiovascular Disease (CVD) and death rate in US

Type of CVD	Death rate (%)	Remarks
Coronary heart disease	43.2	Average age of first **heart** attack: Male—65.6 years Female—72.0 years
Stroke	16.9	Disability following by stroke: Male—3% Female—2%
Peripheral artery disease	3.0	Individual with total cholesterol levels of 200 mg/dL or higher: Male—35.4% Female—41.8%
Hypertension	9.8	–
Heart failure	9.3	–
Other CVDs	17.7	–

Taken from the report by American Heart Association, 2019 (https://www.heart.org/en/about-us/heart-and-stroke-association-statistics)

Sex Differences in Response to Calorie Restriction and Cardiovascular Disease

After nearly a century of research observing the role of reduced energy intake on life extension in model organisms, calorie restriction and related diets continue to draw serious interest not only for their roles as mitigators of aging but also of age-related disease and pathogenesis [8, 9]. Clinical trials and observational studies have also revealed that calorie restriction (CR) has ameliorating effects on metabolic and hormonal homeostasis in both humans and experimental models, and is a promising, inexpensive mode of intervention for the treatment and prevention of cardiovascular disease [9–12]. Furthermore, there is ample evidence from animal and human studies to suggest that there are notable sex-dependent responses to these interventions, possibly influenced by a variety of sex differences including fat distribution in the body and the interaction of steroid hormones [12–15]. The lack of comprehensive analysis of possible sexual dimorphism in response to calorie restriction warrants the need for more rigorous research that considers sex and age as major variables.

Prior to discussing the sex-based differences in the effects of calorie restriction, it is necessary to introduce its implications on cardiovascular disease risk. It should be noted that calorie restriction, as practiced chronically or intermittently, is often defined as a physiological state of reduced energy intake while avoiding malnutrition [11]. Common protocols in rodent studies is 25–40% calorie reduction from usual ad libitum daily intake, with effects dependent not solely on sex but also on dose, age, strain, and nutrient composition of the diet [13, 15]. However, in a pilot study of the CALERIE (Comprehensive Assessment of Long-term Effects of Reducing Intake of Energy) randomized controlled trials, weight loss induced by an average calorie reduction of just 11.5 (±2)% for 1 year in overweight and middle-aged individuals contributed to significant reductions in plasma concentrations of LDL-cholesterol, triglycerides, and C-reactive protein, along with reductions in total cholesterol:HDL ratio and insulin resistance which are the major risk factors of coronary heart disease [16–18]. The phase-2 multicenter trial, CALERIE-2, spanning two years, also demonstrated that prolonged mild calorie restriction significantly improved blood lipid profiles and insulin sensitivity while significantly reducing intramyocellular lipids as well as visceral and subcutaneous abdominal adipose tissue [18–22]. Another 6-month calorie restriction study found a reduction in oxidative stress, as measured by reduced ROS production, higher glutathione levels, and increased catalase activity [23]. Short-term interventions spanning 10 weeks with 20% calorie reduction have also been proven to be effective in reducing systolic and diastolic blood pressure, glucose concentration, as well as reversing fibrinolytic dysfunction, which has been associated with increased cardiovascular disease risk [18, 21, 22].

A cross-sectional look at members of the Calorie Restriction Optimal Nutrition (CRON) Society, who have been practicing, on average, 30% calorie restriction for 3–15 years while maintaining a highly nutrient-dense and low glycemic diet, demonstrated very low risk factors for atherosclerosis. Average serum total cholesterol and LDL-cholesterol concentrations were in the bottom 10% of their age bracket and triglyceride concentrations were comparable to values at the 5th percentile for average 20-year-olds [19]. Carotid intima-media thickness (IMT) in CR practitioners also measured 40% less than age-matched controls on a typical Western diet. Lastly, heart rate variability in the calorie restriction group was comparable to healthy males and females at ages 20 years younger [19]. A considerable number of these studies did not observe or quantify the effects of calorie restriction in relation to sex. Despite limited analysis of sex differences in long-term clinical trials, substantial evidence exists in animal models, short-term calorie restriction and related interventions that may explain some fundamental physiological differences between males and females beyond the context of reproduction.

Effects of Calorie Restriction and Sex Differences on Endothelial Function

Maintaining healthy endothelial function is essential for the management of cardiovascular disease risk in males and females as endothelial impairment has been shown to precede hypertension, atherosclerosis, and other acute disease manifestations [12, 27]. One distinguishing characteristic of healthy endothelial function is estradiol (E_2) -mediated increase in nitric oxide (NO) bioavailability in relation to decreased oxidative stress [11, 27]. Estradiol (E_2) binding to estrogen receptors (ER) expressed on endothelial cells and vascular smooth muscle cells has been shown to stimulate production of NO through the activation of endothelial nitric oxide synthase (eNOS) [27]. And with accelerated NO production comes the need to suppress oxidative stress, particularly superoxide (O_2^-) whose reactivity with NO reduces its bioavailability [24]. In a 2-week study observing rats subjected to 40% calorie reduction (compared to controls) and vascular reactivity tests, caloric restriction significantly increased NO concentrations in males only and reduced NADPH-sensitive O_2^- production with greater effect in males than females [24]. While caloric restriction proved to be more effective towards males in this experiment, it is important to note that female controls had higher relative NO production and markedly lower O_2^- production than their male counterparts. This discrepancy in sex reveals high levels of E_2 in young female rats serves a cardioprotective advantage to which caloric restriction showed no significant effect [24].

Further testing to see if caloric restriction altered serum levels of estrogen and testosterone in animals demonstrated no change in serum level as well as mRNA expression of ERs. This observation indicates that calorie restriction benefits noted in males were not primarily the result of estrogens and ER activity but its mechanisms still remains to be investigated [25]. One known means of activating eNOS, noted when binding to subtypes ERα and ERβ, involves the induction of the phosphatidylinositol 3-kinase (PI3K)/Akt pathway [28, 29]. Independent of ERs, calorie restriction—induced AMPK activation also triggers a downstream cascade of the

PI3K/Akt/eNOS pathway. Conversely, overconsumption of food has been shown to dysregulate this pathway, further demonstrating the potential benefits of calorie restriction in males and postmenopausal females in maintaining healthy vasculature [11, 28].

Effect of Calorie Restriction and Sex Differences in Fat Loss

Sex differences in visceral and subcutaneous fat distribution have largely been investigated in association with sex steroid characteristics. Analysis of participants from the Multi-Ethnic Study of Atherosclerosis revealed that higher levels of bioavailable testosterone was associated with lower visceral and subcutaneous fat composition in males while the reverse was seen in females. Other androgens at high levels, such as dehydroepiandrosterone, were again associated with higher visceral fat content in females but not in males [30]. Another study looking at the role of sex steroid profiles on fat loss from lifestyle modifications showed the amount of visceral and subcutaneous fat was inversely associated with high testosterone and sex-hormones binding globulin (SHBG) levels in males, while females only found correlation with high SHBG levels [31]. Considering the protective functions provided by high testosterone levels in males only, a 12-week 800 kcal/day caloric restriction intervention has been shown to significantly increase testosterone and SHBG levels in obese male subjects through improved testicular function and reduced aromatase conversion of testosterone to estradiol in adipose tissue [32]. Postmenopausal females in the SHAPE-2 weight reduction trial demonstrated that calorie restriction did not elevate serum testosterone and androstenedione levels in females and maintained cardioprotective testosterone to estradiol ratios [33]. These findings demonstrate the role of calorie restriction in ameliorating age-associated decline in cardioprotective sex steroid profiles.

As shown in Fig. 12.1, the observation of calorie restriction effects on fat loss in mice studies reveals male mice tend to lose both visceral and subcutaneous fat while females predominantly lose more visceral fat [34]. A direct comparison of visceral adipose tissue loss after 8-weeks of 800–1000 kcal/day consumption in adult human males and females, revealed males experienced greater loss than females after adjusting for relative changes in fat mass [35]. However, when sample sizes were reduced to compare males and females with matched baseline ratios of visceral-to-subcutaneous fat, the significant difference was no longer evident [35]. Despite evident sexually dimorphic patterns of fat loss between sexes in mice studies, human studies still require further examination with larger sample sizes matched in baseline characteristics. Lastly, an analysis of adipose tissue genes from males and females subjected to 12 weeks of a high protein-low calorie-intermittent fasting protocol showed higher expressions of 158 sex-biased genes, except one, in female adipose tissue. Even after controlling for fat mass (which is generally higher in females), the discrepancy remained consistent in 80% of the genes [36]. While it is difficult to draw conclusions relative to cardiovascular disease risk from these findings, the

Fig. 1 Fat Distribution of male and female mice subjected to Ad-lib feeding (sham control) and caloric restriction (CR) Reprinted from Shi H, Strader AD, Woods SC, Seeley R. Am J Physiol Endocrinol Metab 293, E316–E326, 2007. Subcu—subcutaneous

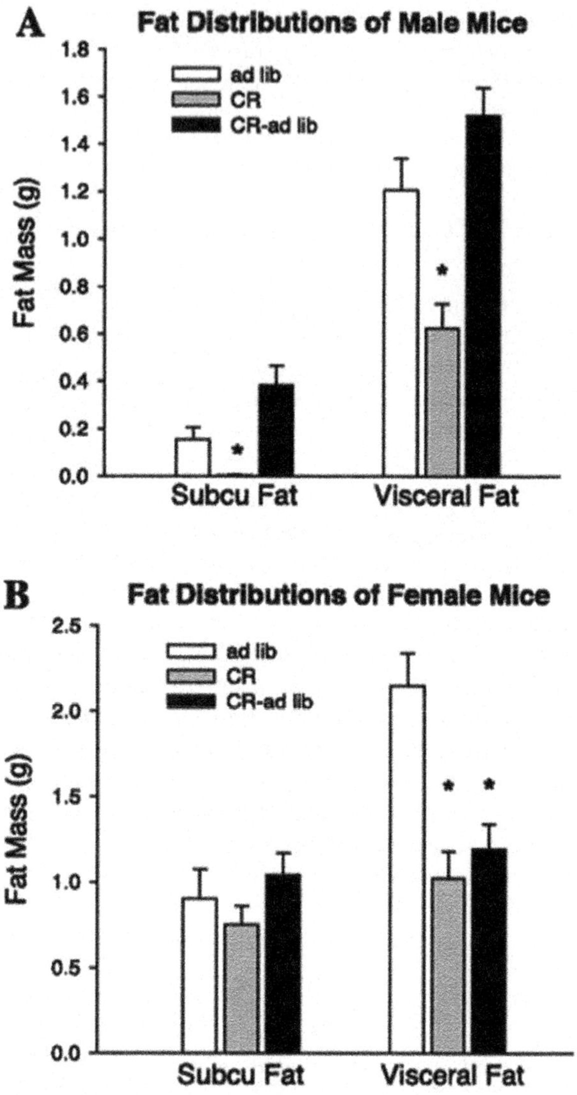

evident sexual dimorphism in adipose tissue gene expression further implicates the possibly greater role of sex on metabolic health [36].

Sex Differences in Response to Hormones and Predisposition to Cardiovascular Diseases

It is now well documented that heart disease is the primary cause of morbidity and mortality in both males and females. While males are at higher risk than females during early adulthood, risk appears to increase in females after menopause. The connection between lower levels of estrogen and higher levels of androgens and their role in cardiac health has been widely studied, although results remain conflicting [37]. Androgens and estrogens are typically thought of, respectively, as the "male" and "female" sex hormones. However, both classes of hormones are found in each sex and proper regulation may be necessary in order to prevent health risk. In addition to the development of reproductive organs and secondary sex characteristics, sex hormones participate in the regulation of a variety of other systems within the body. Testosterone assists in regulation of muscle mass and function, bone mass, and fat distribution [38, 39]. As for estrogen, there are three primary types; Estrone (E1), Estradiol (E2), and Estriol (E3). 17β-Estradiol is the primary estrogen found in the human body. It has been shown to impact cardiac health through stimulating vasodilation, cell differentiation, and regulation of cholesterol production in the liver [40–42].

Obesity and Metabolic Syndrome

Obesity has been significantly linked to increased risk of health-related conditions such as insulin resistance, diabetes, and cardiovascular disease [43, 44]. Measurements of abdominal obesity such as waist circumference, which measure abdominal obesity, verses body mass index which measures general obesity, has shown to indicate greater health risk especially when considering cardiovascular disease [45]. In addition to individual risk, obesity is part of the collection of conditions that make up metabolic syndrome which, together, increase risk for development of cardiovascular disease, diabetes, or stroke. Metabolic syndrome is diagnosed when a person has three or more of the following conditions: abdominal obesity/40 inch waist measurement (male) and 35 inch (female), blood pressure of >/=130/85 mmHg, triglycerides of >150 mg/dL, fasting blood glucose > 100 mg/dL, and HDL < 40 mg/dL (male) or <50 mg/dL (female) [44–46]. It had been largely understood that males had a higher rate of risk of cardiovascular disease than females of the same age. However, continuing research has shown that cardiovascular disease risk increases for peri and post-menopausal females [46]. Understanding the role of sex hormones in obesity, diabetes, cardiovascular disease, and its risk is imperative for treatment and prevention of such conditions [47–54].

Several studies have shown a link between diminished testosterone levels/elevated estrogen levels in males and elevated testosterone/diminished estrogen in females and metabolic syndrome [55, 56]. There are a variety of factors that could affect circulating hormone levels. Both the aging process and increased adiposity have been shown to have a negative association in relation to testosterone and estrogen levels [39, 57]. In addition to serving as a storage site for energy, adipose tissue is metabolically active. Free testosterone is metabolized by adipose tissue into estradiol. Therefore, excess adipose tissue could result in depleted levels of circulating testosterone. In males, this increase in estrogen exasperates testosterone levels further by suppressing testosterone production from the testes [40]. In addition to metabolic factors, adipose tissue could affect circulation and delivery of testosterone to target tissues. While some testosterone circulates in free form, most is bound to either albumin or SHBG. Studies are not clear on the relationship between obesity and SHBG levels, however, there could be a relationship between the effect of visceral fat on SHBG synthesis in the liver [41]. Adipose tissue distribution is also affected by estrogen as it has shown to promote subcutaneous fat accumulation in premenopausal females with normal estrogen levels. During and post-menopause, the decline in estrogen has been linked to a decrease in subcutaneous fat accumulation and an increase in abdominal visceral fat [57, 58]. It is pointed out that sex hormones can increase risk of metabolic syndrome in obese and normal weight persons alike. In males, regardless of weight, low testosterone and androgen receptor deficiency have been linked to hyperinsulinemia and type II diabetes risk. In females, risk is also increased when free testosterone levels are elevated, and SHBG and estrogen levels are diminished [57, 59]. This is of concern as type II diabetes increases risk for cardiovascular disease by contributing to arterial stiffness, hypertension, endothelial dysfunction, inflammation, and atherosclerosis 60].

Hypertension

Males have higher rates of hypertension in relation to age-matched females up until menopause. After menopause, both systolic and diastolic blood pressure increase in females beyond that of comparative aged males [37]. This noted change could point to a connection with decreased estrogen levels after menopause and the effect estrogen has on the regulation of overall blood pressure. Considering estrogen is found in both males and females, it has shown to have cardioprotective effects in both sexes. Estrogen affects blood pressure directly by impacting vascular tone and endothelial function through receptor dependent genomic and non-genomic actions [47]. Endothelial function is primarily regulated by endothelial nitric oxide synthases (eNOS) for the secretion of nitric oxide (NO). Estrogen has also been shown to have a positive effect on blood pressure through rapid non-genomic action of NO. Through genetic transcription, estrogen binding to estrogen receptors (ER-α and ER-β) is integral in the synthesis of eNOS. While this process takes time, endothelial production of nitric oxide appears to be stimulated by estrogen action [37, 53]. In addition to regulating NO production, ER-α and ER-β are observed to protect against vascular injury and atherosclerosis as well as control blood pressure and arterial tone, respectively [47]. Testosterone has also shown to have a possible relationship with NO and eNOS. While reports on this relationship have been conflicting, many studies have reported that deficiency in testosterone is connected to decreased eNOS expression and low NO production [48].

While appropriate endothelial function and vascular tone is important in cardiovascular health, the primary regulator of arterial blood pressure is the renin-angiotensin-aldosterone system (RAAS). Continuing research has examined how sex hormones impact the regulation of RAAS. Estrogen appears to participate in all aspects of RAAS. Ways in which estrogen participates in reduction of blood pressure may include decreasing the synthesis of renin and ACE, decreasing the expression of angiotensin II receptor type 1 (AT_1R) and attenuating the expression of angiotensin II receptor type 2 (AT_2R) [50]. AT_1R is the primary receptor that participates in the regulation of blood pressure. AT_2R is upregulated after vascular injury and appears to assist in cardioprotective activities [49]. While estrogen's role in the RAAS appears to be cardioprotective in nature, testosterone has shown to have an impact on increased arterial pressure by participating in the expression of angiotensinogen, renin, and AT_1R system [50].

Hormone Replacement Therapy

Considering the noted increase of cardiovascular disease risk as associated with sex hormone imbalance, hormone replacement therapy has been used as the primary way to minimize risk of cardiovascular disease development. While estrogen and testosterone hormone replacement therapy are common treatments for the reduction of risk, studies remain conflicting in results [43, 51, 52]. Regarding post-menopausal hormone replacement therapy (HRT), treatment became controversial after the Women's Health Initiative released research findings that conflicted with research that showed the benefits of estrogen HRT. It was found that HRT increased risk of coronary heart disease (CHD) and stroke [61]. In a follow-up study, published in 2013, it was observed that timing of initiation of HRT could play a role in risk of CHD. Starting HRT closer to the point of menopausal onset, in association with decreased independent biomarkers for CHD risk, has shown to reduce HRT CHD risk. However, more research needs to be conducted to determine increased stroke risk and timing of HRT [61]. For testosterone replacement therapy, while a long-term study comparable to the Women's Health Initiative has not been conducted, many studies have pointed to improvement in cardiovascular risk following testosterone replacement [38].

Sex Differences in Response to Genes Involving Cholesterol Biosynthesis

The cholesterol biosynthetic pathway also known as the mevalonate pathway is an important mechanism involved in a vast array of cellular functions. This pathway has been very closely observed by various researchers to study genes affecting heart function. HMG-Coa Reductase (HMGCR) is the rate limiting enzyme of this pathway and catalyze the conversion of HMG-CoA to mevalonate. This pathway provides sterols for membrane structure and non-sterol intermediates for the post-translational modifications and membrane anchorage of growth-related proteins, including Ras, Rac, and Rho GTPase families. In the past years, the regulation of HMGCR has been thoroughly investigated because of its prime involvement in cholesterol and isoprenoid biosynthesis. As all cells require a steady supply of mevalonate, both the sterol (i.e. cholesterol) and non-sterol (i.e. isoprenoid) as products of mevalonate metabolism exert coordinated feedback regulation on HMGCR through different mechanisms. The right functioning of HMGCR as the rate limiting enzyme in the mevalonate pathway is of prime importance under both normal physiologic conditions and in many diseases. HMGCR has diverse roles in various cellular pathways involving cell proliferation, cellular metabolism and cholesterol biosynthesis, in maintaining cytoskeletal stability and dynamics, cellular structure, it's fluidity, mitochondrial function and eventually in regulating overall fate of the cell [62]. Due its implication with endogenous cholesterol metabolism and heart diseases, this pathway has been studied very closely to determine the effect of vitamins and micronutrients targeting the regulation of HMGCR and its downstream effectors. Several studies suggested vitamin isomers mimic the role of statins, a class of drugs used in patients to lower their cholesterol by targeting HMGCR [63]. Earlier studies have substantially found the sex difference in metabolism of mevalonate in rats [64]. The two known pathways of mevalonate metabolism—the shunt pathway and the cholesterol synthesis pathway, both have been demonstrated to exhibit the sex difference. Firstly, the shunt pathway in female rats has shown to exhibit twice the ability to metabolize circulating mevalonate accounted by greater ability of the female kidney to convert mevalonate to CO_2. Secondly, male rats have significantly greater ability to convert circulating mevalonate to cholesterol as compared to the females [64, 65]. However, there is no significant data to extrapolate these findings in humans and further research is needed to justify these findings. It may be a little premature to relate these findings to lower cholesterol levels and less occurrence of atherosclerosis in females in spite of the major findings reported in the studies indicated above.

Conclusion

From the foregoing discussion, it is evident that nutrition and incidence of CV disease go hand in hand. Although, calorie restriction, sex hormones and cholesterol metabolism play a significant role, there is still a void in concise understanding of the exact role of sex with respect to diet and heart disease. There is a lower risk of heart disease, hypertension and atherosclerosis in females before menopause then that in the male; however, this sex-dependent difference is not evident after menopause. Such a switch in the susceptibility of females may be attributed to alterations in the levels of estrogen hormones and/or the sensitivity of its receptors. Differences in the lipid profile, particularly in the cholesterol biosynthetic pathway has also been suggested to explain the differences in the development of cardiovascular disease in male and female population. Nutritional status is considered to have a profound impact on the sex-specific differences in the sensitivity of various factors involved in the induction of heart disease. Caloric restriction has been observed to modify the sex-dependent difference in the incidence of cardiovascular disease by changing the distribution of body fats, altering the difference in the endothelial function, reducing the oxidative stress and regulating the levels of sex hormones. Thus, it is suggested that the nutrition plays a critical role in modifying the incidence of heart disease in both males and females.

References

1. Women's Health Research (2019). https://swhr.org/about/womens-health-research/
2. Kane AE, Sinclair DA, Mitchell JR, Mitchell SJ (2018) Sex differences in the response to dietary restriction in rodents. Curr Opinion Physiol 6:28–34
3. Baar EL, Carbajal KA, Ong IM, Lamming DW (2016) Sex- and tissue-specific changes in mTOR signaling with age in C57BL/6 J mice. Aging Cell 15:155–66
4. Junnila RK, Duran-Ortiz S, Suer O et al (2016) Disruption of the GH receptor gene in adult mice increases maximal lifespan in females. Endocrinology 157:4502–4513
5. Sun LY, Fang Y, Patki A et al (2017) Longevity is impacted by growth hormone action during early postnatal period. Elife 6:e24059
6. Sharples AP, Hughes DC, Deane CS et al (2015) Longevity and skeletal muscle mass: the role of IGF signalling, the sirtuins, dietary restriction and protein intake. Aging Cell 14:511–523
7. Ashpole NM, Logan S, Yabluchanskiy A et al (2017) IGF-1 has sexually dimorphic, pleiotropic, and time-dependent effects on healthspan, pathology, and lifespan. Geroscience 39:129–145
8. McCay CM, Crowell MF, Maynard LA (1989) The effect of retarded growth upon the length of life span and upon the ultimate body size. Nutrition 5:155–171
9. Golbidi S, Daiber A, Korac B et al (2017) Health benefits of fasting and caloric restriction. Curr Diab Rep 7:123
10. Mattson MP, Wan R (2005) Beneficial effects of intermittent fasting and caloric restriction on the cardiovascular and cerebrovascular systems. J Nutr Biochem 16:129–137
11. García-Prieto CF, Fernández-Alfonso MS (2016) Caloric restriction as a strategy to improve vascular dysfunction in metabolic disorders. Nutrients 8:370
12. Nicoll R, Henein MY (2018) Caloric restriction and its effect on blood pressure, heart rate variability and arterial stiffness and dilatation: a review of the evidence. Int J Mol Sci 19:751

13. Austad SN, Bartke A (2015) Sex differences in longevity and in responses to anti-aging interventions: a mini-review. Gerontology 62:40–46
14. Mitchell SJ, Madrigal-Matute J, Scheibye-Knudsen M et al (2016) Effects of sex, strain, and energy intake on hallmarks of aging in mice. Cell Metab 23:1093–1112
15. Fontana L, Villareal DT, Weiss EP, et al. Calorie restriction or exercise: effects on coronary heart disease risk factors. A randomized, controlled trial. Am J Physiol Endocrinol Metab, 293: E197–202, 2007
16. Racette SB, Weiss EP, Villareal DT et al (2006) One year of caloric restriction in humans: feasibility and effects on body composition and abdominal adipose tissue. J Gerontol A Biol Sci Med Sci 61:943–950
17. Most J, Tosti V, Redman LM, Fontana L (2017) Calorie restriction in humans: An update. Ageing Res Rev 39:36–45
18. Ravussin E, Redman LM, Rochon J et al (2015) A 2-Year randomized controlled trial of human caloric restriction: feasibility and effects on predictors of health span and longevity. J Gerontol Series A, Biol Sci Medical Sci 70:1097–1104
19. Fontana L, Meyer TE, Klein S, Holloszy JO (2004) Long-term calorie restriction is highly effective in reducing the risk for atherosclerosis in humans. Proc Natl Acad Sci U S A 101:6659–6663
20. Velthuis-te Wierik EJ, Meijer P, Kluft C, van den Berg H (1995) Beneficial effect of a moderately energy restricted diet on fibrinolytic factors in non-obese men. Metabolism 44:1548–1552
21. Tofler GH, Massaro J, O'Donnell CJ et al (2016) Plasminogen activator inhibitor and the risk of cardiovascular disease: The Framingham heart study. Thromb Res 140:30–35
22. Most J, Gilmore LA, Smith SR et al (2017) Significant improvement in cardiometabolic health in healthy non-obese individuals during caloric restriction-induced weight loss and weight loss maintenance. Am J Physiol Endocrinol Metab 314:E396–E405
23. López-Domènech S, Martínez-Herrera M, Abad-Jiménez Z et al (2019) Dietary weight loss intervention improves subclinical atherosclerosis and oxidative stress markers in leukocytes of obese humans. Int J Obes 43:2200–2209
24. Chou SH, Lee YC, Huang CF et al (2010) Gender-specific effects of caloric restriction on the balance of vascular nitric oxide and superoxide radical. Cardiovasc Res 87:751–759
25. Korybalska K, Swora-Cwynar E, Łuczak J et al (2016) Association of endothelial proliferation with the magnitude of weight loss during calorie restriction. Angiogenesis 19:407–419
26. Stanhewicz AE, Wenner MM, Stachenfeld NS (2018) Sex differences in endothelial function important to vascular health and overall cardiovascular disease risk across the lifespan. Am J Physiol Heart Circ Physiol 31:H1569–H1588
27. García-Prieto CF, Pulido-Olmo H, Ruiz-Hurtado G et al (2015) Mild caloric restriction reduces blood pressure and activates endothelial AMPK-PI3K-Akt-eNOS pathway in obese Zucker rats. Vascul Pharmacol 65–66:3–12
28. Zhang Y, Wang SJ, Han ZH et al (2014) PI3K/AKT signaling pathway plays a role in enhancement of eNOS activity by recombinant human angiotensin converting enzyme 2 in human umbilical vein endothelial cells. Int J Clin Exp Pathol 7:8112–8117
29. García-Prieto CF, Gil-Ortega M, Aránguez I et al (2015) Vascular AMPK as an attractive target in the treatment of vascular complications of obesity. Vascul Pharmacol 67–69:10–20
30. Mongraw-Chaffin ML, Anderson CA, Allison MA et al (2015) Association between sex hormones and adiposity: qualitative differences in women and men in the multi-ethnic study of atherosclerosis. J Clin Endocrinol Metab 100:E596–E600
31. Kim C, Dabelea D, Kalyani RR et al (2017) Changes in visceral adiposity, subcutaneous adiposity, and sex hormones in the diabetes prevention program. J Clin Endocrinol Metab 102:3381–3389
32. Schulte DM, Hahn M, Oberhäuser F et al (2014) Caloric restriction increases serum testosterone concentrations in obese male subjects by two distinct mechanisms. Horm Metab Res 46:283–286
33. van Gemert WA, Schuit AJ, van der Palen J et al (2015) Effect of weight loss, with or without exercise, on body composition and sex hormones in postmenopausal women: the SHAPE-2 trial. Breast Cancer Res 17:120

34. Shi H, Strader AD, Woods SC, Seeley RJ (2007) Sexually dimorphic responses to fat loss after caloric restriction or surgical lipectomy. Am J Physiol Endocrinol Metab 293:E316–E326
35. Gasteyger C, Larsen TM, Vercruysse F et al (2009) Visceral fat loss induced by a low-calorie diet: a direct comparison between women and men. Diabetes Obes Metab 11:596–602
36. Viguerie N, Montastier E, Maoret JJ et al (2012) Determinants of human adipose tissue gene expression: impact of diet, sex, metabolic status, and cis genetic regulation. PLoS Genet 8:e1002959
37. Zhao D, Guallar E, Ouyang P et al (2018) Endogenous sex hormones and incident cardiovascular disease in post-menopausal women. J Am Coll Cardiol 71:2555–2566
38. Goodale T, Sadhu A, Petak S, Robbins R (2017) Testosterone and the heart. Method DeBakey Cardiovasc J 13:68–72
39. Cooper LA, Page ST, Amory JK et al (2015) The association of obesity with sex hormone-binding globulin is stronger than the association with ageing-implications for the interpretation of total testosterone measurements. Clin Endocrinol 83:828–833
40. Bellastella G, Menafra D, Puliani G et al (2019) Obesity programs of nutrition, education, research and assessment (OPERA) group How much does obesity affect the male reproductive function? Int J Obesity supplements 9:50–64
41. Peter A, Kantartzis K, Machann J et al (2010) Relationships of circulating sex hormone-binding globulin with metabolic traits in humans. Diabetes 59:3167–3173
42. Bjornstrom L, Sjoberg M (2005) Mechanisms of estrogen receptor signaling: Convergence of genomic and nongenomic actions on target genes. Mol Endocrinol 19:833–842
43. Simoncini T, Mannella P, Fornari L et al (2004) Genomic and non-genomic effects of estrogens on endothelial cells. Steroids 69:537–542
44. Mokdad AH, Ford ES, Bowman BA et al (2003) Prevalence of obesity, diabetes, and obesity-related health risk factors 2001. JAMA 289:76–79
45. Recio-Rodriguez JI, Gomez-Marcos MA, Patino-Alonso MC et al (2012) Abdominal obesity vs general obesity for identifying arterial stiffness, subclinical atherosclerosis and wave reflection in healthy, diabetics and hypertensive. BMC cardiovas disord 12:3
46. Regnault V, Thomas F, Safar ME et al (2012) Sex difference in cardiovascular risk: role of pulse pressure amplification. J Am Coll Cardiol 59:1771–1777
47. Vitale C, Mendelsohn M, Rosano G (2009) Gender differences in the cardiovascular effect of sex hormones. Nat Rev Cardiol 6:532–542
48. Hotta Y, Kataoka T, Kimura K (2019) Testosterone deficiency and endothelial dysfunction: nitric oxide, asymmetric dimethylarginine, and endothelial progenitor cells. Sex Med Rev 7:661–668
49. Pernomian L, Pernomian L, Gomes MS, Da Silva C (2015) Pharmacological significance of the interplay between angiotensin receptors: MAS receptors as putative final mediators of the effects elicited by angiotensin AT1 receptors antagonists. Eur J Pharmacol 769:143–146
50. Hilliard L, Sampson M, Brown A, Denton K (2013) The, "His and Hers" of the renin-angiotensin system. Curr Hypertens Rep 15:71–79
51. Wong WW, Taylor AA, Smith EO et al (2012) Effect of soy isoflavone supplementation on nitric oxide metabolism and blood pressure in menopausal women. Am J Clin Nutr 95:1487–1494
52. Del Principe D, Ruggieri A, Pietraforte D et al (2015) The relevance of estrogen/estrogen receptor system on the gender difference in cardiovascular risk. Int J Cardiol 187:291–298
53. Guo X, Razandi M, Pedram A et al (2005) Estrogen induces vascular wall dilation: mediation through kinase signaling to nitric oxide and estrogen receptors alpha and beta. The J Biol Chem 280:19704–19710
54. O'Donnell E, Floras J, Harvey P (2014) Estrogen status and the renin angiotensin aldosterone system. Am J Physiol Regul Integr Comp Physiol 307:R498–R500
55. Wang C, Zhang W, Wang Y et al (2019) Novel associations between sex hormones and diabetic vascular complications in men and postmenopausal women: A cross-sectional study. Cardiovasc Diabetol 18:97
56. Ahmed I, Mahdy M, El Oraby H, Abdelazeem E (2017) Association of sex hormones with metabolic syndrome among Egyptian males. Diabetes Metab Syndr 11:S1059–S1064

57. Faulkner J, De Chantemele E (2019) Sex hormones, aging and cardiometabolic syndrome. Biol Sex Differ 10:1–9
58. Brown L, Gent L, Davis K, Clegg D (2010) Metabolic impact of sex hormones on obesity. Brain Res 1350:77–85
59. Navarro G, Allard C, Xu W, Mauvais-Jarvis F (2015) The role of androgens in metabolism, obesity, and diabetes in males and females. Obesity 23:713–719
60. Lathief S, Inzucchi S (2016) Approach to diabetes management in patients with CVD. Trends Cardiovasc Med 26:165–179
61. Miller V, Manson M (2013) Women's health initiative hormone therapy trials: new insights on cardiovascular disease from additional years of follow up. Curr Cardiovasc Risk Rep 7:196–202
62. Yeganeh B, Wiechec E, Ande SR et al (2014) Targeting the mevalonate cascade as a new therapeutic approach in heart disease, cancer and pulmonary disease. Pharmacol Ther 143:87–110
63. Gong L, Xiao Y, Xia F et al (2019) The mevalonate coordinates energy input and cell proliferation. Cell Death Dis 10:327
64. Wiley MH, Howton MM, Siperstein MD (1979) Sex differences in the sterol and nonsterol metabolism of mevalonate. J Biol Chem 254:837–842
65. Feingold KR, Wiley MH, Searle GL et al (1980) Sex difference in human mevalonate metabolism. J Clin Invest 66:361–366

Chapter 13
Cardio-Oncology: Preventing Broken Hearts in Women with Breast Cancer

Hilary Bews, Cameron Eekhoudt, Sonu Varghese, and Davinder S. Jassal

Abstract Breast cancer is the leading cause of cancer in Canadian women and as a result constitutes a significant cause of morbidity and mortality in this patient population. Although overall survival is significantly improved by treatment with anti-cancer drugs including Doxorubicin and Trastuzumab, approximately 1 in 4 Canadian women are at risk of developing chemotherapy induced cardiotoxicity. Current guidelines recommend serial cardiac monitoring using transthoracic echocardiography (TTE) or multigated acquisition scan (MUGA) for the non-invasive assessment of left ventricular ejection fraction (LVEF). Recent studies suggest that echocardiographic deformation measures including tissue velocity imaging and strain parameters can detect adverse cardiac changes prior to quantifiable differences in LVEF. Early detection of chemotherapy mediated cardiotoxicity would allow for early intervention with the use of cardioprotective agents in women with breast cancer. The prophylactic use of angiotensin converting enzyme inhibitors (ACEi), angiotensin receptor blockers (ARBs), β-blockers, statins, anti-oxidants, and nutraceutical agents as cardioprotective agents have been studied in the Cardio-Oncology setting. While animal models and initial clinical trials prove promising, further studies are required to prevent the devastating consequences of chemotherapy mediated cardiotoxicity in women with breast cancer.

Keywords Breast cancer · Doxorubicin · Trastuzumab · Cardiotoxicity · Cardiovascular imaging · Prevention

---

H. Bews · D. S. Jassal
Section of Cardiology, Department of Internal Medicine, Max Rady College of Medicine, Rady Faculty of Health Sciences, University of Manitoba, Winnipeg, Manitoba, Canada

C. Eekhoudt · S. Varghese · D. S. Jassal (✉)
Institute of Cardiovascular Sciences, Department of Physiology and Pathophysiology, Max Rady College of Medicine, Rady Faculty of Health Sciences, University of Manitoba, Winnipeg, Manitoba, Canada
e-mail: djassal@sbgh.mb.ca

D. S. Jassal
Department of Radiology, Max Rady College of Medicine, Rady Faculty of Health Sciences, University of Manitoba, Winnipeg, Manitoba, Canada

B. Ostadal and N. S. Dhalla (eds.), *Sex Differences in Heart Disease*, Advances in Biochemistry in Health and Disease 21,
https://doi.org/10.1007/978-3-030-58677-5_13

Breast Cancer: An Overview

Breast cancer is a major public health concern as the leading cause of cancer in women [1]. Approximately 1 in 8 Canadian women will develop breast cancer throughout their lifetime and 1 in 33 will die from breast cancer [1]. It is estimated that in 2019 alone, 26 900 Canadian women will be diagnosed with breast cancer and over 5 000 women will die as a consequence of the disease [1]. Breast cancer is primarily a disease of females as illustrated by the projected 2019 statistics for Canadian males. It is expected that only 230 males in Canada will be diagnosed with breast cancer and 55 will die of the disease [1]. The incidence of breast cancer increased by 2%/year between 1984 and 1991, largely attributable to increased detection through the implementation of screening programs [1]. A slight decrease in incidence after 2002 is credited to a decline in the use of hormone replacement therapy in post-menopausal women [1, 2]. The overall 5-year survival from breast cancer also differs between males and females at 80% and 88%, respectively [1].

In 2018, the Canadian Task Force on Preventative Health Care published new guidelines for breast cancer screening in women of average risk. The guidelines recommend screening mammography every 2–3 years in women aged 50–69 years and 70–74 years [3]. The implementation of these recommendations is expected to decrease the absolute rate of death from breast cancer by 0.75–0.92/1000 and 1.55/1000 women screened, respectively, for the two age cohorts [3]. Self and clinical breast exams are no longer recommended due to conflicting evidence for benefit [3]. As with any screening program, mammography programs will lead to over diagnosis of early stage breast cancer. Approximately 25% of invasive and in situ cancers identified over five years of screening may be overdiagnosed among women aged 50–59 years [3].

Heritable factors account for approximately 25–30% of breast cancers, of which mutations in the *BRCA* genes are most strongly implicated [4]. *BRCA* genes are postulated to function as tumor suppressor genes, such that mutations provide the "first hit" of genetic susceptibility in the two-hit hypothesis to explain the development of breast cancer [4]. Breast cancer develops in women with a *BRCA* mutation if a second allele becomes mutated [4]. It is estimated that 0.1–0.5% of women carry a mutation in either the *BRCA1* or *BRCA2* genes, associated with a lifetime risk for developing breast cancer of approximately 72% and 69%, respectively [4, 5]. Screening via magnetic resonance imaging (MRI) provides superior sensitivity to detect cancer in these high risk individuals when compared to mammography or ultrasound [6].

Treatment of breast cancer involves a multimodal approach involving surgery, radiotherapy, chemotherapy, hormone modulation, and immunotherapy depending on the stage and pathological features of the malignant cells. Surgical treatment involves either mastectomy or lumpectomy with breast irradiation. A randomized control trial conducted in 1976 compared mastectomy versus lumpectomy with radiation in women with resectable breast cancer and demonstrated no difference in disease-free, distant-disease-free, or overall survival at a follow-up of 20 years [7].

Thus, if negative margins are achievable, breast conserving therapy is favored to minimize cosmetic and functional consequences.

Adjuvant chemotherapy is administered to women with early stage breast cancer, defined as cancer confined to the breast or spread to loco-regional nodes, following resection to control microscopic metastatic spread [8]. A meta-analysis performed by the Early Breast Cancer Trialists' Collaborative Group demonstrated that adjuvant chemotherapy with an anthracycline decreases annual breast cancer death by 20–38% in this setting [8]. The two most commonly employed regimens include either Adriamycin (doxorubicin; DOX) and Cyclophosphamide (AC) or 5-Flurouroucil, Epirubicin, and Cyclophosphamide (FEC).

Anthracycline based chemotherapy constitutes the foundation of breast cancer chemotherapy. The first two anthracyclines developed, DOX and daunorubicin, were isolated from *Streptomyces peucetius* in the 1960s [9]. While DOX is widely utilized in breast cancer treatment, daunorubicin demonstrates activity in hematologic malignancies [9]. Epirubicin, commonly used in breast cancer chemotherapy regimens, is a semi-synthetic derivative with an increased volume of distribution and reduced half-life [9]. A number of proposed mechanisms exist to explain the cytotoxic actions of anthracyclines including: (1) inhibition of DNA synthesis through intercalation and inhibition of DNA polymerase; (2) free-radical generation resulting in DNA damage and lipid peroxidation; (3) DNA alkylation; (4) interference of DNA unwinding through helicase modulation; (5) DNA topoisomerase inhibition; and (6) induction of apoptotic pathways [10]. A number of these mechanisms are only observed at concentrations well above those achieved in the clinical setting [9, 10]. As such, the three most clinically relevant mechanisms of cytotoxicity appear to include inhibition of DNA topoisomerase II, apoptosis induction, and generation of free-radicals [9]. Topoisomerases induce single- or double-strand breaks in DNA which are then resealed, to ultimately modulate DNA helical supercoiling [9]. Anthracyclines stabilize an intermediate structure by creating a covalent bond between the cut DNA strands and the tyrosine moiety of topoisomerase II which prevents the resealing of DNA [9]. A minor contribution to cell death through the inhibition of topoisomerase I has also been proposed [9]. Anthracyclines induce apoptosis through the activation of p53-DNA binding [9]. This ultimately leads to the inhibition of cyclin-dependent kinases involved in the G1 to S cell cycle transition [9]. Alternatively, there is some evidence to suggest that anthracyclines may stimulate the direct release of cytochrome c from the mitochondria, triggering a caspase cascade culminating in cell death [11]. Finally, anthracyclines are strong inducers of free-radical damage and oxidative stress as a consequence of participation in one electron redox cycling to generate by-products such as the superoxide anion and hydrogen peroxide [9]. The free-radicals generated from this process induce harmful post-translational modification of cell constituents and possibly serve as aberrant transmitter signals in cell cycle or apoptotic pathways [9].

In addition to chemotherapy, endocrine therapies were developed to treat hormone receptor positive breast cancers. Hormone receptor positivity, representing 70% of breast cancers, is defined pathologically when >1% of cells stain positive for the estrogen or progesterone receptor on surgical specimens [12]. Estrogen receptor alpha (ERα) in malignant tissue is activated by endogenous estrogen to promote oncogenic growth pathways [12]. The expression of the progesterone receptor is also a marker of ERα signaling [12]. All women with hormone positive breast cancer are treated with endocrine therapy for five years [12]. Specific therapy includes selective estrogen receptor modulators including tamoxifen, which competitively bind to the estrogen receptor and block ERα growth pathways [12]. Aromatase inhibitors, such as anastrozole, exemestane and letrozole, are employed in post-menopausal women to prevent the conversion of androgens to estrogen, thereby reducing estrogen levels [12]. The addition of systemic endocrine therapy results in an approximately 50% relative reduction in cancer recurrence rates for the first 5 years on therapy and a 30% reduction in cancer recurrence for the subsequent five years [13]. Continuing hormonal therapy beyond 5 years may be associated with a small benefit in regards to both recurrence and mortality at the expense of increased side effects such as endometrial cancer, osteoporosis, and thromboembolic events [14, 15].

Targeted immunotherapy is the most recent advance in breast cancer which has drastically improved overall prognosis [12]. The human epidermal growth factor receptors (HER) are a family of tyrosine kinase receptors important in cell proliferation, survival, and migration [16]. HER subsequently activates both mitogen-activated protein kinase (MAPK/MEK/Erk) and phosphatidylinositol-3-kinase (PI3K/PKB/Akt) signaling pathways, to induce mitosis and survival, respectively [16]. Erk activation through HER signaling results in downstream cell migration and proliferation, transformation, and resistance to apoptosis [17, 18]. Atk is a serine/threonine kinase with activity in nutrient metabolism, cell growth, transcriptional regulation, and cell survival [11]. Downstream regulators include YAP, MDM2, and the Forkhead family of transcription factors [11]. Approximately 20–30% of breast cancers overexpress HER2, leading to a more aggressive disease with a worse prognosis [19]. Trastuzumab (TRZ), a recombinant humanized IgG monoclonal antibody, was developed in the 1990s to selectively bind the extracellular domain of HER2 [12, 19]. Proposed anti-cancer mechanisms include internalization and degradation of HER2 receptors, recruitment of immune cells to target tumor cells, and the inhibition of metalloproteinase-induced cleavage of the extracellular HER2 domain which is necessary for signaling [19]. TRZ further inhibits PI3K/Akt and MAPK signaling, as well as modulates other transcription pathways to promote apoptosis and anti-angiogenesis [19]. In 2001, a multi-centre randomized control trial was conducted to compare the efficacy of chemotherapy (consisting of an anthracycline and cyclophosphamide or paclitaxel) versus chemotherapy plus TRZ in women with HER2 positive metastatic breast cancer [20]. The addition of TRZ resulted in a longer median time to disease progression of 7.4 months versus 4.6 months, as well as a longer median survival of 25.1 months versus 20.3 months [20]. More recently, TRZ has been approved for use in early breast cancer in the adjuvant setting [21, 22]. The HERceptin Adjuvant trial randomized 5102 females with early HER2 positive breast cancer following initial treatment with surgery, chemotherapy, and radiation to TRZ for 1 or 2 years, or placebo [21]. After 11 years of follow-up, treatment with TRZ for 1 year was found to decrease disease-free survival with a hazard ratio of 0.76 (95%CI 0.68–0.86) and death with a hazard ratio of 0.74 (95%CI 0.64–0.86) [21].

Treatment with TRZ for 2 years did not significantly alter outcomes when compared to the 1 year cohort [21].

Although treatment for breast cancer, particularly chemotherapy and immunotherapy, is potentially live saving, it is also associated with negative sequelae, specifically cardiotoxicity. As breast cancer is overwhelmingly a disease affecting females, this creates a unique risk factor for the development of cardiovascular disease.

Chemotherapy Mediated Cardiac Dysfunction

Chemotherapy induced cardiotoxicity is a leading cause of morbidity and mortality in cancer survivors [23]. Patients with pre-existing cardiovascular disease, additional cardiovascular risk factors, increasing age, and exposure to multiple cardiotoxic agents are at a particularly high risk of developing cardiac dysfunction [23]. In addition to left ventricular (LV) systolic dysfunction, cancer therapies may promote the development of hypertension, myocardial ischemia, arterial thrombosis, and/or arrhythmias [23]. The American Society of Echocardiography defines cancer therapeutics—related cardiac dysfunction (CTRCD) as a decrease in the LV ejection fraction (LVEF) of >10% to an absolute value of <53% [24]. This definition requires confirmation of a reduced LVEF on repeat cardiovascular imaging, performed 2–3 weeks after the baseline study [24]. CTRCD is further stratified as reversible, partially reversible, or irreversible. Cardiotoxicity is considered reversible if the LVEF recovers to within 5% of the baseline value, partially reversible if the LVEF improves by ≥10% from the nadir but remains >5% from the baseline value, and irreversible if the LVEF improves by <10% from the nadir and remains >5% from the baseline value [24].

CTRCD is further classified by two distinct patterns of toxicity as shown in Table 13.1. Type 1 cardiotoxicity, of which DOX serves as a prototype, is characterized

Table 13.1 Comparison between the features of type 1 and type 2 cardiotoxicity

	Type 1 Cardiotoxicity	Type 2 Cardiotoxicity
Prototypical agent	Anthracycline based chemotherapy	Trastuzumab
Dose dependency	Yes	No
Mechanism	Oxidative stress, topoisomerase inhibition	Variable
Reversibility	Irreversible	Functional recovery after discontinuation is common
Ultrastructural change	Vacuolar swelling, myofibrillar disarray, and cell death	May not lead to apoptosis in isolation
Rechallenge after cardiotoxicity	Not safe	May not result in further cardiotoxicity

by dose-dependent toxicity and microscopic cardiomyocyte damage represented by vacuolar swelling, myofibrillar disarray, and cell death [24]. Damage at the cellular level is permanent and drug re-challenge is associated with a high risk of further damage [24]. In contrast, type 2 CTRCD including agents such as TRZ, does not demonstrate a dose-dependent toxicity and is not believed to be associated with abnormalities at the cellular level [24, 25]. As such, type 2 cardiotoxicity is believed to be potentially reversible and re-challenge with the anti-cancer drug such as TRZ may be safe. Despite this, some evidence exists to challenge the reversibility of type 2 toxicity as ultrastructural changes have been observed in murine models [24, 25].

The dose-dependent relationship between the cumulative DOX dose and development of cardiotoxicity has been well established. Cardiotoxicity develops at an incidence of >4% when the cumulative DOX dose is 500–550 mg/m^2, >18% at a cumulative dose of 551–600 mg/m^2, and increases significantly to 36% at a cumulative dose of >601 mg/m^2 [26]. Furthermore, the accumulation of DOX in cardiac tissue also correlates with the administered concentration [9]. As such, a maximum cumulative dose of 500 mg/m^2 is recommended to avoid these adverse consequences [26]. Specific risk factors that predispose an individual to develop DOX mediated cardiac dysfunction include age >70 years old, combination therapy with agents including cyclophosphamide, concurrent mediastinal irradiation, pre-existing cardiac disease, hypertension, and/or liver disease [26]. DOX induced cardiotoxicity develops at varying time points following treatment, occurring as remotely as 20 years after DOX administration [27]. Rarely, a myocarditis-pericarditis syndrome or myocardial infarction can occur after the first day and during the first month of DOX therapy [26].

DOX mediated cardiotoxicity develops primarily as a consequence of free-radical production and oxidative stress culminating in programmed cell death (Fig. 13.1) [26]. Cell death is promoted by both extrinsic and intrinsic pathways. In a neonatal rat cardiomyocyte model, DOX was shown to induce apoptosis through an extrinsic receptor mediated pathway [28]. DOX induces the intrinsic apoptotic pathway through the up-regulation of Bax. The participation of DOX in redox cycling generates reactive oxygen species such as the superoxide ion and hydrogen peroxide which induce p53 to transcriptionally activate Bax [9]. Bax then functions to promote mitochondrial pore opening, cytochrome c release, and the formation of the apoptosome complex [9]. The reactive oxygen species generated from DOX administration may also directly stimulate pore formation in the mitochondria to promote apoptotic pathways [9]. Anti-apoptotic mediators such as Bcl-xL, responsible for blocking cytochrome c release from the mitochondria through the zinc finger transcription factor GATA-4, are also modulated by DOX to further induce cell death [9]. In addition to promoting the generation of reactive oxygen species, DOX renders cardiomyocytes less resilient to oxidative stress by depleting anti-oxidant levels [26]. Cardiomyocytes are particularly susceptible to this phenomenon as selenium dependent GSH-peroxidase-1 is readily inactivated in these cells following DOX administration, leaving cardiac tissue vulnerable to oxidative stress [9]. Additional pro-apoptotic mechanisms include the activation of NF-kB and p38 MAPK, as well as the dysregulation of iron to catalyze DOX redox reactions, subsequently disrupting

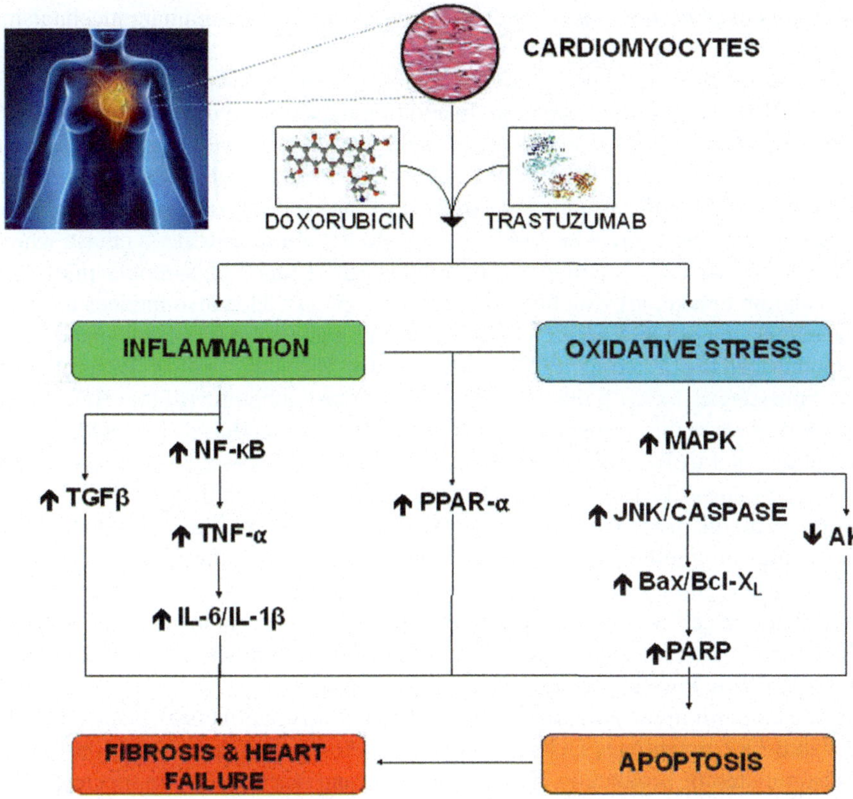

Fig. 13.1 The mechanism of doxorubicin (DOX) + trastuzumab (TRZ) mediated cardiotoxicity

iron dependent cellular processes [9]. While these apoptotic mechanisms prevail in in vitro studies and several animal models, non-apoptotic mechanisms may also contribute to toxicity in the in vivo setting [9, 29, 30]. These include membrane lipid peroxidation, inhibition of nucleic acid and protein synthesis, the release of vasoactive amines, changes in adrenergic function and adenylate cyclase, abnormalities in calcium handling, transcriptional changes, mitochondrial creatine kinase (CK) aberrancies, and induction of nitric oxide synthase [9, 27].

TRZ induced cardiotoxicity is much less understood when compared to the adverse cardiovascular effects associated with DOX. HER2 is present in cardiac tissue and plays an important role in embryonic cardiogenesis and cardiac hypertrophy [31]. Additionally, disruption of the HER2 receptor in adult murine models results in a dilated cardiomyopathy [25]. The cardiotoxic risk of TRZ is associated with increased age and concurrent treatment with anthracyclines [32]. Pre-existing cardiac dysfunction, prior chest wall irradiation, HTN, diabetes, and a post-AC LVEF of 50–54% are proposed risk factors [32, 33]. Although the pathogenesis by which

TRZ results in cardiomyopathy is not well understood, three dominant mechanisms likely explain the majority of damage.

The first mechanism by which TRZ promotes cardiotoxicity is through the inhibition of HER heterodimer formation. In addition to its other elucidated roles, HER2 forms heterodimers with other HER, to facilitate more global signal transduction [25]. For example, HER3 and HER4 activity is impaired when the assembly of heterodimers with HER2 is interrupted [25]. HER heterodimerization initiates downstream signaling by PI3K/Akt, MAPK and endothelial nitric oxide synthase which serve pivotal functions in cell survival, mitochondrial function, sarcoplasmic reticulum calcium uptake, growth, and proliferation [25, 34]. TRZ also induces changes in the expression of DNA repair and stress response genes [25]. Secondly, apoptosis is hypothesized to play a role in TRZ cardiotoxicity. The addition of HER2 antibodies to cardiomyocytes for 24 h increased levels of the pro-apoptotic protein Bcl-xS and decreased levels of the corresponding anti-apoptotic protein Bcl-xL [35]. Ultimately, these effects led to Bax oligomerization, cytochrome c release, caspase activation, and mitochrondrial dysfunction [35]. Finally, TRZ may contribute to cardiac damage through the modulation of the renin-angiotensin system (RAS). TRZ inhibits NADPH oxidase, thus up-regulating angiotensin II and subsequently inhibiting neuregulin-1 [33, 36]. Neuregulin-1 promotes sarcomere stability and relieves oxidative stress contributing to cell survival [33]. Angiotensin II can also bind the angiotensin I receptor leading to the production of oxidative stress and activation of apoptosis signal-regulating kinase-1 to stimulate apoptosis [34, 37].

While the cardiotoxic potential of TRZ in isolation is controversial, it is well established that TRZ potentiates DOX mediated cardiotoxicity. As previously described, DOX stimulates the production of oxidative stress and reactive oxygen species, while the modulation of HER2 signaling by TRZ impedes survival pathways to potentiate DOX toxicity, ultimately leading to apoptosis and cardiomyocyte death [25]. To support this, animal models have demonstrated that HER signaling protects myocytes from anthracycline induced apoptosis. For example, mice were protected from daunorubicin induced apoptosis by treatment with neuregulin-1β and subsequent HER4 activation of the PI3K/Akt pathway [38].

The cardiotoxicity of TRZ in the adjuvant setting following anthracycline chemotherapy has been evaluated in a number of clinical trials [21, 39, 40]. After 11 years of follow-up, only 4.4% of individuals in the chemotherapy + TRZ treatment arm of the HERceptin Adjuvant trial met the secondary cardiac endpoint criteria of a decrease in LVEF by >10% to an absolute value of <50% [21]. Even fewer patients (1%) were severely symptomatic as defined by New York Heart Association (NYHA) class 3 or 4 symptoms [21]. Similar rates of cardiotoxicity were observed in the National Surgical Adjuvant Breast and Bowel Project (NSABP) B-31 trial. At a median follow-up of 8.8 years, 3.4% of patients in the chemotherapy + TRZ arm demonstrated a >10% decrease in LVEF to an absolute value <50% [39]. The exclusion of patients with cardiac risk factors may explain the favorable outcomes observed in trials such as these [25, 39].

Despite the low rates of DOX + TRZ mediated cardiotoxicity initially reported in clinical trials, subsequent real-world observational studies have revealed the adverse event to be far more prevalent. A 2007 observational study conducted in British Columbia followed 155 females treated with adjuvant TRZ in the setting of breast cancer [41]. A total of 21.6% of patients in the study experienced a cardiac event requiring discontinuation of TRZ, of which 4 cases represented clinical heart failure [41]. Of a total of 18 patients who experienced an asymptomatic decline in LVEF, 14 were re-challenged with TRZ and only 2 patients demonstrated further impairment in LVEF [41]. Similar findings were reported in a 2009 retrospective analysis of 152 patients with breast cancer treated with anthracycline based chemotherapy + TRZ in Winnipeg, Manitoba by our group [42]. A total of 24% of patients in this study developed TRZ mediated cardiotoxicity, the majority of which were asymptomatic [42]. As 1 in 4 females treated with DOX + TRZ will develop cardiotoxicity, early detection of adverse cardiac changes is important in the prevention of morbidity and mortality in this setting.

Detection of Chemotherapy Induced Cardiotoxicity

The high prevalence of chemotherapy induced cardiotoxicity in women with breast cancer necessitates early detection to prevent subsequent morbidity and mortality in this vulnerable patient population. Serial cardiac imaging to detect adverse changes constitutes the mainstay of monitoring in women with breast cancer undergoing DOX + TRZ therapy [43]. Unfortunately, the optimal surveillance schedule is unknown and current practice follows research protocols and consensus opinion [23]. Most commonly, LVEF is quantified every 3 months during DOX + TRZ therapy by radionuclide angiography (multiple gated acquisition scan; MUGA), transthoracic echocardiography (TTE), or cardiac MRI (CMR; Table 13.2) [23].

Table 13.2 Comparison of the cardiac modalities used to monitor for chemotherapy mediated cardiac dysfunction: multiple gated acquisition scan (MUGA), transthoracic echocardiography (TTE), and cardiac magnetic resonance imaging (CMR)

	MUGA	TTE	CMR
Spatial Resolution	6–8 mm	3–4 mm	1–2 mm
Temporal Resolution	8–16 frames/s	30–120 frames/s (2D) 1000–5000 frames/s (M mode)	10 frames/s
Portable	No	Yes	No
Radiation	Yes	No	No
mSv	5	0	0
CXR equivalents	10–15	0	0
Potential for renal impairment	No	No	Yes

MUGA scans are commonly employed to detect adverse cardiac changes while breast cancer patients are receiving chemotherapy due to the favorable reproducibility and accuracy of this imaging modality [44]. Despite this, MUGA scans are limited by high radiation exposure, with each study delivering roughly the same radiation as computed tomography (CT) of the chest; approximately equivalent to 10–15 chest x-rays per scan [45]. Furthermore, MUGA scans lack detailed structural information on valvular function, wall motion abnormalities, and pericardial disease and demonstrate limited accuracy in patients with underlying cardiac arrhythmias [43].

TTE is endorsed by the Canadian Cardiovascular Society (CCS) as the imaging modality of choice to detect chemotherapy induced cardiomyopathy, as it is widely available and lacks ionizing radiation [23, 43]. Further benefits include visualization of valvular structure and function along with the non-invasive assessment of the pericardium [23]. Despite these favorable characteristics, the accuracy of two dimensional (2-D) echocardiography is limited. A study of 54 patients with cancer undergoing chemotherapy evaluated the efficacy of both 2-D and 3-D echocardiography [46]. Study patients were selected on the basis of having stable cardiac function (as evaluated by stable global longitudinal strain over 12 months) and underwent LVEF assessment by both echocardiographic modalities over time [46]. As it was assumed that cardiac function remained stable, any variability was attributed to echocardiographic inaccuracy. From this study, the variability was reported to be 6% for 3-D echocardiography and >10% for 2-D echocardiography [46]. A similar study conducted in survivors of childhood cancer compared the efficacy of 2-D and 3-D TTE to CMR as the gold standard to assess cardiac structure and function [47]. As compared to CMR, the sensitivity to detect a LVEF <50% was 25% and 53% for 2-D and 3-D echocardiography, respectively [47]. The efficacy of serial cardiac imaging was also assessed by our group in 50 females with HER2 positive breast cancer treated with DOX + TRZ in the adjuvant setting [44]. Imaging by MUGA, 2-D TTE, and real time 3-D TTE was conducted at baseline, 6, and 12 months after TRZ therapy and compared to CMR as the gold standard [44]. Whereas 2-D TTE demonstrated a weak correlation with CMR ($r = 0.42$), MUGA and 3-D TTE correlated strongly with CMR at 12 months ($r = 0.90$ and $r = 0.91$, respectively) [44]. As such, 3-D TTE has been suggested as superior to 2-D TTE for the detection of chemotherapy induced cardiotoxicity [23]. The use of contrast may further improve LVEF assessment by TTE and should be employed when more than 2 contiguous LV segments are not well visualized on any apical view [23]. A study of 51 patients referred for MUGA because of poor echocardiographic windows, compared 2-D TTE with and without contrast [48]. A LVEF assessment was only possible in 31% of studies without contrast as compared to 100% of cases in which contrast was administered [48]. Furthermore, LVEF as determined by 2-D TTE with contrast strongly correlated with radionucleotide derived values [48].

CMR is considered to be the gold standard of non-invasive cardiac imaging modalities, as the sample size required to detect changes in LVEF is considerably smaller than with TTE [23, 49]. Additional benefits include the ability to detect myocardial edema, perfusion abnormalities, and cardiac fibrosis. Characteristically, TRZ induced cardiomyopathy demonstrates subepicardial linear late gadolinium enhancement on CMR [50]. In a case study performed by our group including 10 women with DOX + TRZ mediated cardiotoxicity detected by MUGA, all patients demonstrated late gadolinium enhancement on CMR [50]. Additionally, the delayed enhancement findings persisted at 6 months despite an improvement of LV systolic function in 6 patients [50]. Despite the low intra- and inter-observer variability of LVEF assessment by CMR, it is limited by universal access and cost, often precluding its routine use for monitoring cardiac function in the setting of breast cancer.

The imaging modality employed for the non-invasive monitoring of cardiac function during chemotherapy may vary as a result of patient specific factors. However, LVEF should be followed serially using the same modality over time as measurements by TTE, MUGA, and CMR may differ. For example, the mean LVEF in 52 patients with chronic stable heart failure was 31 ± 10%, 24 ± 9%, and 30 ± 11% by 2-D echocardiography, MUGA, and CMR respectively [51].

Current monitoring programs to detect chemotherapy induced cardiac dysfunction follow LVEF as a parameter. Unfortunately, a decrease in LVEF is a late finding in chemotherapy induced cardiotoxicity, and is detected only after global cardiomyocyte damage has occurred [23]. Ideally, early identification of adverse changes before structural damage would allow for the discontinuation of chemotherapy and the implementation of cardioprotective strategies to ultimately prevent chemotherapy related cardiotoxicity. Tissue velocity imaging and strain imaging are sensitive echocardiographic techniques that allow for early detection of LV systolic dysfunction. Tissue velocity imaging is a modification of conventional Doppler to detect high amplitude, low velocity Doppler signals derived from tissue [43]. Strain quantifies the deformation of a region of myocardium and can describe the change in length in one direction (i.e. longitudinal) or describe LV function globally by applying speckle tracking [52]. Tissue velocity imaging has been shown to detect early adverse cardiac changes in a murine model of DOX + TRZ mediated cardiotoxicity [53]. In this study, progressive LV systolic dysfunction was observed in the DOX + TRZ treatment arm, as compared to a preserved LVEF observed in control mice [53]. Tissue velocity imaging was adversely affected as early as 24 h in DOX + TRZ treated mice, well before LV dilatation and LV systolic dysfunction were observed on day 4 [53].

Similar findings have been replicated in the clinical setting of DOX + TRZ mediated cardiotoxicity. A study by our group involved 42 women with HER2 positive breast cancer treated with TRZ in the adjuvant setting after anthracycline chemotherapy [54]. A total of 24% of study participants met criteria for DOX + TRZ mediated cardiomyopathy, defined as a decrease in LVEF of at least 10% to an absolute value of <55% and symptoms consistent with congestive heart failure [54]. A decline in LVEF was only detected at 6 months in the subset of women who developed a cardiomyopathy (mean LVEF of 42 ± 9% compared to a mean LVEF of 64 ± 4% in the cohort without cardiac dysfunction) [54]. A further decrease in mean LVEF to 39 ± 5% was observed at 9 months in the cardiomyopathy group, despite discontinuation of TRZ [54]. At 12 months, partial functional recovery was observed (mean LVEF of 49 ± 4%) [54]. In contrast, tissue velocity imaging and strain parameters detected impending cardiac dysfunction as early as 3 months after TRZ therapy [54]. A threshold of 0.6 cm/s for the lateral S' of the mitral valve (difference between baseline and 3 months) and 2% for longitudinal strain (difference between baseline and 3 months) captured the majority of patients who developed chemotherapy induced cardiomyopathy [54]. Additionally, there were no false positive results by tissue velocity imaging and only 3 by strain parameters [54]. Similar findings were reproduced in a study of 35 women with HER2 positive breast cancer undergoing adjuvant TRZ therapy [55]. At 12 months of follow-up, a reduction in tissue velocity imaging and strain rate were observed in a subset of patients, despite all participants demonstrating a preserved LVEF [55]. Of the 18 patients with reduced longitudinal strain rate, 5 women developed chemotherapy mediated cardiac dysfunction, with a reduction in LVEF at 20 months [55]. Finally, a systematic review described the utility of myocardial deformation parameters in 1504 patients treated with chemotherapy [56]. Changes in deformation, including tissue velocity imaging and speckle tracking echocardiography, were uniformly detected well in advance to a measurable decline in LVEF [56]. In this review, a 10–15% decrease in global longitudinal strain was the most useful parameter for predicting future systolic dysfunction [56]. Several additional studies corroborate the potential for tissue velocity imaging and strain rate to detect early changes consistent with chemotherapy induced cardiomyopathy [57–59]. In addition to early detection using non-invasive cardiovascular imaging, preventative strategies are a novel area of research with the potential to decrease the morbidity and mortality associated with chemotherapy mediated cardiotoxicity.

Prevention of DOX + TRZ Mediated Cardiotoxicity

The morbidity and mortality associated with chemotherapy mediated cardiotoxicity has necessitated the search for a prophylactic agent to prevent these devastating consequences. Potential agents investigated thus far include angiotensin converting enzyme inhibitors (ACEi)/angiotensin receptor blockers (ARB), β-blockers, statins, anti-oxidants, and nutraceuticals.

ACEi and ARBs constitute the mainstay of heart failure therapy. The 2017 CCS heart failure guidelines strongly recommend ACEi/ARB therapy in the treatment of heart failure with reduced EF (HFrEF), as large randomized control trials have consistently demonstrated improvement in both patient morbidity and mortality [60]. These favorable effects are largely attributed to the blockade of negative downstream effects of angiotensin II, such as fibroblast hypertrophy, collagen deposition, and subsequent detrimental cardiovascular remodeling [61]. The beneficial effects of ACEi/ARB therapy in established HFrEF and the potential implication of the RAS in chemotherapy mediated toxicity have led to the study of ACEi/ARB in the prevention of DOX + TRZ cardiotoxicity [33, 36]. In a rat model of DOX induced cardiomyopathy, co-administration of an ACEi (captopril) or ARB (telmisartan) reduced biochemical markers of cardiotoxicity (lactate dehydrogenase; LDH and creatine kinase; CK) and attenuated oxidative stress [62]. A similar study was conducted in 240 mice randomized to prophylactic placebo, direct renin inhibition (aliskiren), ACEi (perindopril) or ARB (valsartan) prior to administration of chemotherapy with DOX, TRZ, or the combination of DOX + TRZ [63]. Echocardiography was conducted weekly to monitor cardiac structure and function for the duration of the 13 week study [63]. Mice in the placebo arm demonstrated an increase in LV end diastolic diameter (LVEDD) from 3.1 ± 0.2 mm at baseline to 4.6 ± 0.3 mm at week 13 ($p < 0.05$) [63]. Prophylactic treatment with aliskiren, perindopril, or valsartan attenuated the degree of LV cavity dilatation with LVEDD dimensions of 3.9 ± 0.2 mm, 4.1 ± 0.2 mm, and 4.2 ± 0.1 mm at week 13, respectively ($p < 0.05$) [63]. The LV fractional shortening (LVFS) values corroborated these findings. At week 13, LVFS significantly decreased to 26 ± 2% in placebo treated mice, but this effect was attenuated by the prophylactic administration of aliskiren (40 ± 1%), perindopril (32 ± 1%), or valsartan (33 ± 2%) ($p < 0.05$) [63].

In addition to ACEi/ARB therapy, β-blockers have also consistently demonstrated decreased hospitalization rates with improved long-term survival in the setting of HFrEF [60]. Sympathetic stimulation of β-adrenergic receptors in heart failure leads to further decreases in LVEF, arrhythmias, and tachycardia [64]. Harmful peripheral consequences include activation of the RAS and subsequent fluid retention [64]. In the setting of DOX induced cardiomyopathy, β2-receptors are up-regulated, and as such serve as a potential target for cardioprotection [65]. In a murine model, the beta blocker carvedilol attenuated DOX induced changes in LVEDD and LVEF after 35 days [66]. Furthermore, markers of fibrosis, apoptosis, oxidative stress, and mitochondrial damage were decreased in carvedilol treated mice [66]. In a small clinical trial, 50 patients undergoing anthracycline chemotherapy were randomized to placebo or prophylactic treatment with carvedilol [67]. After 6 months of follow-up, 1 patient died in the carvedilol arm as compared to 4 patients in the control group [67]. Additionally, the mean LVEF remained similar to baseline value in the carvedilol group, while patients in the control group demonstrated a significantly lower mean LVEF by the end of the study period [67]. Only 1 patient in the carvedilol arm demonstrated a LVEF <50%, as compared to 5 patients in the control group [67].

Several clinical studies have investigated the prophylactic use of ACEi/ARB therapy and β-blockade in the prevention of chemotherapy induced cardiac dysfunction with differing results. In the OVERCOME trial, 90 patients with acute leukemia or another hematologic malignancy undergoing hematopoietic stem cell transplantation were randomized to enalapril (ACEi) + carvedilol (β-blocker) or placebo [68]. Cardiac imaging by TTE or CMR was conducted at baseline and after 6 months of follow-up [68]. At the conclusion of the study, the LVEF did not significantly change in the intervention group, but did significantly decrease in the control group (−3.1% by TTE and −3.4% by CMR) [68]. Additionally, the composite endpoint of heart failure and death was significantly reduced in the enalapril + carvedilol arm at 6.7%, compared to 22% in the placebo group [68]. The PRADA trial was a placebo-controlled double blind trial in which 121 women with breast cancer treated with adjuvant epirubicin were randomized to concomitant metoprolol (β-blocker), candesartan (ARB), metoprolol + candesartan, or placebo [69]. Troponin levels were followed as a biomarker of cardiac damage after anthracycline administration and were attenuated by prophylactic treatment with metoprolol but not candesartan [69]. Additionally, no correlation was observed between troponin levels and changes in LV function by cardiac imaging, raising question to the relevance of the findings [69]. More recently, the MANTICORE study randomized 94 women with HER2 positive breast cancer to perindopril (ACEi), bisoprolol (β-blocker), or placebo during adjuvant therapy with TRZ [70]. CMR imaging was conducted at baseline and after 17 cycles of TRZ [70]. Unfortunately, no improvement in the primary outcome of LV end diastolic volume (LVEDV) was observed in the intervention groups. After 17 cycles of TRZ, LVEDV was 7 ± 14 mL/m^2 for the perindopril arm, 8 ± 9 mL/m^2 for the bisoprolol arm, and 4 ± 11 mL/m^2 for the placebo arm [70]. Bisoprolol was partially cardioprotective, a decrease in LVEF by only $-1 \pm 5\%$ was observed compared to a decrease by $-3 \pm 4\%$ and $-5 \pm 5\%$ in the perindopril and placebo arms, respectively [70]. Whether the prophylactic administration of ACEi/ARB and β-blockers can prevent chemotherapy mediated cardiotoxicity warrants further study.

Statins are 3-hydroxy-3-methylglutaryl-coenzyme A (HMG-CoA) reductase inhibitors that are widely prescribed to treat hypercholesterolemia [71]. Statins lower cholesterol by decreasing hepatic cholesterol synthesis and increasing hepatic uptake of low-density lipoprotein (LDL) [71]. At lower doses, statins demonstrate anti-inflammatory and anti-oxidant properties, such as improved endothelial function, cardioprotection, and anti-tumor action [71]. For example, in mice treated with atorvastatin, DOX induced oxidative stress, DNA, and cellular damage were reduced [71]. An observational study involving 628 women with newly diagnosed breast cancer treated with anthracycline chemotherapy compared heart failure hospitalization in patients receiving statin therapy to controls [72]. A hazard ratio of 0.3 was calculated for patients on uninterrupted statin therapy [72]. Similar findings were reported for 51 patients treated with anthracycline based chemotherapy for breast cancer, leukemia or lymphoma followed by CMR imaging for 6 months.[773] The 14 individuals on statin therapy, who were more likely to be older and comorbid with hypertension or hyperlipidemia, demonstrated a preserved LVEF at study completion ($56.6\% \pm 1.4\%$ at baseline, $54.1\% \pm 1.3\%$ at 6 months, $p = 0.15$) as compared to those not on statin therapy ($57.5\% \pm 1.4\%$ at baseline, $52.4\% \pm 1.2\%$ at 6 months; $p = 0.0003$) [73]. More recently, women with HER2 positive breast cancer treated with TRZ ± anthracycline chemotherapy on a statin were retrospectively identified and matched to 2 patients of the same age and anthracycline exposure [74]. The median decrease in LVEF was -6% in the control group, as compared to 0% in the statin group [74]. These studies demonstrate that statins may play an important cardioprotective role in the prevention of chemotherapy mediated cardiotoxicity, however further prospective study is warranted.

Anti-oxidants have been explored as prophylactic agents in the prevention of DOX + TRZ cardiotoxicity given the prominent role of oxidative stress in this process. Although Probucol is a lipid lowering agent used to treat clinical atherosclerosis, its use is limited by an undesirable high-density lipoprotein (HDL) lowering effect [75]. Probucol has demonstrated anti-oxidant properties and may inhibit free-radical lipid peroxidation [75]. In a mouse model of DOX + TRZ cardiotoxicity, 114 mice were treated with Probucol or placebo [76]. Probucol was administered on alternating days for 2 weeks prior to the chemotherapeutic challenge, after which mice were followed by TTE for 10 days [76]. Mortality was significantly decreased from >80% in the placebo group to 40% in the Probucol group at study completion [76]. Prophylactic treatment with Probucol further attenuated adverse LV morphological changes and histological findings of myofibrillar degeneration and vacuolization [76].

N-acetyl-cysteine (NAC), an anti-oxidant commonly used in the setting of Tylenol overdose, was first studied in the setting of DOX induced cardiotoxicity in 1983 [77]. A total of 20 patients were randomized to receive placebo or NAC 1 h prior to the administration of DOX [77]. Endomyocardial biopsies were completed at baseline, 4, and 24 h after chemotherapy [77]. No difference in tubular area or mitochondrial swelling was observed between the two groups and the conclusion was drawn that NAC was not an effective cardioprotective agent in this setting [77]. Contemporary studies have shifted focus to the amide derivative of NAC (NACA). NACA is more bioavailable and readily replenishes intracellular GSH stores in comparison to NAC in a human red blood cell model of tert-butylhydroxyperoxide induced oxidation [78]. In vitro, NACA decreased reactive oxygen species and lipid peroxidation and increased GSH in DOX treated cardiomyocytes [79]. However, NACA did not protect against cell death in this model [79]. In vivo, we conducted a study in which 100 female mice treated with DOX + TRZ received prophylactic placebo or NACA treatment during a 10 day study [80]. Although the LVEF decreased from 72 $\pm$ 3% to 32 $\pm$ 2% in the DOX + TRZ group, it was partially preserved at 55 $\pm$ 3% in those mice pre-treated with NACA [80]. NACA also attenuated the loss of cellular integrity, oxidative stress, and apoptosis in histological and biochemical analyses [80]. To date, no clinical studies have evaluated the efficacy of NACA in the prevention of DOX + TRZ cardiotoxicity.

Finally, nutraceuticals including flaxseed are a prophylactic option in the prevention of DOX + TRZ induced cardiotoxicity. Flaxseed contains plant lignans, naturally occurring chemicals, and the omega-3 fatty acid α-linolenic acid (ALA) which possesses anti-inflammatory and anti-oxidative properties [81, 82]. The lignan secoisolariciresinol diglucoside (SDG) has proven effective against radical-induced DNA damage and liposomal lipid peroxidation in vitro.[83] In a mouse model, ALA attenuated DOX induced increases in cardiac weight, brain natriuretic peptide (BNP), CK, LDH, and troponin levels, and protected against adverse myocardial necrosis, LVEDD, and LVEF changes [84]. Additionally in a study of DOX + TRZ induced cardiotoxicity, 195 mice received daily prophylactic treatment with regular chow, flaxseed, ALA or SDG for 6 weeks [85]. LVEF decreased significantly from 73 $\pm$ 2% to 38 $\pm$ 2% in mice administered regular chow at 6 weeks [85]. Of interest, LVEF was partially preserved at 61 $\pm$ 2%, 60 $\pm$ 3% and 61 $\pm$ 4% in the flaxseed, ALA or SDG treated groups, respectively [85]. Furthermore, flaxseed and ALA attenuated inflammation, as measured by COX-derived oxylipins, and all three treatments attenuated OS and apoptosis, as represented by 8,9-DiHETrE and Bax/Bcl-xL levels, respectively [85].

Conclusion

Chemotherapy mediated cardiotoxicity remains a unique risk to females. As 1 in 4 women with HER2 positive breast cancer may develop cardiotoxicity, early detection and prevention remain important concepts in decreasing the associated morbidity and mortality. The CCS recommends consideration of ACEi/ARB, β-blocker, and/or statin in this high risk population [23]. Future research studies in the evolving field of Cardio-Oncology will continue to focus on developing strategies, programs, and guidelines to prevent cardiovascular complications in women with breast cancer.

References

1. Canadian Cancer Statistics Advisory Committee. Canadian Cancer Statistics 2019. Toronto, ON: Canadian Cancer Society; 2019. cancer.ca/Canadian-Cancer-Statistics-2019-EN. Accessed 15 October 2019
2. Crawford SL, Crandall CJ, Derby CA et al (2018) Menopausal hormone therapy trends before versus after 2002: impact of the women's health initiative study results. Menopause 26:588–597
3. Klarenbach S, Sims-Jones N, Lewin G et al (2018) Recommendations on screening for breast cancer in women aged 40–74 years who are not at increased risk for breast cancer. CMAJ 190:E1441–E1451
4. Pasche B (2008) Recent advances in breast cancer genetics. Cancer Treat Res 141:1–10
5. Kuchenbaecker KB, Hopper JL, Barnes DR et al (2017) Risks of breast, ovarian, and contralateral breast cancer for BRCA1 and BRCA2 mutation carriers. JAMA 317:2402–2416
6. Warner E, Plewes DB, Hill KA (2004) Surveillance of BRCA1 and BRCA2 mutation carriers with magnetic resonance imaging, ultrasound, mammography, and clinical breast examination. JAMA 292:1317–1325
7. Fisher B, Anderson S, Bryant J et al (2002) Twenty-year follow-up of a randomized trial comparing total mastectomy, lumpectomy, and lumpectomy plus irradiation for the treatment of invasive breast cancer. N Engl J Med 347:1233–1241
8. an overview of the randomised trials (2005) Early Breast Cancer Trialists' Collaborative Group (EBCTCG). Effects of chemotherapy and hormonal therapy for early breast cancer on recurrence and 15-year survival. Lancet 365:1687–1717
9. Minotti G, Menna P, Salvatorelli E (2004) Anthracyclines: molecular advances and pharmacologic developments in antitumor activity and cardiotoxicity. Pharmacol Rev 56:185–229
10. Gewirtz DA (1999) A critical evaluation of the mechanisms of action proposed for the antitumor effects of the anthracycline antibiotics adriamycin and daunorubicin. Biochem Pharmacol 57:727–741
11. Song G, Ouyang G, Bao S (2005) The activation of Akt/PKB signaling pathway and cell survival. J Cell Mol Med 9:59–71
12. Waks AG, Winer EP (2019) Breast cancer treatment: a review. JAMA 321:288–300
13. P atient-level meta-analysis of randomised trials (2011) Early Breast Cancer Trialists' Collaborative Group (EBCTCG), Davies C, Godwin J, et al. Relevance of breast cancer hormone receptors and other factors to the efficacy of adjuvant tamoxifen. Lancet 378:771–784
14. Davies C, Pan H, Godwin J et al (2013) Long-term effects of continuing adjuvant tamoxifen to 10 years versus stopping at 5 years after diagnosis of oestrogen receptor-positive breast cancer: ATLAS, a randomised trial. Lancet 381:805–816
15. Goss PE, Ingle JN, Pritchard KI et al (2016) Extending Aromatase-inhibitor adjuvant therapy to 10 years. N Engl J Med 375:209–219

16. Zhang Y, Opresko L, Shankaran H et al (2009) HER/ErbB receptor interactions and signaling patterns in human mammary epithelial cells. BMC Cell Biol 10:78–92
17. Joslin EJ, Opresko LK, Wells A et al (2007) EGF-receptor-mediated mammary epithelial cell migration is driven by sustained ERK signaling from autocrine stimulation. J Cell Sci 120:3688–3699
18. Wu CJ, Qian X, O'Rourke DM (1999) Sustained mitogen-activated protein kinase activation is induced by transforming erbB receptor complexes. DNA Cell Biol 18:731–741
19. Plosker GL, Keam SJ (2006) Trastuzumab: a review of its use in the management of HER2-positive metastatic and early-stage breast cancer. Drugs 66:449–475
20. Slamon DJ, Leyland-Jones B, Shak S et al (2001) Use of chemotherapy plus a monoclonal antibody against HER2 for metastatic breast cancer that overexpresses HER2. N Engl J Med 344:783–792
21. Cameron D, Piccart-Gebhart MJ, Gelber RD et al (2017) 11 years' follow-up of trastuzumab after adjuvant chemotherapy in HER2-positive early breast cancer: final analysis of the HERceptin Adjuvant (HERA) trial. Lancet 389:1195–1205
22. Romond EH, Perez EA, Bryant J et al (2005) Trastuzumab plus adjuvant chemotherapy for operable HER2-positive breast cancer. N Engl J Med 353:1673–1684
23. Virani SA, Dent S, Brezden-Masley C et al (2016) Canadian cardiovascular society guidelines for evaluation and management of cardiovascular complications of cancer therapy. Can J Cardiol 32:831–841
24. Plana JC, Galderisi M, Barac A et al (2014) Expert consensus for multimodality imaging evaluation of adult patients during and after cancer therapy: a report from the American Society of Echocardiography and the European Association of Cardiovascular Imaging. J Am Soc Echocardiogr 27:911–939
25. Nemeth BT, Varga ZV, Wu WJ, Pacher P (2017) TRZ cardiotoxicity: from clinical trials to experimental studies. Br J Pharmacol 174:3727–3748
26. Singal PK, Iliskovic N (1998) Doxorubicin-induced cardiomyopathy. N Engl J Med 339:900–905
27. Singal PK, Iliskovic N, Li T, Kumar D (1997) Adriamycin cardiomyopathy: pathophysiology and prevention. FASEB J 11:931–936
28. Yamaoka M, Yamaguchi S, Suzuki T et al (2000) Apoptosis in rat cardiac myocytes induced by Fas ligand: priming for Fas-mediated apoptosis with doxorubicin. J Mol Cell Cardiol 32:881–889
29. Kumar D, Kirshenbaum LA, Li T et al (2001) Apoptosis in adriamycin cardiomyopathy and its modulation by probucol. Antioxid Redox Signal 3:135–145
30. Childs AC, Phaneuf SL, Dirks AJ et al (2002) Doxorubicin treatment in vivo causes cytochrome C release and cardiomyocyte apoptosis, as well as increased mitochondrial efficiency, superoxide dismutase activity, and Bcl-2: Bax ratio. Cancer Res 62:4592–4598
31. Keefe DL (2002) Trastuzumab-associated cardiotoxicity. Cancer 95:1592–1600
32. Jawa Z, Perez RM, Garlie L (2016) Risk factors of trastuzumab-induced cardiotoxicity in breast cancer: a meta-analysis. Medicine (Baltimore) 95: e5195
33. Mitri Z, Constantine T, O'Regan R (2012) The HER2 receptor in breast cancer: pathophysiology, clinical use, and new advances in therapy. Chemother Res Pract 2012:743193
34. Zeglinski M, Ludke A, Jassal DS, Singal PK (2011) Trastuzumab-induced cardiac dysfunction: a 'dual-hit' . Exp Clin Cardiol 16:70–74
35. Grazette LP, Boecker W, Matsui T et al (2004) Inhibition of ErbB2 causes mitochondrial dysfunction in cardiomyocytes: implications for herceptin-induced cardiomyopathy. J Am Coll Cardiol 44:2231–2238
36. Nakagami H, Takemoto M, Liao JK (2003) NADPH oxidase-derived superoxide anion mediates angiotensin II-induced cardiac hypertrophy. J Mol Cell Cardiol 35:851–859
37. Lemarié CA, Paradis P, Schiffrin EL (2008) New insights on signaling cascades induced by cross-talk between angiotensin II and aldosterone. J Mol Med (Berl) 86:673–678
38. Fukazawa R, Miller TA, Kuramochi Y et al (2003) Neuregulin-1 protects ventricular myocytes from anthracycline-induced apoptosis via erbB4-dependent activation of PI3-kinase/Akt. J Mol Cell Cardiol 35:1473–1479

39. Ganz PA, Romond EH, Cecchini RS et al (2017) Long-term follow-up of cardiac function and quality of life for patients in NSABP Protocol B-31/NRG Oncology: A randomized trial comparing the safety and efficacy of Doxorubicin and Cyclophosphamide (AC) followed by Paclitaxel With AC followed by Paclitaxel and Trastuzumab in patients with node-positive breast cancer with tumors overexpressing human epidermal growth factor receptor 2. J Clin Oncol 35:3942–3948
40. Advani PP, Ballman KV, Dockter TJ et al (2016) Long-term cardiac safety analysis of NCCTG N9831 (Alliance) adjuvant Trastuzumab trial. J Clin Oncol 34:581–587
41. McArthur HL, Chia S (2007) Cardiotoxicity of trastuzumab in clinical practice. N Engl J Med 357: 94–5
42. Wadhwa D, Fallah-Rad N, Grenier D et al (2009) Trastuzumab mediated cardiotoxicity in the setting of adjuvant chemotherapy for breast cancer: a retrospective study. Breast Cancer Res Treat 117:357–364
43. Saeed MF, Premecz S, Goyal V et al (2014) Catching broken hearts: pre-clinical detection of doxorubicin and trastuzumab mediated cardiac dysfunction in the breast cancer setting. Can J Physiol Pharmacol 92:546–550
44. Walker J, Bhullar N, Fallah-Rad N et al (2010) Role of three-dimensional echocardiography in breast cancer: comparison with two-dimensional echocardiography, multiple-gated acquisition scans, and cardiac magnetic resonance imaging. J Clin Oncol 28:3429–3436
45. Kaul P, Medvedev S, Hohmann SF et al (2010) Ionizing radiation exposure to patient s admitted with acute myocardial infarction in the united states. Circulation 122:2160–2169
46. Thavendiranathan P, Grant AD, Negishi T et al (2013) Reproducibility of echocardiographic techniques for sequential assessment of left ventricular ejection fraction and volumes: application to patients undergoing cancer chemotherapy. J Am Coll Cardiol 61:77–84
47. Armstrong GT, Plana JC, Zhang N et al (2012) Screening adult survivors of childhood cancer for cardiomyopathy: comparison of echocardiography and cardiac magnetic resonance imaging. J Clin Oncol 30:2876–2884
48. Yu EH, Sloggett CE, Iwanochko RM et al (2000) Feasibility and accuracy of left ventricular volumes and ejection fraction determination by fundamental, tissue harmonic, and intravenous contrast imaging in difficult-to-image patients. J Am Soc Echocardiogr 13:216–224
49. Strohm O, Schultz-Menger J, Pilz B et al (2001) Measurement of left ventricular dimensions and function in patients with dilated cardiomyopathy. J Magn Reson Imaging 13:367–371
50. Fallah-Rad N, Lytwyn M, Fang T et al (2008) Delayed contrast enhancement cardiac magnetic resonance imaging in trastuzumab induced cardiomyopathy. J Cardiovasc Magn Reson 10:5
51. Bellenger NG, Burgess MI, Ray SG et al (2000) Comparison of left ventricular ejection fraction and volumes in heart failure by echocardiography, radionuclide ventriculography and cardiovascular magnetic resonance; are they interchangeable? Eur Heart J 21:1387–1396
52. Mirea O, Duchenne J, Voigt JU (2016) Recent advances in echocardiography: strain and strain rate imaging. F1000Res 29
53. Jassal DS, Han SY, Hans C et al (2009) Utility of tissue Doppler and strain rate imaging in the early detection of trastuzumab and anthracycline mediated cardiomyopathy. J Am Soc Echocardiogr 22:418–424
54. Fallah-Rad N, Walker JR, Wassef A et al (2011) The utility of cardiac biomarkers, tissue velocity and strain imaging, and cardiac magnetic resonance imaging in predicting early left ventricular dysfunction in patients with human epidermal growth factor receptor II-positive breast cancer treated with adjuvant trastuzumab therapy. J Am Coll Cardiol 57:2263–2270
55. Hare JL, Brown JK, Leano R et al (2009) Use of myocardial deformation imaging to detect preclinical myocardial dysfunction before conventional measures in patients undergoing breast cancer treatment with trastuzumab. Am Heart J 158:294–301
56. Thavendiranathan P, Poulin F, Lim KD et al (2014) Use of myocardial strain imaging by echocardiography for the early detection of cardiotoxicity in patients during and after cancer chemotherapy: a systematic review. J Am Coll Cardiol 63:2751–2768

57. Ho E, Brown A, Barrett P et al (2010) Subclinical anthracycline- and trastuzumab-induced cardiotoxicity in the long-term follow-up of asymptomatic breast cancer survivors: a speckle tracking echocardiographic study. Heart 96:701–707
58. Di Lisi D, Bonura F, Macaione F et al (2011) Chemotherapy-induced cardiotoxicity: role of the tissue Doppler in the early diagnosis of left ventricular dysfunction. Anticancer Drugs 22:468–472
59. Sawaya H, Sebag IA, Plana JC et al (2011) Early detection and prediction of cardiotoxicity in chemotherapy-treated patients. Am J Cardiol 107:1375–1380
60. Ezekowitz JA, O'Meara E, McDonald MA et al (2017) Comprehensive update of the canadian cardiovascular society guidelines for the management of heart failure. Can J Cardiol 33:1342–1433
61. Sayer G, Bhat G (2014) The renin-angiotensin-aldosterone system and heart failure. Cardiol Clin 32:21–32
62. Ibrahim MA, Ashour OM, Ibrahim YF et al (2009) Angiotensin-converting enzyme inhibition and angiotensin AT(1)-receptor antagonism equally improve doxorubicin-induced cardiotoxicity and nephrotoxicity. Pharmacol Res 60:373–381
63. Akolkar G, Bhullar N, Bews H et al (2015) The role of renin angiotensin system antagonists in the prevention of DOX+TRZ induced cardiotoxicity. Cardiovasc Ultrasound 13:18
64. Kemp CD, Conte JV (2012) The pathophysiology of heart failure. Cardiovasc Pathol 21:365–371
65. Merlet N, Piriou N, Rozec B et al (2013) Increased beta2-adrenoceptors in doxorubicin-induced cardiomyopathy in rat. PLoS ONE 8:e64711
66. Chen YL, Chung SY, Chai HT et al (2015) Early administration of Carvedilol protected against Doxorubicin-Induced Cardiomyopathy. J Pharmacol Exp Ther 355:516–527
67. Kalay N, Basar E, Ozdogru I et al (2006) Protective effects of carvedilol against anthracycline-induced cardiomyopathy. J Am Coll Cardiol 48:2258–2262
68. Bosch X, Rovira M, Sitges M et al (2013) Enalapril and carvedilol for preventing chemotherapy-induced left ventricular systolic dysfunction in patients with malignant hemopathies: the OVERCOME trial (preventiOn of left Ventricular dysfunction with Enalapril and caRvedilol in patients submitted to intensive ChemOtherapy for the treatment of Malignant hEmopathies). J Am Coll Cardiol 61:2355–2362
69. Gulati G, Heck SL, Røsjø H et al (2017) Neurohormonal blockade and circulating cardiovascular biomarkers during anthracycline therapy in breast cancer patients: results From the PRADA (Prevention of Cardiac Dysfunction During Adjuvant Breast Cancer Therapy) study. J Am Heart Assoc 6
70. Pituskin E, Mackey JR, Koshman S et al (2017) Multidisciplinary approach to novel therapies in cardio-oncology research (MANTICORE 101-Breast): a randomized trial for the prevention of Trastuzumab-associated cardiotoxicity. J Clin Oncol 35:870–877
71. Ramanjaneyulu SV, Trivedi PP, Kushwaha S et al (2013) Protective role of atorvastatin against doxorubicin-induced cardiotoxicity and testicular toxicity in mice. J Physiol Biochem 69:513–525
72. Seicean S, Seicean A, Plana JC et al (2012) Effect of statin therapy on the risk for incident heart failure in patients with breast cancer receiving anthracycline chemotherapy: an observational clinical cohort study. J Am Coll Cardiol 60:2384–2390
73. Chotenimitkhun R, D'Agostino R Jr, Lawrence JA et al (2015) Chronic statin administration may attenuate early anthracycline-associated declines in left ventricular ejection function. Can J Cardiol 31:302–307
74. Calvillo-Argüelles O, Abdel-Qadir H, Michalowska M et al (2019) Cardioprotective effect of statins in patients With HER2-positive breast cancer receiving Trastuzumab therapy. Can J Cardiol 35:153–159
75. Yamashita S, Masuda D, Matsuzawa Y (2015) Did we abandon probucol too soon? Curr Opin Lipidol 26:304–316
76. Walker JR, Sharma A, Lytwyn M et al (2011) The cardioprotective role of probucol against anthracycline and trastuzumab-mediated cardiotoxicity. J Am Soc Echocardiogr 24:699–705

77. Unverferth DV, Jagadeesh JM, Unverferth BJ et al (1983) Attempt to prevent doxorubicin-induced acute human myocardial morphologic damage with acetylcysteine. J Natl Cancer Inst 71:917–920
78. Grinberg L, Fibach E, Amer J, Atlas D (2005) N-acetylcysteine amide, a novel cell-permeating thiol, restores cellular glutathione and protects human red blood cells from oxidative stress. Free Radic Biol Med 38:136–145
79. Shi R, Huang CC, Aronstam RS et al (2009) N-acetylcysteine amide decreases oxidative stress but not cell death induced by doxorubicin in H9c2 cardiomyocytes. BMC Pharmacol 9:7
80. Goyal V, Bews H, Cheung D et al (2016) The cardioprotective role of N-Acetyl Cysteine Amide in the prevention of Doxorubicin and Trastuzumab-Mediated cardiac dysfunction. Can J Cardiol 32:1513–1519
81. Gillingham LG, Gustafson JA, Han SY et al (2011) High-oleic rapeseed (canola) and flaxseed oils modulate serum lipids and inflammatory biomarkers in hypercholesterolaemic subjects. Br J Nutr 105:417–427
82. Newairy AS, Abdou HM (2009) Protective role of flax lignans against lead acetate induced oxidative damage and hyperlipidemia in rats. Food Chem Toxicol 47:813–818
83. Hu C, Yuan YV, Kitts DD (2007) Antioxidant activities of the flaxseed lignan secoisolariciresinol diglucoside, its aglycone secoisolariciresinol and the mammalian lignans enterodiol and enterolactone in vitro. Food Chem Toxicol 45:2219–2227
84. Yu X, Cui L, Zhang Z et al (2013) α-Linolenic acid attenuates doxorubicin-induced cardiotoxicity in rats through suppression of oxidative stress and apoptosis. Acta Biochim Biophys Sin (Shanghai) 45:817–826
85. Asselin CY, Lam A, Cheung DY et al (2019) The cardioprotective role of flaxseed in the prevention of doxorubicin and trastuzumab mediated cardiotoxicity. Can J Cardiol. Submitted

Part IV
Sex Differences in Cardiac Mitochondria in Heart Disease

Chapter 14
Sex-Related Pathophysiological Differences in Cardiac Mitochondria: Role of Estrogens

Natalia Pavón, Francisco Correa-Segura, Elizabeth Lira-Silva, and Alfredo Cabrera-Orefice

Abstract A higher prevalence of cardiovascular diseases has been observed for long time in men; although initially associated with stressful working situations that women supposedly did not deal with. However, the subsequent inclusion of women in all kind of formal jobs proved this idea wrong since cardiac affections still occur more frequently in men. Apart from clinical studies, considerable research has been performed in order to unveil the pathophysiological causes accounting for this sex-related difference. Nowadays, it is known that female hormones, particularly estrogens, have cardioprotective effects; thus pointing them out as key factors in the development of cardiovascular disorders. Estrogens regulate the heart function not only at the organ/cellular levels but also at the mitochondrial level. The effects of estrogens on mitochondria may result from their binding to nuclear and mitochondrial estrogen receptors, which modulate gene expression and signaling pathways or directly by interacting with proteins and modifying physicochemical properties of mitochondrial membranes. In female animal models, estrogens deprivation leads to a number of heart mitochondrial impairments and higher sensitivity to ischemia/reperfusion. Despite we have gained substantial descriptive evidence of these effects in heart, the molecular basis remains unclear. This book chapter aimed to revisit the current knowledge about sex-related differences in cardiac pathophysiology, emphasizing on the role of estrogens in heart mitochondrial function under normal conditions and after ischemia/reperfusion injury.

Keywords Mitochondria · Estrogens · Oophorectomy · Heart · Cardiovascular disease · Gender difference · Ischemia/reperfusion

N. Pavón (✉) · E. Lira-Silva
Departamento de Farmacología, Instituto Nacional de Cardiología, Ciudad de México, Mexico
e-mail: pavitonat@yahoo.com.mx

F. Correa-Segura
Departamento de Biomedicina Cardiovascular, Instituto Nacional de Cardiología, Ciudad de México, Mexico

A. Cabrera-Orefice
Radboud Institute for Molecular Life Sciences, Radboud University Medical Center, Nijmegen, The Netherlands

B. Ostadal and N. S. Dhalla (eds.), *Sex Differences in Heart Disease*, Advances in Biochemistry in Health and Disease 21,
https://doi.org/10.1007/978-3-030-58677-5_14

Introduction: "A Females' Thing"

Until the World War I, it was common to believe that heart stroke was a "men's illness" due to the low prevalence of this condition in women. In those times, cardiovascular affections were mostly related to high stress situations that women, typically staying home, apparently did not deal with. In the course of the mid-twentieth century, the majority of men continued working in the stressful fields, factories and offices; whereas women were still far from such a pressure. Nowadays, in spite of the presence of women in all working fields, their susceptibility to suffer a heart attack is still lower. Thus, a sex-related explanation for this condition becomes essential.

Testosterone and other androgenic steroids have been associated with myocardial ischemia, sudden cardiac death and hypertension in males, particularly athletes, leading to the view that these hormones are detrimental for the cardiovascular system [1–3]. In contrast, different intervention studies rather suggest a beneficial effect of natural circulating androgens on coronary heart disease in males [4]. Indeed, there is substantial evidence that physiological testosterone levels have favorable effects on blood vessels and cardiovascular system due to the activity of aromatase, which catalyzes the interconversion of testosterone to estradiol. Inhibition of this enzyme increases the fatty streak lesions in male mice at the same extent as castration does [5]. It has also been demonstrated that the aromatizable androgen dehydroepiandrosterone (DHEA) prevents atherosclerosis in intact cholesterol-fed rabbits, whereas non-aromatizable androgens have no effects [6]. Cardiovascular diseases have been described to increase in women entering menopause [7]. This female stage is characterized by a decline in hormone production by the ovaries, mainly estrogens [8]. It is also known that cardiovascular diseases show a better prognosis in pre-menopausal women than in men. Several epidemiological studies have also shown that the risk of atherosclerotic disease is low in pre-menopausal women and increases dramatically after menopause. Reasonably, estrogens have been pointed out as the causative factors of this sex-related cardiac difference.

However, current knowledge about testosterone/androgens effects on cardiovascular system remains controversial; therefore, further studies are required to unveil the pathophysiological mechanisms exerted by these hormones, especially in humans.

Estrogens and Heart

Estrogens are considered traditionally as female sex hormones; most likely due to their high concentrations and the remarkable physical changes they promote in females. Nevertheless, these steroid hormones are also synthesized in males and participate in a number of physiological and reproductive processes; e.g. spermatogenesis and erectile function [9]. There are three main estrogens with hormonal activity: 17β-estradiol, estrone and estriol [10]. Above all, the most potent estrogen in circulation is 17β-estradiol. In general, estrogen hormones not only regulate reproductive system functions, but also participate in other physiological processes such as cognition [11], cardiovascular function [12], immunity [13] and mineral and bone metabolism [14].

The pleiotropic role of estrogens arises essentially from their binding to intracellular estrogen receptors (ERs). These hormone-receptor complexes are translocated into the nucleus to regulate gene expression [15]. Two isoforms of ERs have been identified: estrogen receptor α (ERα) and β (ERβ). These receptors are encoded respectively by the *ESR1* and *ESR2* genes in humans. Both isoforms belong to the superfamily of nuclear receptor transcription factors [16]. Estrogen hormones bind similarly to the two isoforms since the DNA- and ligand-binding domains are highly conserved [17]. In this regard, it has been shown that ERs may trigger signaling via both genomic and non-genomic pathways [18]. Recent studies have also evidenced that a G protein-coupled receptor with a high affinity for estrogens, known as GPER or GPR30, is present in heart and may modulate rapid signaling events [19–21], even in the presence of the so-called weak estrogens; i.e. estrone and estriol.

The presence of functional ERs has been demonstrated in cardiac myocytes and fibroblasts from both male and female rats [22]. Interestingly, the group of Regitz-Zagrosek has shown that ERα is localized in the intercalated discs, adjacent to the sarcolemma and in some but not all nuclei; besides, it exhibits a striated pattern with troponin T [23]. In rats with dilated cardiomyopathy, ERs are up regulated in both genders; the expression of ERs is remarkably higher in females though. Furthermore, the intercalated discs, which are critical sites for the normal propagation of electrical stimuli in cardiac myocytes, undergo significant changes in injured hearts. The intercalated discs are composed by gap junctions, desmosomes and adherens junctions. The gap junctions regulate electrical coupling, while desmosomes and adherens junctions integrate intermediate filaments and anchor myofibrils to the sarcolemma. Taken together, these data are particularly relevant and suggest that estrogens and their receptors could modulate not only the function, but also maintain normal cell–cell interactions between myocytes.

Role of Estrogens in Heart Mitochondria

A new perspective of estrogenic regulation emerged after the discovery of estrogen receptors α and β within mitochondria [24, 25]. This finding suggests that estrogens may modify directly the mitochondrial function and energy metabolism, although depending on the cell type. Prior to describe those effects, we have included a brief description of mitochondrial energy metabolism in cardiac cells for a basic biochemical understanding.

In aerobic eukaryotes, mitochondria are the major source of adenosine triphosphate (ATP), which is synthesized in the oxidative phosphorylation (OXPHOS) pathway. Four membrane oxidoreductases transfer electrons from different reduced electron carriers to the oxygen (O_2). These multi-subunit complexes (I, II, III and

IV) constitute the mitochondrial respiratory chain; a.k.a. electron transport chain. Complexes I, III and IV use the energy released during the electron transfer to translocate protons from the mitochondrial matrix to the intermembrane space; hence generating an electrochemical gradient of protons that is used by the F_1F_O-ATP synthase (complex V) to drive the ATP synthesis [26]. The establishment of the mitochondrial transmembrane potential ($\Delta\Psi_m$) by the respiratory chain is also important for other processes, such as metabolite transport, protein import, calcium (Ca^{2+}) uptake and fusion/fission [27, 28].

In heart, the excitation–contraction coupling and optimal myocardial performance has to be supported by the OXPHOS system, which keeps providing ATP in both systolic and diastolic periods. During maximal demand, cardiac muscle uses up to 90% of its oxidative capacity; i.e. cardiac cell function relies mainly on mitochondrial energy metabolism [29]. Finely tuning of ATP synthesis [30] is achieved when all OXPHOS components work tightly coupled; otherwise ATP yield is compromised leading to energy deficits and cardiac impairment [31]. Even though cardiac mitochondria are essentially dedicated for aerobic energy metabolism, these organelles are also involved in steroidogenesis, ion homeostasis and biosynthesis of heme groups, lipids, aminoacids and nucleotides [32–34].

The presence of ERs allows regulation of gene expression by estrogens not only in the nuclear genome but also in the mitochondrial genome [35]. It should be noted that the majority of mitochondrial proteins are encoded in the nuclear DNA (nDNA), synthesized in the cytosol and later imported into mitochondria. Nonetheless, few subunits of the OXPHOS complexes, as well as tRNAs and rRNAs are encoded in the mitochondrial DNA (mtDNA) and synthesized in the matrix [36]. In particular, estrogens regulate the mitochondrial metabolism, biogenesis and apoptosis by modulating the expression of a number of nDNA- and mtDNA-encoded proteins; e.g. subunit II of cytochrome *c* oxidase, PGC1α, Bax and Bcl-2 [37–41].

On the other hand, estrogen effects on mitochondrial functions are not only limited to gene regulation. For example, in the brain, estrogens may bind directly with complex V inhibiting ATP synthesis; whereas in heart, interaction of estrogens with this enzyme increases its activity suggesting a direct role in mitochondrial ATP production [42]. Importantly, estrogens have other effects that do not involve interactions with proteins or genes; instead they can directly modify the biophysical properties of mitochondrial membranes, such as microviscosity [43].

Sex-Related Differences in Cardiac Sensitivity to Ischemia/reperfusion Damage

Ischemia/Reperfusion (I/R) in Heart

Myocardial infarction, stroke and peripheral vascular disease occur when myocardial tissue receives insufficient blood supply; i.e. ischemia. The lack of oxygen and nutrients results in lower mitochondrial ATP production, increased lactate production, cytosolic pH acidification, Ca^{2+} overload and multiple ion and redox imbalances [44]. The level of cardiac tissue damage depends directly on the extent of blood supply reduction and the length of the ischemic interval [45]. Paradoxically, restoration of blood supply to ischemic tissues may cause additional cell damage; i.e. reperfusion injury [46]. It should be noted that, during ischemia, redox centers and the pools of electron carriers, such as NAD(H), NADP(H), FAD(H_2), ubiquinone (Q/QH_2) and glutathione (GSSG/GSH), become highly reduced. When O_2 is again available, the electron transport chain flow is restored; however, excessive reduction promotes the release of free radicals, which results in overproduction of reactive oxygen species (ROS) [47]. On the other hand, during reperfusion, the Ca^{2+} is accumulated within the injured tissue triggering additional overload, thus causing fatal cardiac arrhythmias and cell death [48]. The restored blood supply can also exacerbate the inflammation response, promoting white blood cells infiltration leading to necrosis and tissue destruction [49, 50].

Lack of Estrogens Results in Cardiac Mitochondrial Dysfunction, Particularly After I/R

In an interesting work performed in 2000 [51], Zhai and co-workers reported morphological alterations in mitochondria from hearts subjected to ischemia/reperfusion, that were obtained from ovariectomized (OVX) female rats, fed with a diet without phytoestrogens and another group supplemented with 17β-estradiol. Such alterations included abnormal cristae and matrix cleared out (simulating vacuoles), in clear contraposition with the normal matrix and slight changes in cristae and shape observed in the OVX rats group that received a diet supplemented with 17β-estradiol [51]. These results reinforced the idea that estrogens have an important role in preserving mitochondrial structure and consequently protecting the myocardium against I/R injury.

In this respect, our group has established a clear association between the heart electrical activity, inflammation processes and mitochondrial function with the levels of sexual hormones in both female and male rats. We have observed that in male rats, castration does not affect the heart function, showing better electrocardiogram records; whereas in female rats the opposite was observed; i.e. removal of ovaries led to irreversible cardiac damage [52]. This sex-related difference was more evident when cardiac mitochondrial function was explored. In female rats, oophorectomy resulted in ~50% loss of mitochondrial OXPHOS coupling, characterized by a low respiratory control (RC) ratio (Intact RC $= 6 \pm 0.5$ vs OVX RC $= 3 \pm 0.3$); whereas in castrated male, the RC did not change significantly (Intact RC $= 5 \pm 0.8$ vs Cast RC $= 4 \pm 0.9$). The RC value not only reflects the intactness of the mitochondrial inner membrane after isolation but also the coupling between ATP synthesis and the respiratory chain activity [53]. It is calculated as the ratio between oxygen consumption rates

(OCR) in phosphorylating state (plus adenosine diphosphate (ADP) and phosphate; state 3) and in resting state (state 4); i.e. $RC = OCR_{state3}/OCR_{state4}$. Since the RC is a substrate-, tissue- and organism-dependent ratio, there is no absolute diagnostic value for mitochondrial dysfunction [54]. Nevertheless, it is widely accepted that higher RC values are representative of a more intact and "coupled"/functional preparation of isolated mitochondria. These results suggested that the OXPHOS pathway was altered in heart mitochondria from OVX female rats, but it is mandatory to explore these functional abnormalities further.

We have also found that mitochondria from OVX female rats were not capable to accumulate Ca^{2+}, in contrast with mitochondria from castrated males (Fig. 14.1) and that this effect was indeed time-dependent (Fig. 14.2) [55]. The mitochondrial Ca^{2+}-handling capacity was gradually lost after the second month post-oophorectomy (Fig. 14.2B). In general, mitochondrial Ca^{2+} accumulation is an important process regulating a variety of metabolic and pathological processes within cells [56].

In mammalian cardiac and skeletal muscle tissues, mitochondria occur as individual organelles, situated either in clusters beneath the sarcolemma (subsarcolemmal mitochondria, SSM) or in longitudinal rows within the contractile apparatus (interfibrillar mitochondria, IFM); thus occupying the entire space between Z-lines with usually one mitochondrion per sarcomere [57]. Overall, perinuclear mitochondria are smaller than interfibrillar and have a more rounded shape [58]. IFM are mainly situated in close proximity to the sarcoplasmic reticulum, which

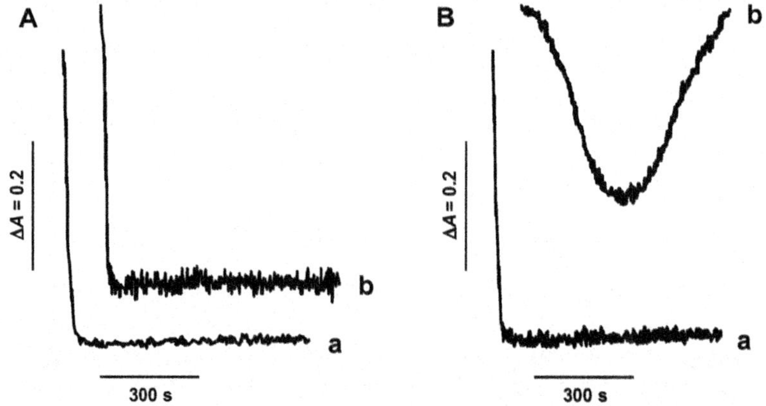

Fig. 14.1 Effect of castration/oophorectomy on Ca^{2+} uptake by heart mitochondria isolated from male and female rats. Mitochondrial Ca^{2+} uptake was monitored spectrophotometrically at 675–685 nm using the indicator Arsenazo III at 25 °C. To start the assay, mitochondrial protein (2 mg) was added into 3 mL of a medium containing 125 mM KCl, 10 mM succinate, 10 mM HEPES, 3 mM phosphate, 100 μM ADP, 5 μg rotenone, 50 μM Arsenazo III and 100 μM $CaCl_2$. (**A**); trace *a* shows Ca^{2+} handling in mitochondria from intact male rats; trace *b*, shows behavior in mitochondria from castrated male rats. (**B**); trace *a*, shows Ca^{2+} handling by intact female heart mitochondria; trace *b* shows failure in Ca^{2+} retention capacity in mitochondria isolated from OVX female rats. Representative traces from 10 independent experiments

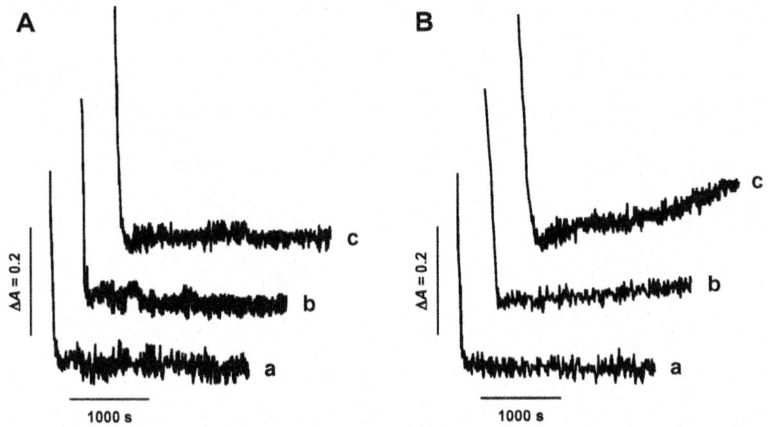

Fig. 14.2 Effect of oophorectomy on Ca^{2+} retention capacity by heart mitochondria isolated from female rats at different times post-surgery. Reaction mixture and assay conditions as described in Fig. 14.1. (**A**); traces *a, b, c* show Ca^{2+} uptake in mitochondria isolated from control female rats after 1, 2 and 3 months of sham surgery, respectively. (**B**); traces *a, b, c* show Ca^{2+} uptake in mitochondria isolated from OVX rats after 1, 2 and 3 months of ovaries removal, respectively. Note the clear impairment in Ca^{2+} retention that takes place at the third month post-oophorectomy (trace c). Representative traces from 10 independent experiments

stands between the IFM and T-tubule. This arrangement creates microdomains of high Ca^{2+} release; a.k.a. Ca^{2+} hotspots [59]. In other words, the correct function and location of mitochondria are critical for accurate Ca^{2+} cycling. The Ca^{2+} uptake in mitochondria is driven by the $\Delta\Psi_m$; thereby, both Ca^{2+} handling and OXPHOS processes should be tightly interconnected [60, 61]. In a related study, Ribeiro Junior and co-workers reported that the IFM and SSM exhibited a different susceptibility to estrogen deprivation [62]. These authors have also shown a lower ADP/O ratio in both subpopulations and a poor Ca^{2+} retention capacity in IFM from OVX rats. In addition, their studies showed that estrogen replacement restored the majority of the heart mitochondrial alterations induced by the ovariectomy.

The restoring effect of estrogens has been also demonstrated in human cell lines. After growing these cells in the presence of 17β-estradiol during 4–6 days, an increase in mitochondrial biogenesis and higher oxygen consumption were observed. In fact, a higher expression of different OXPHOS-related proteins occurred in 17β-estradiol-treated cells [63]. Similarly, our group observed a partial recuperation of the Ca^{2+} retention in isolated heart mitochondria from OVX female rats after being incubated with 17β-estradiol [52]. This effect was achieved by incubating the isolated organelles for ~2–3 h in the presence of the hormone (Fig. 14.3, trace c). Our results hence illustrate the direct effect estrogens have on mitochondrial function; however, further studies are essential in order to explain this phenomenon at the molecular level.

We have also shown that estrogens deprivation resulted in a progressive decrease in the expression and function of several aerobic metabolism-related proteins (Fig. 14.4) [55]. This situation was not dependent on testosterone. Protein content and enzyme

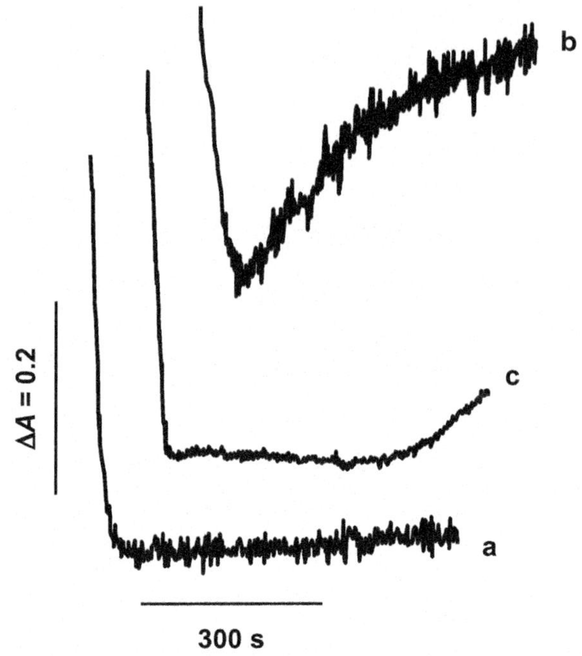

Fig. 14.3 Effect of 17β-estradiol on Ca^{2+} uptake by heart mitochondria isolated from OVX rats. Reaction mixture and assay conditions as described in Fig. 14.1. Trace *a*; Ca^{2+} retention by heart mitochondria isolated from female rats; trace *b* shows failure in Ca^{2+} retention capacity in mitochondria isolated from OVX female rats; trace *c* shows restoration of Ca^{2+} uptake in mitochondria from OVX females rats after being incubated with 17β-estradiol (100–300 nM) for ~2–3 h before the assay

activities of respiratory complexes I (CI) and IV (CIV), pyruvate dehydrogenase (PDH) and 2-oxoglutarate dehydrogenase (OGDH) were clearly diminished in OVX female rats after the second month post-oophorectomy (Fig. 14.4). Although ~ 50% of CIV activity was lost (Fig. 14.4C), complex II (CII)-linked respiration and coupling did not change after estrogens deprivation. Conversely, both CI-linked respiration and coupling were severely affected. When electrons enter the respiratory chain through either CI or CII, the major flux control relies on complexes I/III or III/IV, respectively [64]. The stoichiometry for the respiratory complexes I:II:III:IV in heart mitochondria has been reported as 1:1.5:3:6–7 [65]; therefore, a partial decrease in CII and CIV contents would not be expected to modify the oxygen consumption activity as a CI deficiency would. It has also been described that the expression of genes encoding CI subunits ND1, NDUFS7 and NDUFS8 might be regulated by estrogens [66]. The latter may explain the decrease in cardiac CI content and activity in the OVX rats. Unexpectedly, expression of subunit ND1 of complex I did not change after oophorectomy (Fig. 14.4D). Subunit ND1 is a mtDNA-encoded protein, whereas subunits NDUFS7 and NDUFS8 are codified by nuclear genes. In this regard, the expression of nuclear-encoded subunits seemed to be more susceptible to estrogenic regulation since SDHC, COX4, glutaminase and PDH-E1α were markedly downregulated after oophorectomy. Moreover, it has been described that the regulation of glutaminase expression involves the estrogen-related receptor alpha (ERRα) during cell differentiation [67].

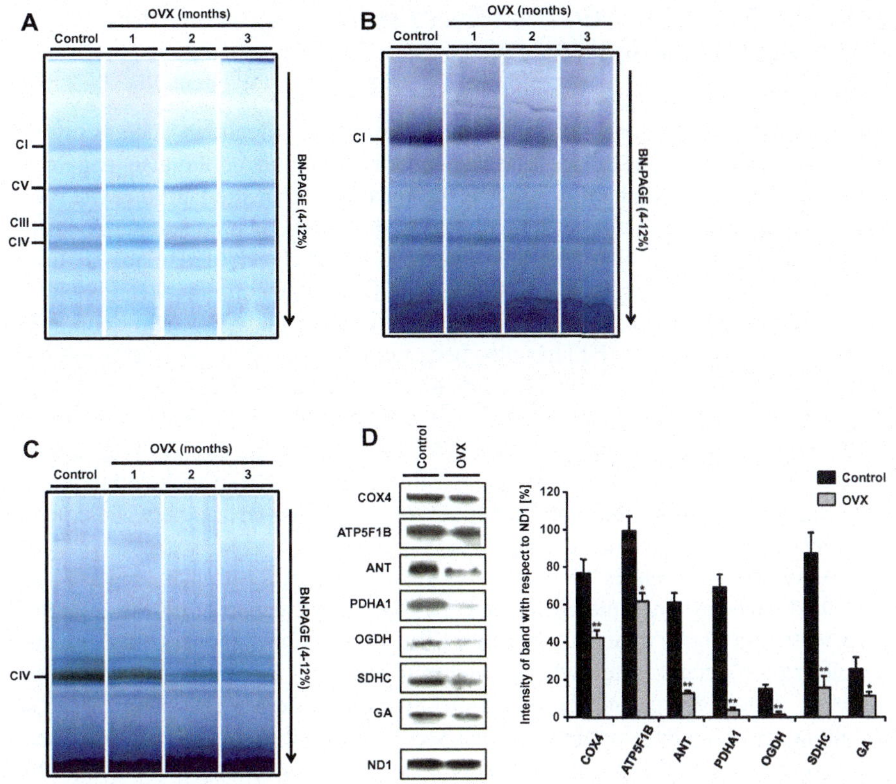

Fig. 14.4 Oophorectomy results in progressive decrease of several cardiac mitochondrial proteins and OXPHOS complexes. Isolated heart mitochondria from control and OVX (1-, 2- and 3-months post-surgery) rats were solubilized with dodecyl-maltoside (2 g/g protein) and separated by Blue Native-PAGE in 4–12% polyacrylamide gradient gels. (**A**); after electrophoretic separation, the proteins were stained with Coomassie blue dye. (**B**); In-gel NADH dehydrogenase activity staining for complex I was performed by incubating the gel in 10 mM Tris/HCl pH 7.0 supplemented with 1 mM NADH and 0.5 mg/ml Nitrotetrazolium blue chloride. (**C**); In-gel cytochrome *c* oxidase activity staining for complex IV was carried out by incubating the gel in sodium phosphate buffer pH 7.4 supplemented with 0.04% diaminobenzidine and 0.02% horse heart cytochrome *c*. OXPHOS complexes are marked as CI (complex I), CIII (complex III), CIV (complex IV) and CV (ATP synthase). (**D**); *Left panel*, western blot analysis of different mitochondrial proteins from control and OVX (3 months post-oophorectomy) female rat mitochondria. Analyzed proteins: ND1, subunit ND1 from complex I; COX4, cytochrome *c* oxidase subunit 4; ATP5F1B, ATP synthase subunit 5B (beta); ANT, adenine nucleotide translocase; PDHA1, pyruvate dehydrogenase subunit E1; OGDH, 2-oxoglutarate dehydrogenase; SDHC, succinate dehydrogenase subunit B; GA, glutaminase. *Right panel*, variations in relative content of analyzed proteins compared to the loading control (ND1). Representative blots and data from three independent experiments; $*p < 0.05$ and $**p < 0.01$

Estrogen Effects on Mitochondrial Permeability Transition and ROS Production After I/R

Mitochondrial impairments occurring in cardiovascular affections, such as myocardial hypertrophy, heart failure and ischemia/reperfusion could also trigger the opening of the so-called mitochondrial permeability transition pore (MPTP). As a result, molecules with masses up to 1.5 kDa can diffuse in a non-selective fashion across the mitochondrial membranes causing a high level of damage including large amplitude swelling, $\Delta\Psi_m$ dissipation, OXPHOS uncoupling, ATP depletion, exhaustion of glycolytic substrates, acidosis and cell death [68]. The mitochondrial permeability transition (MPT) has also been recognized as a key mechanism underlying both necrotic and apoptotic cell death [69]. However, the structure of MPTP remains controversial. On the one hand, it has been described that it is composed by the voltage-dependent anion channel (VDAC), phosphate carrier (PiC), adenine nucleotide translocase (ANT) and cyclophilin D (CypD) [70]. On the other hand, it has been recently proposed that the MPTP forms at the interface between complex V dimers [71].

Sex-related differences in mitochondrial permeability transition have been described in heart. In female rats, the Ca^{2+} retaining capacity of the mitochondrial matrix is higher than in male rats [72]. In addition, recovery of $\Delta\Psi_m$ after depolarization occurs faster, whereas the mitochondrial swelling after Ca^{2+} uptake is lower; hence, decreasing the possibility to trigger the MPTP opening [73, 74]. Remarkably, in hearts of male mice that were subjected to I/R, a brief exposure to 17β-estradiol favored the Ca^{2+} retention capacity by inhibiting the MPTP, which diminished the infarct size [75]. This cardio-protective effect might result from the interactions between CypD and the MPTP modulated by 17β-estradiol and ERβ [76]. Although the molecular mechanism is unclear, the combined effect of estrogens on the Ca^{2+} homeostasis and OXPHOS seems to prevent the MPTP opening after I/R. Since complex V and CypD levels are similar in both male and female hearts, it has been recently proposed that the lower sensitivity of the MPTP to the Ca^{2+}-induced swelling is also related to a higher ischemic tolerance of the female myocardium [77].

Besides, mitochondria are the major cellular source of ROS, i.e. H_2O_2, superoxide anion ($O_2^{\bullet-}$) and hydroxyl free radical ($^{\bullet}OH$). It has been estimated that 2% of the respiratory chain flux ends in formation of H_2O_2 [78]. Within mitochondria, the manganese-dependent superoxide dismutase (MnSOD or SOD2) converts $O_2^{\bullet-}$ to H_2O_2. Mitochondrial ROS may diffuse to the cytosol, where they participate in redox signaling pathways and other cell processes. However, ROS overproduction and failure of the antioxidant systems lead to oxidative stress and cell death. It is important to mention that estrogens control the expression of several antioxidant enzymes too, among them MnSOD, which has a higher expression in females [79]. Estrogens might upregulate ROS handling systems, thus protecting the female heart against oxidative stress. Until now two major ROS-forming sites at the OXPHOS system have been recognized: (*a*) the flavin mononucleotide (FMN) group of CI and (*b*) the Q_O site in CIII [80]. Likewise, FAD-containing proteins, such as CII, PDH and OGDH, have been potentially suggested as ROS sources [81]. It has also been described that 17β-estradiol is capable to inhibit complex I at the FMN site [82]. Hence, it is proposed that this hormone may block the superoxide generation by CI, preferentially in females.

Other Cardiac Impairments Related to Estrogens Deprivation

The estrogen treatment has been shown to ameliorate and even prevent different levels of cardiac damage in several models of infarct, especially in vivo. Treating adult OVX mice subjected to coronary artery ligation with 17β-estradiol reduced the cardiomyocytes apoptosis, which is otherwise observed in untreated animals [83]. Similarly, OVX rabbits treated with estrogens prior to the induction of myocardial infarction, exhibited smaller heart stroke dimensions [84]. The estrogens effects observed in vitro are also remarkable; for instance, the presence of estrogens in cultures of adult rat myocytes obtained from infarcted hearts led to a lower degree of apoptosis and higher viability in comparison with the untreated controls [85]. Therefore, these observations deserve special attention since estrogens are able to activate different molecules involved in survival signaling pathways; e.g. Akt.

Akt (protein kinase B, PKB) is an important kinase which regulates not only apoptosis and cell proliferation but also a number of other cellular processes; such as, carbohydrates metabolism, motility and transcription. A higher activation of this kinase has been described in neonatal rat cardiomyocytes supplemented with 17β-estradiol [83]. The same effect has been also observed using an in vivo murine model; i.e. administration of 17β-estradiol before blocking the coronary artery promoted a high activation of Akt. In contrast, cardiac damage situations, such as pressure overload, result in a lower activation of Akt in OVX rats in comparison to the sham controls [86]. The latter reinforces the notion of a crucial connection among estrogens and Akt. In fact, it has been reported that cardiomyocytes from pre-menopausal women have a higher Akt activity than the one displayed by the same cells from men or post-menopausal women [87]. Certainly, all these findings strongly suggest that the cardioprotective effects of estrogens observed in the different cardiovascular disease models may also have a common molecular basis involving the regulation of the Akt activity/expression.

Moreover, estrogens and ERs have effects on the heart contractile function by controlling ion channels and indirectly the excitation–contraction (EC) coupling. The presence of 17β-estradiol in cultures of guinea-pig isolated ventricular myocytes reduces the peak inward Ca^{2+} current (I_{Ca}), which means that the hormone might have a Ca^{2+} channel inhibitory feature [88]. In another study using estrogen receptor-knockout (ERKO) mice [89], it has been observed an elongation in the action potential duration (APD), as well as a clear electrocardiographic difference; i.e., an increase in the QT interval (~70%). In addition, the authors found an increased expression of the L-type Ca^{2+} channels resulting in a cardiac repolarization delay in ERKO mice [89]. When female rats were ovariectomized, it has been observed a higher expression of subunit Cav1.2 from L-type Ca^{2+} channels in ventricular myocytes [90].

Even so, estrogens not only affect calcium channels but also potassium (K^+) channels. It has been also noticed a downregulation of the subunits Kv1.5 and Kv4.3 in estrogen-treated OVX mice resulting in lower transient outward and delayed K^+ currents; i.e. APD is prolonged [91, 92]. These findings confirm again the existence of sex-related differences in the female heart and thus suggest that the signals triggered by estrogens are indeed important for Ca^{2+} and K^+ channels expression and optimal ion currents.

Perspectives for Elucidating the Molecular Basis of Sex Differences in Cardiovascular Disorders

Sex-related differences in cardiovascular diseases come about from a complex interaction of genetic, hormonal and environmental factors resulting in a profile of individual risk and phenotypic presentation of the disorder. Therefore, there is an enormous interest in elucidating all these interactions in order to optimize the patient treatments and outcomes. If estrogens were the only factors accounting for sex-related differences in the prevalence of ischemic heart disease, it might be expected an increase of this affection after the menopause and that estrogenic hormone replacement therapy would prevent it. The Women's Health Initiative study, however, found that the hormone replacement therapy did not reduce the risk of ischemic heart disease in postmenopausal women [93]. A recent large population-based prospective study found only a minor increase in the risk of ischemic heart disease after the menopause in 55–65 years old women [94]. These findings suggest that ovarian hormones are not the only molecular basis for the sex differences observed in heart; accordingly, we should continue looking for other factors in both males and females and their mechanistic roles in cardiac diseases.

The studies summarized on this chapter reveal pervasive sex-related differences in cardiovascular diseases (CVD) and document diverse mechanisms that contribute to their development. Since most of the studies have only focused on the role of gonadal hormones in cardiac disorders, the biological variable of sex has to be taken apart into its components to better understand the molecular mechanisms underlying those differences. As it has been reviewed within the previous specific sections, estrogens are the most frequently studied factors due to their cardioprotective effects. The shortage of studies about the roles of androgens in cardiac pathophysiology offers, however, an important opportunity for further investigations.

Interestingly, recent studies in gonadectomized mice with altered sex chromosomes have shown differences in I/R injured hearts related to the number and type of sex chromosomes [95, 96]. The cardiac infarct size after I/R in gonadectomized adult mice with two X chromosomes (XX or XXY) is ~40–50%, larger than those containing only one X chromosome (XY or XO) [95]. In this case, the absence or presence of the Y chromosome had no effect; i.e. genes encoded in Y chromosome do not seem to contribute for sex-related differences in I/R injury. The presence of a second X chromosome was associated with adiposity which is a risk factor in the coronary heart disease [95]. Nevertheless, the effects on I/R injury were not related with this condition, since the damage occurred before the signs of adiposity become evident. In any case, it has been suggested that some putative genes of the second X chromosome encoding for proteins with metabolic function could be related to account for the sex differences in different metabolic phenotypes [97]. Therefore, this subset of genes escape inactivation since their expression is higher [98].

The effect of an additional X chromosome in female mice is paradoxical in I/R injury. In OVX female mice, it promotes higher damage instead of preventing it. This observation has a potential clinical significance in humans since female-specific factors may interact with one another resulting in a negative effect in women. In other words, the cardioprotective effect of 17β-estradiol could be mitigated by the presence of the second X chromosome. In women, the heart might be protected by the estrogens at early ages; but later, after the menopause, it would become more susceptible to I/R injury due to the deleterious effects of the second X chromosome [98].

The potential implication of sex chromosomes, especially X chromosome copy number, has only been reported in fundamental research studies. The particular genes involved, as well as their sites and mechanisms of action are still unidentified. In addition, both hormonal and sex chromosome functions clearly influence the same disease outcomes, but sometimes in opposite directions; again, the molecular basis for these interactions remains unclear. The specific sex-biasing factors in animal research have yet to be fully translated into the clinical field. However, the current evidence suggests that the effectiveness of specific drug therapies may depend on sex or the levels of sex-biasing factors, for instance estrogens [99].

Another research field that is emerging is related with the transsexual community, where men use estrogens by different ways of administration (transdermal, oral, injected, etc.) and in unknown doses; even up to 20 times the recommended doses for post-menopausal women. On the other hand, a similar situation happens in women who have chosen to change their gender and take supraphysiological doses of testosterone. It is urgent to perform more studies including these groups in order to prevent possible physiological alterations and damage. In this regard, a study from 2011 has found cardiovascular diseases, particularly ischemic heart disease, as one of the most frequent death causes in this community [100]. Accordingly, it is probable that we will observe more sex hormone-related effects in members of this group, which have not been described or even predicted from the laboratory research so far.

It is very interesting to notice that significant sex-related differences do exist not only in cardiovascular diseases but also in the basal heart function. For example, healthy women have higher ejection phase indices compared to healthy age-matched men. The effects of estrogens on mitochondrial and heart functions are unquestionably important and result from a combination of several genetic, metabolic and signaling factors, which have a tremendous therapeutic potential in ameliorating the complications of cardiovascular diseases. Yet, administration of estrogens for this purpose is far from being safe and it must be still extensively tested in agreement with individual needs.

Acknowledgments Authors thank to Dr. Cecilia Zazueta and Alicia Minor for carefully reading this manuscript and helpful suggestions.

References

1. Dickerman RD, Schaller F, Prather I, McConathy WJ (1995) Sudden cardiac death in a 20-year-old bodybuilder using anabolic steroids. Cardiology 86:172–173
2. Fontana K, Oliveira HC, Leonardo MB et al (2008) Adverse effect of the anabolic-androgenic steroid mesterolone on cardiac remodelling and lipoprotein profile is attenuated by aerobicz exercise training. Int J Exp Pathol 89:358–366
3. Wang M, Tsai BM, Kher A et al (2005) Role of endogenous testosterone in myocardial proinflammatory and proapoptotic signaling after acute ischemia-reperfusion. Am J Physiol Heart Circ Physiol 288:H221-226
4. Kirby M, Hackett G, Ramachandran S (2019) Testosterone and the heart. Eur Cardiol 14:103–110
5. Nathan L, Shi W, Dinh H et al (2001) Testosterone inhibits early atherogenesis by conversion to estradiol: critical role of aromatase. Proc Natl Acad Sci U S a 98:3589–3593
6. Vitale C, Mendelsohn ME, Rosano GM (2009) Gender differences in the cardiovascular effect of sex hormones. Nat Rev Cardiol 6:532–542
7. Hayward CS, Kelly RP, Collins P (2000) The roles of gender, the menopause and hormone replacement on cardiovascular function. Cardiovasc Res 46:28–49
8. Takahashi TA, Johnson KM (2015) Menopause. Med Clin North Am 99:521–534
9. Schulster M, Bernie AM, Ramasamy R (2016) The role of estradiol in male reproductive function. Asian J Androl 18:435–440
10. Ryan KJ (1982) Biochemistry of aromatase: significance to female reproductive physiology. Cancer Res 42:3342s–3344s
11. Sherwin BB (1999) Can estrogen keep you smart? Evidence from clinical studies. J Psychiatry Neurosci 24:315–321
12. Stevenson JC (2000) Cardiovascular effects of oestrogens. J Steroid Biochem Mol Biol 74:387–393
13. Ahmed SA, Hissong BD, Verthelyi D et al (1999) Gender and risk of autoimmune diseases: possible role of estrogenic compounds. Environ Health Perspect 107(Suppl 5):681–686
14. Compston JE (2001) Sex steroids and bone. Physiol Rev 81:419–447
15. Flouriot G, Brand H, Denger S et al (2000) Identification of a new isoform of the human estrogen receptor-alpha (hER-alpha) that is encoded by distinct transcripts and that is able to repress hER-alpha activation function 1. EMBO J 19:4688–4700
16. Yasar P, Ayaz G, User SD et al (2017) Molecular mechanism of estrogen-estrogen receptor signaling. Reprod Med Biol 16:4–20
17. Murphy E (2011) Estrogen signaling and cardiovascular disease. Circ Res 109:687–696
18. Czubryt MP, Espira L, Lamoureux L, Abrenica B (2006) The role of sex in cardiac function and disease. Can J Physiol Pharmacol 84:93–109
19. Kadokawa H, Pandey K, Onalenna K, Nahar A (2018) Reconsidering the roles of endogenous estrogens and xenoestrogens: the membrane estradiol receptor G protein-coupled receptor 30 (GPR30) mediates the effects of various estrogens. J Reprod Dev 64:203–208

20. Menazza S, Sun J, Appachi S et al (2017) Non-nuclear estrogen receptor alpha activation in endothelium reduces cardiac ischemia-reperfusion injury in mice. J Mol Cell Cardiol 107:41–51
21. Deschamps AM, Murphy E (2009) Activation of a novel estrogen receptor, GPER, is cardioprotective in male and female rats. Am J Physiol Heart Circ Physiol 297:H1806-1813
22. Grohe C, Kahlert S, Lobbert K et al (1997) Cardiac myocytes and fibroblasts contain functional estrogen receptors. FEBS Lett 416:107–112
23. Mahmoodzadeh S, Eder S, Nordmeyer J et al (2006) Estrogen receptor alpha up-regulation and redistribution in human heart failure. FASEB J 20:926–934
24. Cammarata PR, Chu S, Moor A et al (2004) Subcellular distribution of native estrogen receptor alpha and beta subtypes in cultured human lens epithelial cells. Exp Eye Res 78:861–871
25. Monje P, Boland R (2001) Subcellular distribution of native estrogen receptor alpha and beta isoforms in rabbit uterus and ovary. J Cell Biochem 82:467–479
26. Saraste M (1999) Oxidative phosphorylation at the fin de siecle. Science 283:1488–1493
27. Zorova LD, Popkov VA, Plotnikov EY et al (2018) Mitochondrial membrane potential. Anal Biochem 552:50–59
28. van der Bliek AM, Shen Q, Kawajiri S (2013) Mechanisms of mitochondrial fission and fusion. Cold Spring Harb Perspect Biol 5
29. Harris DA, Das AM (1991) Control of mitochondrial ATP synthesis in the heart. Biochem J 280(Pt 3):561–573
30. Rosca MG, Hoppel CL (2013) Mitochondrial dysfunction in heart failure. Heart Fail Rev 18:607–622
31. Stanley WC, Recchia FA, Lopaschuk GD (2005) Myocardial substrate metabolism in the normal and failing heart. Physiol Rev 85:1093–1129
32. Boengler K, Kosiol M, Mayr M et al (2017) Mitochondria and ageing: role in heart, skeletal muscle and adipose tissue. J Cachexia Sarcopenia Muscle 8:349–369
33. Anuka E, Yivgi-Ohana N, Eimerl S et al (2013) Infarct-induced steroidogenic acute regulatory protein: a survival role in cardiac fibroblasts. Mol Endocrinol 27:1502–1517
34. Nguyen BY, Ruiz-Velasco A, Bui T, et al (2018) Mitochondrial function in the heart: the insight into mechanisms and therapeutic potentials. Br J Pharmacol
35. Sanchez MI, Shearwood AM, Chia T et al (2015) Estrogen-mediated regulation of mitochondrial gene expression. Mol Endocrinol 29:14–27
36. Chen XJ, Butow RA (2005) The organization and inheritance of the mitochondrial genome. Nat Rev Genet 6:815–825
37. Van Itallie CM, Dannies PS (1988) Estrogen induces accumulation of the mitochondrial ribonucleic acid for subunit II of cytochrome oxidase in pituitary tumor cells. Mol Endocrinol 2:332–337
38. Schreiber SN, Emter R, Hock MB et al (2004) The estrogen-related receptor alpha (ERRalpha) functions in PPARgamma coactivator 1alpha (PGC-1alpha)-induced mitochondrial biogenesis. Proc Natl Acad Sci USa 101:6472–6477
39. Subramanian M, Shaha C (2007) Up-regulation of Bcl-2 through ERK phosphorylation is associated with human macrophage survival in an estrogen microenvironment. J Immunol 179:2330–2338
40. Nilsen J, Diaz BR (2003) Mechanism of estrogen-mediated neuroprotection: regulation of mitochondrial calcium and Bcl-2 expression. Proc Natl Acad Sci USA 100:2842–2847
41. Liang J, Xie Q, Li P et al (2015) Mitochondrial estrogen receptor beta inhibits cell apoptosis via interaction with Bad in a ligand-independent manner. Mol Cell Biochem 401:71–86
42. Zheng J, Ramirez VD (1999) Purification and identification of an estrogen binding protein from rat brain: oligomycin sensitivity-conferring protein (OSCP), a subunit of mitochondrial F0F1-ATP synthase/ATPase. J Steroid Biochem Mol Biol 68:65–75
43. Torres MJ, Kew KA, Ryan TE, et al (2018) 17beta-estradiol directly lowers mitochondrial membrane microviscosity and improves bioenergetic function in skeletal muscle. Cell Metab 27:167–179 e167

44. Korge P, Calmettes G, Weiss JN (1847) Increased reactive oxygen species production during reductive stress: The roles of mitochondrial glutathione and thioredoxin reductases. Biochim Biophys Acta 514–525:2015
45. Kalogeris T, Baines CP, Krenz M, Korthuis RJ (2016) Ischemia/Reperfusion. Compr Physiol 7:113–170
46. Yellon DM, Hausenloy DJ (2007) Myocardial reperfusion injury. N Engl J Med 357:1121–1135
47. Cadenas S (2018) ROS and redox signaling in myocardial ischemia-reperfusion injury and cardioprotection. Free Radic Biol Med 117:76–89
48. Consolini AE, Ragone MI, Bonazzola P, Colareda GA (2017) Mitochondrial Bioenergetics During Ischemia and Reperfusion. Adv Exp Med Biol 982:141–167
49. Zuidema MY, Zhang C (2010) Ischemia/reperfusion injury: the role of immune cells. World J Cardiol 2:325–332
50. Van Linthout S, Tschope C (2017) Inflammation—cause or consequence of heart failure or both? Curr Heart Fail Rep 14:251–265
51. Zhai P, Eurell TE, Cotthaus R et al (2000) Effect of estrogen on global myocardial ischemia-reperfusion injury in female rats. Am J Physiol Heart Circ Physiol 279:H2766-2775
52. Pavon N, Martinez-Abundis E, Hernandez L, et al. Sexual hormones: effects on cardiac and mitochondrial activity after ischemia-reperfusion in adult rats. Gender difference. J Steroid Biochem Mol Biol 132: 135–146, 2012.
53. Nicholls DG, Ferguson SJ (2003) 4 - The chemiosmotic proton circuit. In: Nicholls DG, Ferguson SJ (eds) Bioenergetics (Third Edition). Academic Press, London, pp 57–88
54. Brand MD, Nicholls DG (2011) Assessing mitochondrial dysfunction in cells. Biochem J 435:297–312
55. Pavon N, Cabrera-Orefice A, Gallardo-Perez JC et al (2017) In female rat heart mitochondria, oophorectomy results in loss of oxidative phosphorylation. J Endocrinol 232:221–235
56. Granatiero V, De Stefani D, Rizzuto R (2017) Mitochondrial calcium handling in physiology and disease. Adv Exp Med Biol 982:25–47
57. Palmer JW, Tandler B, Hoppel CL (1985) Biochemical differences between subsarcolemmal and interfibrillar mitochondria from rat cardiac muscle: effects of procedural manipulations. Arch Biochem Biophys 236:691–702
58. Lukyanenko V, Chikando A, Lederer WJ (2009) Mitochondria in cardiomyocyte Ca2+ signaling. Int J Biochem Cell Biol 41:1957–1971
59. Giacomello M, Drago I, Bortolozzi M et al (2010) Ca2+ hot spots on the mitochondrial surface are generated by Ca2+ mobilization from stores, but not by activation of store-operated Ca2+ channels. Mol Cell 38:280–290
60. Griffiths EJ, Rutter GA (2009) Mitochondrial calcium as a key regulator of mitochondrial ATP production in mammalian cells. Biochim Biophys Acta 1787:1324–1333
61. McKenzie M, Duchen MR (2016) Impaired cellular bioenergetics causes mitochondrial calcium handling defects in MT-ND5 mutant cybrids. PLoS ONE 11:e0154371
62. Ribeiro Junior RF, Rodrigues PL, Morra EA et al (2017) Estrogen regulates spatially distinct cardiac mitochondrial subpopulations. Mitochondrion 35:87–96
63. Mattingly KA, Ivanova MM, Riggs KA et al (2008) Estradiol stimulates transcription of nuclear respiratory factor-1 and increases mitochondrial biogenesis. Mol Endocrinol 22:609–622
64. Bianchi C, Genova ML, Parenti Castelli G, Lenaz G (2004) The mitochondrial respiratory chain is partially organized in a supercomplex assembly: kinetic evidence using flux control analysis. J Biol Chem 279:36562–36569
65. Schagger H, Pfeiffer K (2001) The ratio of oxidative phosphorylation complexes I-V in bovine heart mitochondria and the composition of respiratory chain supercomplexes. J Biol Chem 276:37861–37867
66. Chen JQ, Cammarata PR, Baines CP, Yager JD (2009) Regulation of mitochondrial respiratory chain biogenesis by estrogens/estrogen receptors and physiological, pathological and pharmacological implications. Biochim Biophys Acta 1793:1540–1570

67. Huang T, Liu R, Fu X et al (2017) Aging reduces an ERRalpha-directed mitochondrial glutaminase expression suppressing glutamine anaplerosis and osteogenic differentiation of mesenchymal stem cells. Stem Cells 35:411–424
68. Crompton M (1999) The mitochondrial permeability transition pore and its role in cell death. Biochem J 341(Pt 2):233–249
69. Petronilli V, Penzo D, Scorrano L, et al (2001) The mitochondrial permeability transition, release of cytochrome c and cell death. Correlation with the duration of pore openings in situ. J Biol Chem 276:12030–12034
70. Crompton M, Virji S, Ward JM (1998) Cyclophilin-D binds strongly to complexes of the voltage-dependent anion channel and the adenine nucleotide translocase to form the permeability transition pore. Eur J Biochem 258:729–735
71. Giorgio V, von Stockum S, Antoniel M et al (2013) Dimers of mitochondrial ATP synthase form the permeability transition pore. Proc Natl Acad Sci USA 110:5887–5892
72. Ribeiro RF Jr, Ronconi KS, Morra EA et al (2016) Sex differences in the regulation of spatially distinct cardiac mitochondrial subpopulations. Mol Cell Biochem 419:41–51
73. Milerova M, Drahota Z, Chytilova A et al (2016) Sex difference in the sensitivity of cardiac mitochondrial permeability transition pore to calcium load. Mol Cell Biochem 412:147–154
74. Arieli Y, Gursahani H, Eaton MM et al (2004) Gender modulation of Ca(2+) uptake in cardiac mitochondria. J Mol Cell Cardiol 37:507–513
75. Bopassa JC, Ciobotaru A, Eghbali M et al (2009) Estrogen-induced Protection of Heart Ischemia-reperfusion Injury by the Inhibition of the Mitochondrial Permeability Transition Pore (mPTP) in Isolated Heart Mouse. Biophys J 96:530a
76. Fels JA, Manfredi G (2019) Sex differences in ischemia/reperfusion injury: the role of mitochondrial permeability transition. Neurochem Res
77. Ostadal B, Drahota Z, Houstek J et al (2019) Developmental and sex differences in cardiac tolerance to ischemia-reperfusion injury: the role of mitochondria (1). Can J Physiol Pharmacol 97:808–814
78. Grivennikova VG, Vinogradov AD (1827) Partitioning of superoxide and hydrogen peroxide production by mitochondrial respiratory complex I. Biochim Biophys Acta 446–454:2013
79. Borras C, Sastre J, Garcia-Sala D et al (2003) Mitochondria from females exhibit higher antioxidant gene expression and lower oxidative damage than males. Free Radic Biol Med 34:546–552
80. Zorov DB, Juhaszova M, Sollott SJ (2014) Mitochondrial reactive oxygen species (ROS) and ROS-induced ROS release. Physiol Rev 94:909–950
81. Wong HS, Dighe PA, Mezera V et al (2017) Production of superoxide and hydrogen peroxide from specific mitochondrial sites under different bioenergetic conditions. J Biol Chem 292:16804–16809
82. Moreira PI, Custodio J, Moreno A et al (2006) Tamoxifen and estradiol interact with the flavin mononucleotide site of complex I leading to mitochondrial failure. J Biol Chem 281:10143–10152
83. Patten RD, Pourati I, Aronovitz MJ et al (2004) 17beta-estradiol reduces cardiomyocyte apoptosis in vivo and in vitro via activation of phospho-inositide-3 kinase/Akt signaling. Circ Res 95:692–699
84. Booth EA, Marchesi M, Kilbourne EJ, Lucchesi BR (2003) 17Beta-estradiol as a receptor-mediated cardioprotective agent. J Pharmacol Exp Ther 307:395–401
85. Brinckmann M, Kaschina E, Altarche-Xifro W et al (2009) Estrogen receptor alpha supports cardiomyocytes indirectly through post-infarct cardiac c-kit+ cells. J Mol Cell Cardiol 47:66–75
86. Bhuiyan MS, Shioda N, Fukunaga K (2007) Ovariectomy augments pressure overload-induced hypertrophy associated with changes in Akt and nitric oxide synthase signaling pathways in female rats. Am J Physiol Endocrinol Metab 293:E1606-1614
87. Camper-Kirby D, Welch S, Walker A et al (2001) Myocardial Akt activation and gender: increased nuclear activity in females versus males. Circ Res 88:1020–1027

88. Jiang C, Poole-Wilson PA, Sarrel PM et al (1992) Effect of 17 beta-oestradiol on contraction, Ca2+ current and intracellular free Ca2+ in guinea-pig isolated cardiac myocytes. Br J Pharmacol 106:739–745
89. Johnson BD, Zheng W, Korach KS et al (1997) Increased expression of the cardiac L-type calcium channel in estrogen receptor-deficient mice. J Gen Physiol 110:135–140
90. Chu SH, Goldspink P, Kowalski J et al (2006) Effect of estrogen on calcium-handling proteins, β-adrenergic receptors, and function in rat heart. Life Sci 79:1257–1267
91. Saito T, Ciobotaru A, Bopassa JC et al (2009) Estrogen contributes to gender differences in mouse ventricular repolarization. Circ Res 105:343–352
92. Trepanier-Boulay V, St-Michel C, Tremblay A, Fiset C (2001) Gender-based differences in cardiac repolarization in mouse ventricle. Circ Res 89:437–444
93. Rossouw JE, Anderson GL, Prentice RL et al (2002) Risks and benefits of estrogen plus progestin in healthy postmenopausal women: principal results From the Women's Health Initiative randomized controlled trial. JAMA 288:321–333
94. Mosca L, Barrett-Connor E, Wenger NK (2011) Sex/gender differences in cardiovascular disease prevention: what a difference a decade makes. Circulation 124:2145–2154
95. Arnold AP, Reue K, Eghbali M et al (2016) The importance of having two X chromosomes. Philos Trans R Soc Lond B Biol Sci 371:20150113
96. Li J, Chen X, McClusky R et al (2014) The number of X chromosomes influences protection from cardiac ischaemia/reperfusion injury in mice: one X is better than two. Cardiovasc Res 102:375–384
97. Arnold AP, Cassis LA, Eghbali M et al (2017) Sex hormones and sex chromosomes cause sex differences in the development of cardiovascular diseases. Arterioscler Thromb Vasc Biol 37:746–756
98. Chen X, McClusky R, Chen J et al (2012) The number of x chromosomes causes sex differences in adiposity in mice. PLoS Genet 8:e1002709–e1002709
99. Arnold AP, Chen X, Link JC et al (2013) Cell-autonomous sex determination outside of the gonad. Dev Dyn 242:371–379
100. Asscheman H, Giltay EJ, Megens JA et al (2011) A long-term follow-up study of mortality in transsexuals receiving treatment with cross-sex hormones. Eur J Endocrinol 164:635–642

Chapter 15
Cardiac Mitochondria and Ischemia/Reperfusion Injury—Sex Differences

Zdenek Drahota, Marketa Hlavackova, and Bohuslav Ostadal

Abstract It is now widely accepted that function of cardiac mitochondria is sex-dependent. Female mitochondria better utilize lipids and exhibit higher oxidative capacity in comparison with males. Our results have revealed that female cardiac mitochondria are more resistant to calcium–induced swelling as compared with the male myocardium; this suggests their better protection against mitochondrial permeability transition pore (PTP) opening. It seems to us that sex-dependent specificity of the PTP function is not the result of differences in its protein composition, since the male and female rat heart contains comparable amount of ATP synthase and its regulatory protein cyclophilinD (CypD). The higher hypoxic tolerance of the female cardiac mitochondria thus rather reflects sex differences in the regulation of PTP function, probably together with regulation of CypD by post-translational modifications. The precise knowledge of the composition of the PTP complex and regulation of pore opening are essential conditions for the development of new drugs targeting the function of PTP.

Keywords Sex · Heart · Mitochondria · Ischemia/reperfusion injury · Mitochondrial permeability transition pore

Introduction

Mitochondria are essential for the function of the cardiovascular system owing to their central position in the oxygen and calcium handling. Why are mitochondria unique? (i) They arose through an endosymbiotic process more than 1.45 billion years ago and (ii) they are exclusively maternally inherited [1]. They consist of an external membrane and the inner membrane formed by multiple invaginations or cristae, connected to the boundary region at narrow circular junctions, delineating a matrix and an intermembrane space. Mitochondria contain their own circular DNA (mtDNA); the mitochondrial genome encodes 13 subunits of the mitochondrial

Z. Drahota · M. Hlavackova · B. Ostadal (✉)
Institute of Physiology, Czech Academy of Sciences, Prague, Czech Republic
e-mail: ostadal@biomed.cas.cz

B. Ostadal and N. S. Dhalla (eds.), *Sex Differences in Heart Disease*,
Advances in Biochemistry in Health and Disease 21,
https://doi.org/10.1007/978-3-030-58677-5_15

respiratory chain, 2 ribosomal and 22 transfer RNAs. The mitochondrial proteome contains more than 1000 proteins; each cell contains multiple mitochondria (7000–10,000 in cardiac cells) and their number and activity are highly dependent on the cell type and energy requirements [2]. Mitochondria proliferate by division of preexisting organelles through a process called mitochondrial biogenesis. In most cells mitochondria are present as a network undergoing frequent fusion and fission events. This process, called mitochondrial dynamics, is an important factor in several disease conditions. The final step of mitochondrial biogenesis needs the biosynthesis of membrane lipids, including cardiolipin [3]. Poorly functioning mitochondria can be eliminated through mitochondrial autophagy (mitophagy), which not only allows elimination of defective mitochondria but may also accelerate the mitochondrial turnover [4].

Mitochondrial Function

Mitochondria are the site of oxidative phosphorylation (OXPHOS); they produce the majority of energy-rich bonds in the form of ATP (>90% for the heart). They use mainly fatty acids and carbohydrates to produce reducing equivalents that are oxidized along the respiratory chain, providing a proton gradient across the inner membrane. A by-product of electron transport is the generation of reactive oxygen species (ROS), essentially at complex I and III of the respiratory chain. Naturally low levels of ROS represent key signaling molecule to regulate biological processes. However, when mitochondria are dysfunctional, higher production of ROS can cause irreversible damage to lipids, DNA and proteins [1].

Increase of the permeability of the inner mitochondrial membrane is mediated by the permeability transition pore (PTP). PTP opening may be induced by calcium and ROS, and inhibited by adenine nucleotides, and acidic pH. The final effect depends on time: whereas short-term opening may participate in Ca^{2+} and ROS homeostasis, long-lasting opening induces different types of adverse effects, e.g. mitochondrial swelling, collapse of membrane potential, and cessation of OXPHOS and ATP synthesis [5]. Whereas the discovery of the existence of PTP is more than fifty years old [6], its molecular identity is still unclear. Genetic analysis [7] ruled out the original concept of multi-component PTP. New potential candidate is F_0F_1 – ATP synthase (complex V), which consists of two protein domains, F_1 and F_0 with 17 different types of subunits, responsible for stabilization of the ATP synthase structure. The composition of these two domains is different: while the catalytic domain F_1 consists of five different subunits ($\alpha,\beta,\gamma,\delta,\varepsilon$), domain F_0 includes the regular subunits c,a,b,d, F_6, oligomycin sensitivity-conferring protein and the accessory subunits e,f,g and A6L. Bonora et al. [8] identified the c subunit as a fundamental regulator of PTP activity; this original finding was a few months later confirmed by Giorgio et al. [9]. ATP synthase has a capacity to form an ATP synthasome through its dimerization [10]. Increasing evidence suggest that PTP formation is regulated by cyclophilin D

(CypD). It represents the primary binding targets of immunosuppressive agents such as cyclosporine [11, 12].

Important is the role of mitochondria in calcium homeostasis. They accumulate calcium along its electrochemical gradient; this process may favor OXPHOS by stimulating several mitochondrial dehydrogenases. Excess calcium leads, however, to an opening of the PTP and finally cell death [5].

Mitochondria are involved in sex steroid hormone biosynthesis and in turn these hormones like estrogen, progesterone and testosterone may regulate mitochondrial function. Cells contain functional estrogen, androgen and progesterone receptors that are targets for sex hormone action. Estrogen regulation of mitochondrial capacity and function has been shown to participate in vascular, cardiac, and neuronal protection [13].

Another very important by-product of energy transformation process is energy dissipation as heat. This portion of dissipated energy plays an important role in maintenance of thermal homeostasis of mammalian organisms. It may be increased by so called uncoupling of the oxidative phosphorylation process by various endogenous or exogenous factors that discharge the proton gradient across the inner mitochondrial membrane formed by the mitochondrial respiratory enzymes and inhibit its utilization for generation of ATP. Recently, it was found that mitochondria under physiological conditions operate at temperature about 10 degrees higher than is the cell and tissue temperature [14] and that activity of mitochondrial respiratory chain enzymes is maximal at this temperature. However, mitochondria possess also specific regulated proton channels (uncoupling proteins) that can dissipate membrane potential and further increase heat production. There exists a special tissue in newborn mammals and hibernating animals - brown adipose tissue - with high amount of mitochondria with highly expressed uncoupling proteins in their inner membrane. This tissue produces heat by oxidation of fatty acids [15], protects newborn animals against cold stress after birth, and helps hibernating animals to recover body temperature during arousal [16, 17].

It should be also mentioned that mitochondria show big diversity in various mammalian tissues. They all have the same enzyme equipment for energy transformation—oxidative phosphorylation system—but they contain many different enzymes related to specific functions of particular tissue. They differ in quantity, morphology and localization inside the cell; in adult cardiomyocytes are mitochondria regularly localized between myofibrils and under the cell membrane. These fractions can be isolated as subsarcolemmal and myofibrilar mitochondria; they have many differences in quantity, morphology, and enzyme activities [18]. The final localization of these fractions occurs only during postnatal development.

Sex Differences in Cardiac Mitochondria

Mitochondria exhibit sex-specific behaviour (for rev see [19]). Cardiomyocytes from female rats have lower mitochondrial content, but are more efficient and more differentiated than male mitochondria [20]. Moreover, they generate less ROS than male ones and have higher capacity of antioxidant defence [21]. At baseline, no difference in oxygen consumption rate and cardiolipin content is observed between mitochondria from male and female rats [22]. Subsarcolemmal and intermyofibrillar isolated mitochondria from female hearts have the same respiration rates as the male ones except for glutamate-malate-stimulated respiration which is lower in females, while the ADP/O ratio is higher in the female heart [23]. Taken together, these results suggest that female cardiac mitochondria have higher specific activity than the male ones but lower mitochondrial content, explaining the similar oxidative capacity in males and females [1].

It is now generally accepted that mitochondria are major targets of cardioprotective signaling; in addition, females have different cardiac mitochondrial function [24, 25]. Lagranha et al. [21] have observed that mitochondria isolated from the female heart exhibit reduced ROS generation, reoxygenation, and oxidative metabolism. It is interesting to mention that the opening of PTP can be blocked by estrogens, similar to blockade with cyclosporine [26, 27].

Mitochondria from female hearts have lower Ca^{2+} uptake rates and improved recovery of mitochondrial membrane potential from Ca^{2+}-induced depolarization [28]; they cope more successfully with external calcium load by decreasing the rate of calcium influx by the calcium uniporter (MCU). The important role plays the interaction between MCU and calcium uptake regulatory proteins M1CU1, M1CU2, MCUR1, SLC25A23, and EMRE [29]. In addition, it has been found that the concentration threshold for net mitochondrial Ca^{2+} uptake was higher in the female than in male myocardium [30]. These findings suggest that female heart mitochondria are less prone to Ca^{2+} overload upon I/R, suggesting why the female heart suffers less I/R injury in susceptibility to ischemic heart disease [31].

Mitochondria in Cardiac Ischemia/Reperfusion Injury

Degree of I/R injury increases significantly during postnatal ontogeny and the adult female heart is more tolerant than the male myocardium. Some recent data indicate that mitochondria could play an important role in this effect [1, 8].

It was described that mitochondria become leaky, uncoupled, and massively swollen if they are exposed to high Ca^{2+} concentrations, especially when accompanied by oxidative stress [32–36]. The collapse of mitochondrial membrane potential due to opening of PTP has been implicated in the molecular mechanism of cardiac I/R injury [37–39]. PTP is closed during ischemia due to the low pH (<7.0), but it opens during the first minutes of reperfusion, together with normalization of pH, ROS accumulation, and rise in intracellular calcium. PTP opening, accompanied by matrix swelling, leads finally to myocardial cell death [40]. If the I/R injury is minimal, the cell may recover, if the damage is moderate, the cell may undergo apoptosis; and if the damage is severe, the cell may die due to inadequate energy production [41].

The blockade of PTP by cyclosporine and sanglifehrin in perfused heart was cardioprotective in most animal models of cardiac I/R injury [42]. Experimental studies were, however, carried mostly on subsarcolemmal mitochondria in spite of the fact that interfibrillar mitochondria are more resistant to high concentration of calcium [18, 43, 44]. Cyclosporine A was cardioprotective also in a small clinical study with patients suffering from myocardial infarction [45]. However, a large multicenter clinical trial (CIRCUS) revealed no protective effect of this drug [46, 47]. For the explanation of these negative results, severity of infarction, a quite narrow window of protection, timing of administration as well as comorbidities have to be taken into consideration. Nevertheless, these studies challenge the clinical use of blockade of PTP as well as possible CypD inhibitors for cardioprotection [10].

CypD has been shown to interact with the ATP synthase through binding to the lateral stalk; the ATP synthase-CypD interaction could mediate the enzymatic activity of this complex [48]. Hypothetically, the activity of CypD to stimulate or inhibit PTP induction can be regulated by post-translational modifications of CypD by phosphorylation, nitrosylation, oxidation or acetylation.

Sex Differences in the Sensitivity of Cardiac Mitochondrial PTP to Calcium Load

In our previous papers, we have tested the hypothesis that the role of cardiac PTP in I/R injury is sex-dependent [49]. We have observed that female cardiac mitochondria are more resistant to swelling at higher calcium concentrations (Fig. 15.1), suggesting a better protection of these mitochondria against opening of PTP. This observation may at least partly contribute to the explanation of the pathogenetic mechanisms involved in the increased tolerance of the female heart to I/R injury.

It was, therefore, of interest to know whether changes in PTP components can be involved in this process [49]. Our results have revealed that there are no sex differences in substrate oxidation and coupled ATP generation. These findings were confirmed by quantitative immunodetection analysis: female mitochondria contain comparable amount of ATP synthase as well as of PTP regulatory protein CypD. It may be of interest to note that we have obtained similar results also in our previous study comparing the role of PTP in highly hypoxic tolerant neonatal and adult cardiac mitochondria [50], (for rev. see [19]). On the basis of all these observations the high myocardial hypoxic tolerance cannot be explained by changes in the composition of PTP and rather reflects regulation of its function. Until now, we can only speculate on the possible role of differences in calcium interaction with CypD and PTP opening, particularly in accordance with regulation of CypD by post-translational modification

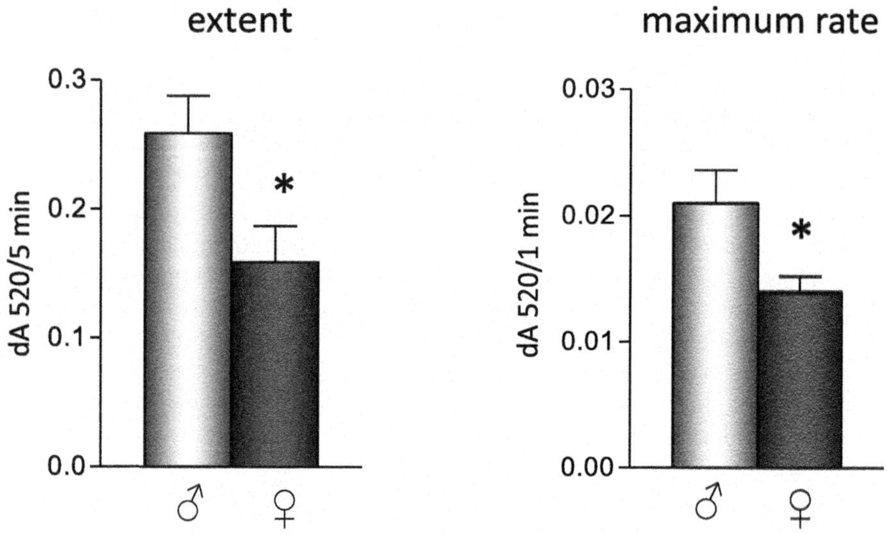

Fig. 15.1 Calcium induced swelling by rat heart mitochondria from male and female rats. Extent of swelling was calculated from the swelling curves and expressed as the decrease of absorbance at 520 nm during 5 min after addition of 200 μM $CaCl_2$. Maximum rate of swelling was calculated from curves obtained after derivatization of data of the extent of swelling. *$P < 0.01$. Data from [50]

[10, 45]. CypD thus remains an attractive target for both experimental studies and clinical application of PTP inhibitory strategies [5].

Conclusions

It may be concluded from our data that cardiac mitochondria reveal sex-dependent sensitivity of PTP to calcium load: female mitochondria are significantly more resistant to swelling at higher calcium concentration. It seems to us that sex-dependent specificity of the PTP function is not the result of differences in its protein composition, since the male and female rat heart mitochondria contain comparable amount of ATP synthase and its regulatory protein CypD. The increased resistance of the female heart mitochondria thus rather reflects sex differences in the regulation of PTP function, probably together with regulation of CypD. The precise knowledge of the protein composition of the PTP complex and regulation of its opening are essential conditions for the development of new drugs targeting the opening of PTP.

Acknowledgments The work was supported by Czech Science Foundation No 19-04790Y.

References

1. Ventura-Clapier, Moulin M, Piquereau J, et al (2017) Mitochondria: a central target for sex differences in pathologies. Clin Sci 131: 803–822
2. Fernandez-Vizarra E, Enriqez JA, Perez-Martos A et al (2011) Tissue-specific differences in mitochondrial activity and biogenesis. Mitochondrion 11:207–213
3. Mejia EM, Nguyen H, Hatch GM (2014) Mammalian cardiolipin biosynthesis. Chem Phys Lipids 179:11–16
4. Ren J, Taegtmeyer H (2015) Too much or not enough of a good thing-the Janus faces of autophagy in cardiac fuel and protein homeostasis. J Mol Cell Cardiol 84:223–226
5. Bernardi P, Di Lisa F (2015) The mitochondrial permeability transition pore: molecular nature and role as a target in cardioprotection. J Mol Cell Cardiol 78:100–106
6. Haworth RA, Hunter DR (1979) The Ca $^{2+}$-induced membrane transition in mitochondria. II. Nature of the Ca $^{2+}$ trigger site. Arch Biochem Biophys 195:460–467
7. Rasola A, Bernardi P (2015) The mitochondrial permeability transition pore and its adaptive responses in tumor cells. Cell Calcium 58:18–26
8. Bonora M, Bononi A, De Marchi E et al (2013) Role of the c subunit of the F_o ATP synthase in mitochondrial permeability transition. Cell Cycle 12:674–683
9. Giorgio V, von Stockum S, Antoniel M et al (2013) Dimers of mitochondrial ATP synthase form the permeability transition pore. Proc Natl Acad Sci USA 110:5887–5892
10. Javadov S, Jang S, Parodi-Rullan R et al (2017) Mitochondrial permeability transition in cardiac ischemia-reperfusion: whether cyclophilin D is a viable target for cardioprotection ? Cell Mol Life Sci 74:2795–2813
11. Galat A (1993) Peptidylproline cis-trans-isomerases: immunophilins. Eur J Biochem 216:689–707
12. Gutierrez-Aguilar M, Baines CP (1850) Structural mechanisms of cyclophilin D-dependent control of the mitochondrial permeability transition pore. Biochim Biophys Acta 2041–2047:2015
13. Velarde MC (2014) Mitochondrial and sex steroid hormone crosstalk during aging. Longev Healthspan 3:2
14. Chretien D, Benit P, Huyung-Ho H et al (2018) Mitochondria are physiologically maintained at close to 50° C. PLOS Biol 16:1–17
15. Bulichev A, Kramar L, Drahota Z, Lindberg O (1972) Exp Cell Res 72:169–187
16. Jansky L (1973) Non-shivering thermogenesis and its thermoregulatory significance. Biol Rev 48:85–132
17. Flatmark T, Pedersen JI (1975) Brown adipose tissue mitochondria. Biochim Biophys Acta 416:53–103
18. Palmer JW, Tandler B, Hoppel CL (1986) Heterogenous response of subsarcolemmal heart mitochondria to calcium. Am J Physiol 250:H741–H748
19. Ostadal B, Drahota Z, Houstek J et al (2019) Developmental and sex differences in cardiac tolerance to ischemia-reperfusion injury: the role of mitochondria. Can J Physiol Pharmacol 97:808–814
20. Colom B, Oliver J, Roca P, Garcia-Palmer FJ (2007) Caloric restriction and gender modulate cardiac muscle mitochondrial $H_2 O_2$ production and oxidative damage. Cardiovasc Res 74:456–465
21. Laghranha CJ, Deschamps A, Aponte A et al (2010) Sex differences in the phosphorylation of mitochondrial proteins result in reduced production of reactive oxygen species and cardioprotection in females. Circ Res 106:1681–1691
22. Moulin M, Piquereau J, Mateo P et al (2015) Sexual dimorphism of doxorubicin-mediated cardiotoxicity: potential role of energy metabolism remodeling. Circ Heart Fail 8:98–108
23. Ribeiro RF Jr, Ronconi KS, Morra EA et al (2016) Sex differences in regulation of spatially distinct cardiac mitochondrial subpopulations. Mol Cell Biochem 419:41–51
24. Murphy E, Steenbergen C (2007) Gender-based differences in mechanisms of protection in myocardial ischemia-reperfusion injury. Cardiovasc Res 75:478–486

25. Stirone C, Duckles SP, Krause DN, Procaccio V (2005) Estrogen increases mitochondrial efficiency and reduces oxidative stress in cerebral blood vessels. Mol Pharmacol 68:959–965
26. Morkuniene R, Arandarcikaite O, Ivanoviene L, Borutaite V (2010) Estradiol-induced protection against ischemia-induced heart mitochondrial damage and caspase activation is mediated by protein kinase G. Biochim Biophys Acta 1797:1012–1017
27. Pavon N, Martinez-Abundis E, Hernandez L, et al (2012) Sexual hormones: effects on cardiac and mitochondrial activity after ischemia-reperfusion in adult rats. Gender difference. J Steroid Biochem Mol Biol 132:135–146
28. Arieli Y, Gursahani H, Eaton MM (2004) Gender modulation of Ca $^{2+}$ uptake in cardiac mitochondria. J Mol Cell Cardiol 37:507–513
29. Williams GS, Boyman L, Lederer WJ (2015) Mitochondrial calcium and the regulation of metabolism in the heart. J Mol Cell Cardiol 78:35–45
30. Chweih H, Castilho RF, Figueira TR (2015) Tissue and sex specificities in Ca $^{2+}$ handling by isolated mitochondria in conditioning avoiding the permeability transition. Exp Physiol 100:1073–1092
31. Vijay V, Han T, Moland CL, et al (2015) Sexual dimorphism in the expression of mitochondria-related genes in rat heart at different ages. PloS One 10: e0117047
32. Halestrap AP (2010) A pore way to die: the role of mitochondria in reperfusion injury and cardioprotection. Biochem Soc Trans 38:841–860
33. Bernardi P. The mitochondrial permeability transition pore: a mystery solved? Front Physiol 4: 95. Doi: 10.3389
34. Drahota Z, Endlicher R, Stankova P et al (2012) Characterization of calcium, phosphate and peroxide interactions in activation of mitochondrial swelling using derivative of the swelling curves. J Bioenerg Biomembr 44:309–315
35. Halestrap AP, Richardson AP (2015) The mitochondrial permeability transition: a current perspective on its identity and role in ischaemia/reperfusion injury. J Mol Cell Cardiol 78:129–141
36. Endlicher R, Drahota Z, Cervinkova Z (2019) Modification of calcium retention capacity of rat liver mitochondria by phosphate and tert-butylhydroperoxide. Physiol Res 68:59–65
37. Di Lisa F, Bernardi P (1998) Mitochondrial function as a determinant of recovery or death in cell response to injury. Mol Cell Biochem 184:379–391
38. Halestrap AP, Clarke SJ, Javadov SA (2004) Mitochondrial permeability transition pore opening during myocardial reperfusion—A target for cardioprotection. Cardiovasc Res 61:372–385
39. Bernardi P, Krauskopf A, Basso E et al (2006) The mitochondrial permeability transition from in vitro artifact to disease target. FEBS J 273:2077–2099
40. Halestrap AP (2006) Calcium, mitochondria and reperfusion injury: a pore way to die. Biochem Soc Trans 34:232–237
41. Morciano G, Giorgi C, Bonora M et al (2015) Molecular identity of the mitochondrial permeability transition pore and its role in ischemia-reperfusion injury. J Mol Cell Cardiol 78:142–153
42. Ong SB, Samangouei P, Kalkhoran SB, Hausenloy DJ (2015) The mitochondrial permeability transition pore and its role in mitochondrial ischemia reperfusion injury. J Mol Cell Cardiol 78:23–34
43. Kuznetsov AV, Margreiter R (2009) Heterogeneity of mitochondria and mitochondrial function within cells as another level of mitochondrial complexity. Int J Mol Sci 10:1911–1929
44. Hollander JM, Thapa D, Shepherd DL (2014) Physiological and structural differences in spatially distinct subpopulations of cardiac mitochondria: influence of cardiac pathologies. Am J Physiol Heart Circ Physiol 307:H1–H14
45. Alam MR, Baetz D, Ovize M (2015) Cyclophilin D and myocardial ischemia-reperfusion injury: a fresh perspective. J Mol Cell Cardiol 78:80–89
46. Cung TT, Morel O, Cayla G et al (2015) Cyclosporine before PCI in patients with acute myocardial infarction. N Engl J Med 373:1021–1031

47. Mewton N, Cung TT, Morel O et al (2015) Rationale and design of the Cyclosporine to ImpRove Clinical oUtcome in ST-elevation myocardial infarction patients (the CIRCUS trial). Am Heart J 169:758–766
48. Giorgio V, Bisetto E, Soriano ME et al (2009) Cyclophilin D modulates mitochondrial F_0F_1-ATP synthase by interacting with the lateral stalk of the complex. J Biol Chem 284:33982–33988
49. Milerova M, Drahota Z, Chytilova A et al (2016) Sex difference in the sensitivity of cardiac mitochondrial permeability transition pore to calcium load. Mol Cell Biochem 412:147–154
50. Milerova M, Charvatova Z, Skarka L et al (2010) Neonatal cardiac mitochondria and ischemia/reperfusion injury. Mol Cell Biochem 335:147–153

Chapter 16
Sex Differences in Mitochondrial Antioxidant Gene Expression

Consuelo Borrás, Cristina Mas-Bargues, Ana B. Paes, and Susana Novella

Abstract Females live longer than males. This could be in part due to the higher levels of estrogens in females, which protect them against aging. Physiological concentrations of estrogens have antioxidant effects as they induce the expression of manganese superoxide dismutase and glutathione peroxidase by stimulating estrogen receptors and the mitogen-activated protein kinase and nuclear factor kappa B pathways. However, estrogens can have undesirable effects such as they are feminizing to males, so other alternatives need to be searched. Phytoestrogens are good candidates as they can also bind to estrogens receptors, and in fact, they are able to mimic the antioxidant properties of estrogens. It is very important to consider that the expression of estrogen receptors is not the same between sexes, organs or that their proportion changes with age. Depending on the organ studied, there are differences in the estrogen receptors involved in the beneficial effects of estrogens.

Keywords Estrogens · Phytoestrogens · ER · Antioxidant · Aging

Introduction

A major aim of gerontology is to find suitable models to understand aging. A very remarkable possibility to study differential aging is the fact that in many species, including humans, females live longer than males [1, 2]. Thus, animals with a very similar genetic background may differ in their average life span by as much as 10%. Oxidative damage, especially that of mitochondria, has been shown to increase with age [3, 4], therefore, one of the reasons for the longer longevity of females could be that they are protected against oxidative stress. This is the case; females overexpress

C. Borrás (✉) · C. Mas-Bargues
Freshage Research Group, Department of Physiology, Faculty of Medicine, University of Valencia, CIBERFES. INCLIVA. Avenida Blasco Ibañez, 15, Valencia, Spain
e-mail: consuelo.borras@uv.es

A. B. Paes · S. Novella
Department of Physiology, Faculty of Medicine, University of Valencia, Av Blasco Ibañez, 15, 46010 Valencia, Spain

B. Ostadal and N. S. Dhalla (eds.), *Sex Differences in Heart Disease*,
Advances in Biochemistry in Health and Disease 21,
https://doi.org/10.1007/978-3-030-58677-5_16

mitochondrial antioxidant enzymes which protect them against oxidative damage [5]. Estrogens are responsible for this up-regulation, as they trigger the expression by binding to estrogen receptors (ER), which in turn are able to activate MAPK/NFkB signaling pathway and finally Mn-superoxide dismutase and glutathione peroxidase, both of them mitochondrial antioxidant enzymes [6]. But estrogens can have undesirable effects such as they are feminizing to males, so other alternatives need to be searched. Phytoestrogens are good candidates as they can bind to estrogens receptors, and in fact, they are able to mimic the antioxidant properties of estrogens [7, 8]. It is very important to consider that the expression of ER is not the same between sexes, organs or that their proportion changes with age. Depending on the organ studied, there are differences in the ER involved in the beneficial effects of estrogens. At the cardiovascular level, ERα is the one that exerts the beneficial cardiovascular effects [9].

Redox Stress and Aging

The free radical theory of aging states that the accumulation of oxidative damage causes aging [1]. Oxidative stress is a cellular phenomenon or condition which occurs as a result of physiological imbalance between the levels of antioxidants and oxidants, in favor of oxidants. These latter ones include free radicals or reactive species that are chemical species with a single unpaired electron which is highly reactive, and a chain reaction of free radicals can occur, leading to more and more damaging reactions. The majority of free radicals are oxygen-derived and are produced in several sites: mitochondrial electron transport, peroxisomal β-oxidation of fatty acids, cytochrome P-450 and phagocytic cells.

Reactive oxygen species (ROS) at low doses play a beneficial role in processes such as cell signaling [2], although their accumulation (when their production exceeds the body's natural antioxidant defense mechanisms) leads to oxidative damage to biomolecules (DNA, proteins and lipids) which in turn is associated with aging and has negative effects on longevity [10–12]. Several studies have found increased levels of 8-hydro-2′-deoxyguanosine (8-OHdG) in mitochondrial and nuclear DNA in post mortem brains of aged subjects [13]. Mitochondrial DNA (mtDNA) is highly susceptible to oxidative damage due to a lack of repair mechanisms, a lack of protection by histones, and the fact that it is located closer to the inner mitochondrial membrane, where ROS are generated. Furthermore, an inverse correlation has been shown between the levels of oxidative damage to mtDNA and maximum longevity in both heart and brain [14]: slowly aging mammals exhibit lower mtDNA damage than those who age faster. Changes in membrane fatty acid composition are also related to aging. For example, peroxidation of arachidonic acid (AA) forms malondialdehyde (MDA), and increased levels of MDA have been found in red blood cells obtained from elderly subjects [15]. An increased accumulation of protein carbonyl groups has also been observed along with age in human brains [16]. Similarly, carbamylated proteins from the peripheral blood of healthy subjects were found to be related to oxidative damage, aging and gender [17].

Lifespan and Antioxidant Enzymes

Eukaryotic cells are provided with important antioxidant enzymes such as superoxide dismutase (SOD), catalase (CAT) and glutathione peroxidase (GPx). Tolmassof et al. found that the higher activity of SOD, the longest-lived species [18]. However, Cutler showed years later that the CAT and GPx activities were negatively correlated with maximum life span potential [10]. Moreover, other studies have found no changes with age of these three enzyme activities. Thus, the decrease of the activity of these enzymes do not explain the aging process by themselves. However, what is true is that free radicals cause damage to the tissues and cells and this damage accumulates with aging [19]. Accordingly, when comparing individuals of the same species, it has been shown that those who enjoy a longer life, also have an increased activity of antioxidant enzymes and lower oxidative damage that the short-lived ones [20]. For example, a recent finding has reported that female flies have higher levels of these three antioxidant enzymes as well as an increased lifespan and reduced levels of ROS [21]. There are many studies in which the levels of antioxidant enzymes have been genetically upregulated to increase their expression for testing the effect on longevity of such manipulation. We found that mice over-expressing regulated p53 and Arf lived longer because p53 increased the expression of the antioxidant enzymes called sestrins and that this effect in longevity was independent of the so-called effect of p53 against cancer [22, 23].

Taken together, these findings are at least consistent with the idea that increased antioxidant protection during the life span is able to slow down physiological deterioration and aging at the individual level [24]. However, we should not forget that there are controversial results which show that increasing oxidative stress with mild stressors, can also increase longevity. We refer to the terms hormesis [25] and mitohormesis [26], which have been shown to be able to increase longevity. Some years ago, we published a complete review of this interesting topic, and formulated a new theory of aging and oxidative stress called The Cell Signaling Disruption Theory of Aging [27].

Females are Protected by Estrogens and Live Longer Than Males

During the twentieth century there has been a dramatic increase in life expectancy in the human population (see Table 16.1). At the beginning of the century, median life span was around 35 years, and nowadays, although depends on the countries

Table 16.1 Life expectancy in Spain between 1900 and 2010 (Data taken from Spanish national Institute for Statistics)

Year	Men	Women	Difference (%)
1900	33.8	35.1	3.8
1960	67.4	72.2	7.1
1980	72.5	78.6	8.1
1992	73.9	81.1	9.1
2000	75.9	82.7	9.2
2010	79.0	85.0	9.3

and many other factors, it is around 80 years [28]. Despite these changes in life expectancy, comparing sex differences in longevity, women have always lived longer than men. A noteworthy detail is that, in addition, just when people get older, that is, when they reach longevity more similar to the current one—which would be from 1960 onwards—this is when the difference in longevity between men and women begins to increase. This not only occurs in Spain, but also occurs in other countries, as has been analyzed in many publications [29].

Of course, there are many sociocultural conditions behind all these differences. In our laboratory we consider whether, in addition to other differences, there could be biological differences that support these sex-associated gap in longevity. For this, we analyzed the longevity of males and females in laboratory animals, in which these sociocultural differences do not exist, since all are housed under the same conditions, *i.e.* they can eat the same diet and have the same light–dark cycles, etc. We found that indeed, females also lived longer than males, therefore we considered that there should be also biological differences underlying their sex-associated longevity gap. The basic biological differences between men and women are sex hormones, and, within them, estrogens have a structure that gives them a certain antioxidant capacity, so they were the chosen candidates. Estrogens, in their action as sex hormones, bind to their ER, and a special feature is that ER are not distributed in the same way throughout the body or between men or women, nor within the same sex, so the effects could be different in some tissues and in others. This will be discussed later in the review.

The beneficial effects of estrogens in women have been widely demonstrated. Estradiol has been shown to have a cardioprotective role [30], a possible beneficial effect in the prevention of Alzheimer's disease [31] or type 2 diabetes [32] and of course a protection against osteoporosis [33], all of these diseases being associated to aging. Certainly, recent studied have demonstrated that they are not only good for females, but also males can benefit from their effects. Garratt et al. published in 2018 that administration of estradiol did increase lifespan in male mice [34].

Role of Mitochondria in Oxidative Stress-Related Sex Differences

Aging is an extremely complex process, so there are many theories that try to explain how and why it occurs [35]. Within the theories of aging, we focused on the theory of free radicals in aging, because as stated before, there is a direct relationship between aging and oxidative stress when it is maintained continuously. Mitochondria has been postulated as the main source and target of free radicals, and it is the best organelle to study when looking for oxidative damage associated to aging [3].

Considering this, we evaluated sex-associated differences in mitochondrial oxidative stress and found that the mitochondrial production of hydrogen peroxide, whether in the liver, whether in the brain (non-synaptic or synaptic mitochondria), was higher in males than in females. Therefore, the mitochondria of males produce more free radicals than those of females [5]. When we analyzed the consequences of this greater production of free radicals in males, we observed that oxidative damage to mitochondrial DNA was much more pronounced in that of males than in that of females, and the oxidation level of this parameter has been related to longevity [14]. In addition, we demonstrated that the protection against oxidative stress of females was mediated by the up-regulation of mitochondrial antioxidant enzymes, such as Mn-SOD and GPx. This is important in the context of previous work by Prof. Orr [36] who showed that the overexpression of SOD and CAT in *Drosophila* increased average life span in the double transgenic flies. This in itself may explain why the average life span of females is higher than that of males.

Further studies demonstrated that estrogens were responsible for the up-regulation of the mitochondrial antioxidant enzymes [5]. It is known that estrogens can have an in vitro antioxidant effect due to its phenolic chemical structure. However, the in vitro dose used is supraphysiological, which prevents the protective antioxidant role of estrogens from being due to its chemical structure. Thus, we reasoned that this effect should be mediated by a cellular process, and we demonstrated that estrogens bind to ER and activate the MAPK-NFkB signaling pathway, which in turn induce the upregulation of the mitochondrial antioxidant enzymes Mn-SOD and GPx, conferring this antioxidant protection to the cells [6].

Therefore, estrogens are protective hormones against oxidative damage, and can be in part responsible for the longer longevity of females when compared with males. The challenge for the future is finding similar molecules with the beneficial effects of estrogens, but without its undesirable effects. Good candidates to meet this challenge may be phytoestrogens.

Phytoestrogens Mimic Antioxidant Beneficial Effects of Estrogens

Phytoestrogens are plant substances or metabolites that induce biological responses in vertebrates and can mimic or modulate the actions of endogenous estrogens, usually by binding to ER [37]. The main phytoestrogens in food are isoflavones, coumestrans, lignans and prenylated flavonoids. The structural similarity of lignans and isoflavones to estrogens was described by induced to Setchell et al. [38] and Adlercreutz et al. [39]; they to proposed their possible effects on cancer prevention. Cross-sectional studies have shown that populations with higher phytoestrogen levels in urine and plasma enjoy lower risk of diseases such as heart disease, osteoporosis, breast cancer, prostate cancer, and menopausal symptoms [40, 41]. Certainly, their beneficial health effects have been reported repeatedly [7, 42] and, to our knowledge, very few, if any, reports have shown unfavorable effects. Some years ago, we studied if phytoestrogens could mimic the beneficial antioxidant properties of estrogens, and we found that indeed, soy consumption by male rats increased the expression of antioxidant enzymes, which was at the same time responsible for an improvement of cardiovascular function [7]. The molecular mechanisms underlying this action were similar to estrogens, i.e. through the activation of the ER (mainly ERβ), and MAPK/NFkB signaling pathway [8]. The concentration of genistein we used can be considered as nutritionally relevant as it is the concentration normally found in plasma of people at the Far East, who usually eat relatively large quantities of soy in their diet.

More recently, we have shown that genistein, as it is able to increase antioxidant defense, is also useful as treatment or in the prevention of age-associated diseases such as Alzheimer's [43] or type 2 diabetes mellitus (unpublished results). Thus, a practical approach of these studies is that administration of phytoestrogens may be very beneficial for longevity (in males and females) because they bind very preferentially to ERβ and promote the upregulation of longevity related genes.

Role of ER in the Beneficial Effects of Estrogens in the Cardiovascular System

In women, the hormonal status achieved after menopause seems to be the main trigger for cardiovascular disease [44–46]. Menopause is characterized by estrogen levels decline including 17β-estradiol (estradiol), which is produced in women mainly by the ovaries, and is the predominant form of circulating estrogen during women reproductive years. At cardiovascular level, actions exerted by estradiol are mainly mediated by genomic and non-genomic mechanisms through ER which are in turn affected by biological processes such as aging and inflammation. In fact, cardiomyocytes, endothelial and vascular smooth muscle cells express different classes of ER (ERα, ERβ and G protein-coupled estrogen receptor, GPER) [47], as well as cells involved in inflammatory processes such as macrophages, monocytes and dendritic cells, which suggest an estradiol modulation of inflammation, a key event in cardiovascular disease development [48, 49]. Estradiol binds to differently located receptors to mediate genomic and non-genomic responses. The binding of estradiol to classical intracellular receptors (ERα and ERβ) induces the dimerization and translocation to the nucleus, binding directly to specific motifs known as estrogen response elements (ERE) in target gene promoters, and thus regulating transcription of estrogen responsive genes [50]. The different distribution of ERα and ERβ throughout the body involves contrasting biological effects, being described opposite [51, 52] or redundant roles [53, 54] on gene expression. Apart from intracellular receptors, estradiol also binds to membrane ERα and ERβ receptors and to G protein-coupled ER (GPER) [55], and activates mainly protein-kinase cascades to trigger faster responses. But the heterogeneity of ER is not only limited to their location on cytosol or membrane but also by the different ER isoforms generated by alternative promoter usage and splicing [56] in a tissue-specific manner [57]. In addition to full-length ERα (ERα66), two truncated splice variants of the ERα, 46 kDa estrogen receptor (ERα46) and 36 kDa estrogen receptor (ERα36) have been identified as membrane ER. The membrane-expressed ERα46, which shares the ligand-binding domain with ERα66 but is deficient in the AF-1 transactivation domain [58], is abundant in human endothelial cells and plays a key role in these endothelial responses to estradiol [56, 59]. ERα46 also preserves domains for palmitoylation, which is required for ERα-protein interaction and caveolin-1 association, and for its plasma membrane localization. Likewise, estrogen binding to membrane ERα46 could also induce eNOS activation more efficiently than the intracellular full-length ERα66 [58]. Another splice variant of ERα66 is the membrane-associated ERα36. It lacks both AF-1 and AF-2 transactivation domains, and present a truncated ligand-binding domain in the C-terminal, explaining its different binding affinity to estradiol. ERα36 mediates the acutely stimulation by estradiol of MAPK pathway and also mobilizes intracellular calcium [58] (different functional domains in ERα is showed in Fig. 16.1). For ERβ, it has been described four isoforms, referred to as ERβcx/2, ERβ3, ERβ4 and ERβ5. All ERβ variants have novel C-terminus, being unable to bind estrogens and other explored ligands [60].

The protective effect of estrogens has often been attributed to modulation of vascular mechanisms such as vascular tone, inflammation, lipid metabolism, and oxidative stress [61–63]. The spread distribution of ERα and ERβ balances estradiol action in different tissues [64], although several studies highlight the importance of ERα in the vasoprotective effects of estradiol [65, 66]. Also, GPER has been linked to the beneficial vascular effects of estradiol on vascular level [67], by reducing the expression of myocardial pro-inflammatory cytokines, inhibiting the VSMC proliferation and inducing vasodilation through a nitric oxide (NO)-dependent mechanism [68]. Estradiol can also affect vascular tone through the regulation of endothelial function [69]. This modulation is accomplished through the release of several endogenous vasoactive substances thereby promoting vasodilation via NO, prostacyclin [70] and angiotensin (Ang) 1–7 [71], or reducing vasoconstriction through the decrease of endothelin-1 [72] and Ang II receptor type 1 expression [73].

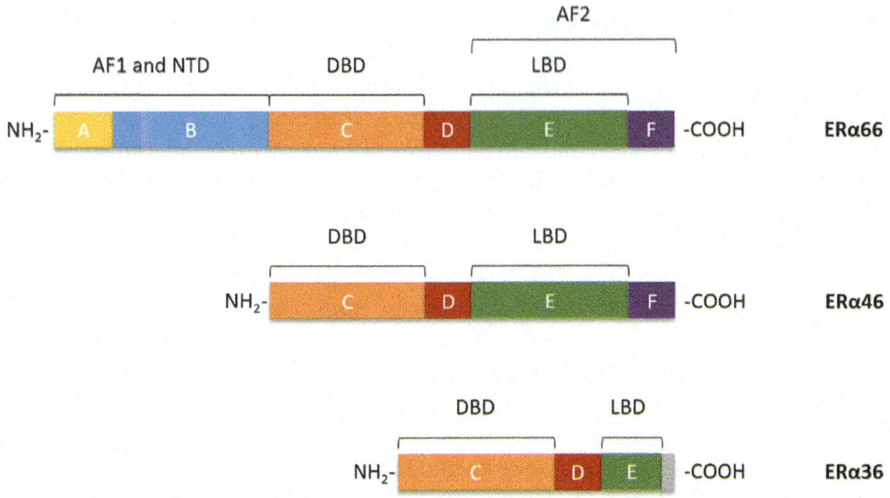

Fig. 16.1 The structures of the ERα isoforms. Briefly, the NH_2-terminal domain (NTD) contains a ligand-independent transactivation domain, activation function (AF1), the region responsible for recruitment of co-regulatory proteins that transactivates gene transcription in the absence of estrogen. DNA-binding domain (DBD) which will bind to the Estrogen Response Elements (ERE) in ERα-regulated genes (orange), the D region (red), known as the hinge region which separates the non-classical signaling pathway (left) with the classical signaling pathway (right), and the COOH-terminal or ligand-binding domain (LBD) (green), which contains the ligand binding cavity and will activate gene transcription. The F domain is variable in length, and its function is not clear [60, 108]

The modulation of NO by estrogen has been extensively studied and is attributed to both genomic and non-genomic effects. Estradiol increases NO bioavailability by increasing NO synthase (eNOS) gene expression [74], the enzyme responsible for NO synthesis in endothelial cells, and non-genomic and rapid activation of enzyme activity via cascades that activate kinases [59]. Findings from cell culture and ERα knockout mice indicate that estrogen may regulate vascular endothelial function via ERα modulation of eNOS [75]. Also, estradiol increases intracellular free Ca^{2+} concentration in endothelial cells [76], regulates endogenous inhibitors and cellular location [9], and attenuates superoxide anion ($O2^-$) concentration, thereby decreasing $O2^-$-mediated NO inactivation [77, 78]. Some of these rapid effects of estradiol on the NO signaling pathway are mediated by ER46, which mediates estrogen-induced eNOS activation in a more efficient manner than ER66 [56]. An increasing number of studies also point to GPER in NO production via c-Src/PI3K signaling pathway [79]. Therefore, either through ERα or GPER, estradiol increases NO release, which through multiple effects on vascular homeostasis promotes vasoprotection [80].

Also, of key importance is the vasoprotective effect that estradiol exerts through the prostanoid pathway. In endothelial cells, estradiol mediates a shift towards prostacyclin production through cyclooxygenase [81] and via ERα [70]. Prostacyclin shows vasodilator and antiaggregatory effects, and the up-regulation of prostacyclin production improves cerebral perfusion and prevents thrombus formation in cerebral blood vessels from ovariectomized rats [78]. In the same way, GPER affects prostanoid pathway by inhibiting vasoconstrictor prostanoids production under in vitro pro-inflammatory conditions, which could lead to inhibition of vascular tone and inflammation [82].

Additionally, estradiol is able to modulate the renin–angiotensin system (RAS) [83, 84], which plays a pivotal role in the physiological regulation of cardiovascular system. Two axes of RAS modulate cardiovascular function. On the one hand, the classical view of RAS is represented by Angiotensin II (Ang II), which promotes vasoconstriction and vascular remodeling through receptor type 1 (AT1R), and exerts opposing effects through receptor type 2 (AT2R). On the other hand, tissue angiotensin-converting enzyme 2 (ACE2) converts Ang II into angiotensin 1–7 [Ang-(1–7)], which acts as vasodilator through Mas receptor. Therefore, an essential balance is established between the two axes in cardiovascular homeostasis [85]. Circulating plasma Ang-(1–7)) concentrations have been related to sex differences, being higher in healthy premenopausal women than in healthy men of a similar age [86]. Again, as in NO and prostacyclin pathways, ERα seems to be responsible for estradiol modulation of RAS and the observed sex differences both in kidney from ovariectomized mice [87] and in endothelial cells [71], where a crosstalk between estradiol and Ang-(1–7)–Mas axis has been described [88]. Accordingly, the lack of estrogen that occurs in menopause could shift the Ang-(1–7) vasoprotective axis towards Ang II pathway, losing the vasoprotection conferred by ACE2 [89–91]. Therefore, ERα appears to play a pivotal role in the cardiovascular protection through its interaction with several fundamental pathways of cardiovascular physiology [59].

Estrogen can thus modulate cardiovascular function through ER-mediated actions, either through genomic or rapid effects. But sex hormones can also regulate gene expression through epigenetics. In this sense, microRNA (miRNA) are emerging as a novel regulatory mechanism involved in several physiological and pathological processes, where the identification of different profile of circulating miRNA raises the possibility of serving as potential biomarkers in CVD [92]. MiRNA are small (approximately 22 nucleotides in length), non-coding, RNA molecules that play a role in post-transcriptional gene regulation through repression and/or mRNA degradation, and are also related to the estrogen levels in women [93]. Also, estradiol changes the expression of miRNA biosynthesis genes in endothelial cells [94]. ER are partially involved in estrogen regulation of miRNA as demonstrated by bioinformatics and pharmacological approaches. ER binding sites close to estrogen-regulated miRNA have been determined in endothelial cells [95] and VSMC [96], and several estradiol-regulated miRNA expression changes by using ER agonists and antagonists. It is striking that the most of the analyzed miRNA are regulated by ERα in endothelial cells [95]. ERα participates in the antiproliferative action of estrogens on VSMC mediated by miR-203 [96], or in the antioxidant effect of estrogen on cardiovascular tissues mediated by miR-22 [97], although the expression of miR-21 involved in cardiac fibrosis [98] is regulated by an ERβ-dependent mechanism [99].

As it mentioned above, ERβ mediates different effects and somehow counteracts the beneficial profile of estradiol through ERα. ERβ is more—highly expressed than ERα in inflammation, hypoxia and oxidative stress [100], and also mediates a protective role in–in injured arteries [101]. The physiological balance between ERα and ERβ expression appears to change during aging that in women is inevitably linked to menopause. In fact, the uterine arteries of postmenopausal women express increasing levels of ERβ after 10 years from menopause, with no significant changes in ERα expression. The increasing level of ERβ expression related to age and menopause is further correlated with a pro-inflammatory profile of estradiol [102]. It is possible that this increased ratio of ERβ/ERα could be responsible of the increase in oxidative stress due to aging, as observed in an experimental model of menopause [103]. These age-dependent effects of estradiol could explain, at least in part, the cardiovascular risk of Hormone Replacement Therapy (HRT) in clinical trials.

HRT is associated with improved endothelial function [104] but as age moderates estrogen's vasodilatory [105] and anti-inflammatory [102] effect on vascular tissue, years since menopause could affect the benefit/risk profile of HRT [106]. This is the so-called timing hypothesis, which assumes that the beneficial impact or HRT in primary prevention of CVD occur only if started before the detrimental aging-associated changes in the cardiovascular system have become established [107].

Therefore, the vasoprotective profile of estrogen, mediated primarily by ERα, involves multiple vascular pathways, including NO, prostanoids, RAS and epigenomics. These intracellular pathways which are also regulated by aging, become important in women after menopause, when increases the risk of developing CVD. However, despite that sex is an important factor in clinical research, there is a paucity of cardiovascular studies that take sex into account. Therefore, there is a need for a more detailed understanding of sex-specific cardiovascular mechanisms underlying the observed sexual differences in cardiovascular disease.

Conclusions

There is a sex-associated difference in mitochondrial antioxidant expression in favor of females, which can explain, in part, the longer longevity of them. Estrogens are able to up-regulate these antioxidant enzymes through a mechanism that phytoestrogens are able to mimic. Therefore, also phytoestrogens can exert health beneficial effects (See summary in Fig. 16.2). The expression of the different ER is not homogeneous throughout among the body and it is also different between males and females, and changes with age. Therefore, it is very important to consider that the effect of estrogens or phytoestrogens is dependent on the organ or cell type studied and the sex and age of the subject studied.

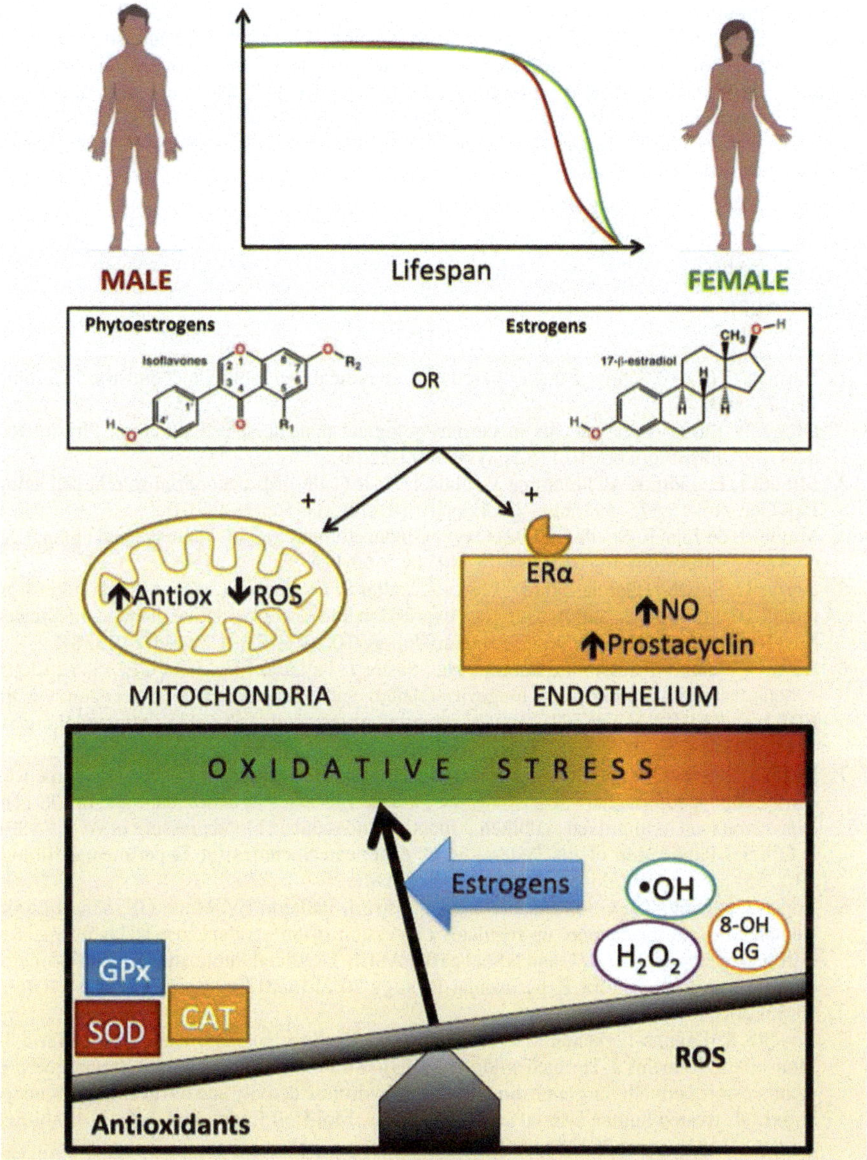

Fig. 16.2 Schematic rendition of the estrogens action as antioxidants

Acknowledgments This work was supported by the following grants: PCIN-2017-117 of the Ministry of Economy and Competitiveness, PI 16/00229 and PI 19/1714 of the Ministry of Science, Innovation and Universities, Health Institute Carlos III—FEDER-ERDF, the EU Joint Programming Initiative 'A Healthy Diet for a Healthy Life' (JPI HDHL INTIMIC-085) to C.B. A.B.P. is recipient of a predoctoral grant financed by Health Institute Carlos III (FI18/00323). Dr. C. M.-B. is recipient of a postdoctoral grant financed by Generalitat Valenciana (APOSTD/2018/230) and FSE (European Social Fund).

References

1. Harman D (1956) Aging: a theory based on free radical and radiation chemistry. J Gerontol 11:298–300
2. Dröge W (2002) Free radicals in the physiological control of cell function. Physiol Rev 82:47–95. https://doi.org/10.1152/physrev.00018.2001
3. Miquel J, Economos AC, Fleming J, Johnson JE Jr (1980) Mitochondrial role in cell aging. Exp Gerontol 15:575–591. https://doi.org/10.1016/0531-5565(80)90010-8
4. Miquel J, de Juan E, Sevila I (1992) Oxygen-induced mitochondrial damage and aging. EXS 62:47–57. https://doi.org/10.1007/978-3-0348-7460-1_5
5. Borras C, Sastre J, Garcia-Sala D, Lloret A, Pallardo FV, Vina J (2003) Mitochondria from females exhibit higher antioxidant gene expression and lower oxidative damage than males. Free Radical Biol Med 34:546–552. https://doi.org/10.1016/S0891-5849(02)01356-4
6. Borras C, Gambini J, Gomez-Cabrera MC, Sastre J, Pallardo FV, Mann GE, Vina J (2005) 17beta-oestradiol up-regulates longevity-related, antioxidant enzyme expression via the ERK1 and ERK2[MAPK]/NFkappaB cascade. Aging Cell 4:113–118. https://doi.org/10.1111/j.1474-9726.2005.00151.x
7. Mahn K, Borras C, Knock GA, Taylor P, Khan IY, Sugden D, Poston L, Ward JP, Sharpe RM, Vina J et al (2005) Dietary soy isoflavone induced increases in antioxidant and eNOS gene expression lead to improved endothelial function and reduced blood pressure in vivo. FASEB J: Official Publication of the Federation of American Societies for Experimental Biology 19:1755–1757. https://doi.org/10.1096/fj.05-4008fje
8. Borras C, Gambini J, Gomez-Cabrera MC, Sastre J, Pallardo FV, Mann GE, Vina J (2006) Genistein, a soy isoflavone, up-regulates expression of antioxidant genes: involvement of estrogen receptors, ERK1/2, and NFkappaB. FASEB J: Official Publication of the Federation of American Societies for Experimental Biology 20:2136–2138. https://doi.org/10.1096/fj.05-5522fje
9. Novella S, Laguna-Fernandez A, Lazaro-Franco M, Sobrino A, Bueno-Beti C, Tarin JJ, Monsalve E, Sanchis J, Hermenegildo C (2013) Estradiol, acting through estrogen receptor alpha, restores dimethylarginine dimethylaminohydrolase activity and nitric oxide production in oxLDL-treated human arterial endothelial cells. Mol Cell Endocrinol 365:11–16. https://doi.org/10.1016/j.mce.2012.08.020
10. Cutler RG (1984) Antioxidants, aging and longevity. In Free radicals in biology, Pryor WA (Ed). Academic Press: Orlando, Vol IV, pp 371–428
11. Cutler RG (1991) Human longevity and aging: possible role of reactive oxygen species. Ann N Y Acad Sci 621:1–28. https://doi.org/10.1111/j.1749-6632.1991.tb16965.x
12. Finkel T, Holbrook NJ (2000) Oxidants, oxidative stress and the biology of ageing. Nature 408:239–247. https://doi.org/10.1038/35041687
13. Richter C, Park JW, Ames BN (1988) Normal oxidative damage to mitochondrial and nuclear DNA is extensive. Proc Natl Acad Sci USA 85:6465–6467
14. Barja G, Herrero A (2000) Oxidative damage to mitochondrial DNA is inversely related to maximum life span in the heart and brain of mammals. FASEB J: Official Publication of the

Federation of American Societies for Experimental Biology 14:312–318. https://doi.org/10.1096/fasebj.14.2.312
15. Gil P, Fariñas F, Casado A, López-Fernández E (2002) Malondialdehyde: a possible marker of ageing. Gerontology 48:209–214. https://doi.org/10.1159/000058352
16. Smith CD, Carney JM, Starke-Reed PE, Oliver CN, Stadtman ER, Floyd RA, Markesbery WR (1991) Excess brain protein oxidation and enzyme dysfunction in normal aging and in Alzheimer disease. Proc Natl Acad Sci USA 88:10540–10543. https://doi.org/10.1073/pnas.88.23.10540
17. Carracedo J, Ramírez-Carracedo R, Martínez de Toda I, Vida C, Alique M, De la Fuente M, Ramírez-Chamond R (2018) Protein Carbamylation: a marker reflecting increased age-related cell oxidation. Int J Mol Sci 19:1495. https://doi.org/10.3390/ijms19051495
18. Tolmasoff JM, Ono T, Cutler RG (1980) Superoxide dismutase: correlation with life-span and specific metabolic rate in primate species. Proc Natl Acad Sci USA 77:2777–2781. https://doi.org/10.1073/pnas.77.5.2777
19. Remacle J, Michiels C, Raes M (1992) The importance of antioxidant enzymes in cellular aging and degeneration. EXS 62:99–108. https://doi.org/10.1007/978-3-0348-7460-1_11
20. Sohal RS, Toy PL, Allen RG (1986) Relationship between life expectancy, endogenous antioxidants and products of oxygen free radical reactions in the housefly, Musca domestica. Mech Ageing Dev 36:71–77. https://doi.org/10.1016/0047-6374(86)90140-5
21. Niveditha S, Deepashree S, Ramesh SR, Shivanandappa T (2017) Sex differences in oxidative stress resistance in relation to longevity in Drosophila melanogaster. J Comp Physiol B 187:899–909. https://doi.org/10.1007/s00360-017-1061-1
22. Sablina AA, Budanov AV, Ilyinskaya GV, Agapova LS, Kravchenko JE, Chumakov PM (2005) The antioxidant function of the p53 tumor suppressor. Nat Med 11:1306–1313. https://doi.org/10.1038/nm1320
23. Budanov AV, Karin M (2008) p53 target genes sestrin1 and sestrin2 connect genotoxic stress and mTOR signaling. Cell 134:451–460. https://doi.org/10.1016/j.cell.2008.06.028
24. Wickens AP (2001) Ageing and the free radical theory. Respir Physiol 128:379–391. https://doi.org/10.1016/S0034-5687(01)00313-9
25. Miwa S, Riyahi K, Partridge L, Brand MD (2004) Lack of correlation between mitochondrial reactive oxygen species production and life span in Drosophila. Ann N Y Acad Sci 1019:388–391. https://doi.org/10.1196/annals.1297.069
26. Ristow M, Zarse K (2010) How increased oxidative stress promotes longevity and metabolic health: the concept of mitochondrial hormesis (mitohormesis). Exp Gerontol 45:410–418. https://doi.org/10.1016/j.exger.2010.03.014
27. Vina J, Borras C, Abdelaziz KM, Garcia-Valles R, Gomez-Cabrera MC (2013) The free radical theory of aging revisited: the cell signaling disruption theory of aging. Antioxid Redox Signal 19:779–787. https://doi.org/10.1089/ars.2012.5111
28. Vina J, Sastre J, Pallardo F, Borras C (2003) Mitochondrial theory of aging: importance to explain why females live longer than males. Antioxid Redox Signal 5:549–556. https://doi.org/10.1089/152308603770310194
29. Austad SN (2006) Why women live longer than men: sex differences in longevity. Gend Med 3:79–92. https://doi.org/10.1016/s1550-8579(06)80198-1
30. Arnal JF, Clamens S, Pechet C, Negre-Salvayre A, Allera C, Girolami JP, Salvayre R, Bayard F (1996) Ethinylestradiol does not enhance the expression of nitric oxide synthase in bovine endothelial cells but increases the release of bioactive nitric oxide by inhibiting superoxide anion production. Proc Natl Acad Sci USA 93:4108–4113. https://doi.org/10.1073/pnas.93.9.4108
31. Russell JK, Jones CK, Newhouse PA (2019) The role of estrogen in brain and cognitive aging. Neurother: J Am Soc Exp NeuroTher 16:649–665. https://doi.org/10.1007/s13311-019-00766-9
32. Diaz A, Lopez-Grueso R, Gambini J, Monleon D, Mas-Bargues C, Abdelaziz KM, Vina J, Borras C (2019) Sex differences in age-associated type 2 diabetes in rats-role of estrogens and oxidative stress. Oxid Med Cell Longev 2019:6734836. https://doi.org/10.1155/2019/6734836

33. Brown C (2017) Osteoporosis: staying strong. Nature 550:S15-s17. https://doi.org/10.1038/550S15a
34. Garratt M, Lagerborg KA, Tsai YM, Galecki A, Jain M, Miller RA (2018) Male lifespan extension with 17-alpha estradiol is linked to a sex-specific metabolomic response modulated by gonadal hormones in mice. Aging Cell 17:e12786. https://doi.org/10.1111/acel.12786
35. Vina J, Borras C, Miquel J (2007) Theories of ageing. IUBMB Life 59:249–254. https://doi.org/10.1080/15216540601178067
36. Orr WC, Sohal RS (1994) Extension of life-span by overexpression of superoxide dismutase and catalase in *Drosophila melanogaster*. Science (New York, N.Y.) 263:1128–1130
37. Madej A, Persson E, Lundh T, Ridderstrale Y (2002) Thyroid gland function in ovariectomized ewes exposed to phytoestrogens. J Chromatogr B Analyt Technol Biomed Life Sci 777:281–287
38. Setchell KD, Borriello SP, Hulme P, Kirk DN, Axelson M (1984) Nonsteroidal estrogens of dietary origin: possible roles in hormone-dependent disease. Am J Clin Nutr 40:569–578. https://doi.org/10.1093/ajcn/40.3.569
39. Adlercreutz H, Hockerstedt K, Bannwart C, Bloigu S, Hamalainen E, Fotsis T, Ollus A (1987) Effect of dietary components, including lignans and phytoestrogens, on enterohepatic circulation and liver metabolism of estrogens and on sex hormone binding globulin (SHBG). J Steroid Biochem 27:1135–1144. https://doi.org/10.1016/0022-4731(87)90200-7
40. Adlercreutz H, Fotsis T, Heikkinen R, Dwyer JT, Woods M, Goldin BR, Gorbach SL (1982) Excretion of the lignans enterolactone and enterodiol and of equol in omnivorous and vegetarian postmenopausal women and in women with breast cancer. Lancet (London, England) 2:1295–1299. https://doi.org/10.1016/s0140-6736(82)91507-0
41. Adlercreutz H, Honjo H, Higashi A, Fotsis T, Hamalainen E, Hasegawa T, Okada H (1991) Urinary excretion of lignans and isoflavonoid phytoestrogens in Japanese men and women consuming a traditional Japanese diet. Am J Clin Nutr 54:1093–1100. https://doi.org/10.1093/ajcn/54.6.1093
42. Munro IC, Harwood M, Hlywka JJ, Stephen AM, Doull J, Flamm WG, Adlercreutz H (2003) Soy isoflavones: a safety review. Nutr Rev 61:1–33. https://doi.org/10.1301/nr.2003.janr.1-33
43. Bonet-Costa V, Herranz-Perez V, Blanco-Gandia M, Mas-Bargues C, Ingles M, Garcia-Tarraga P, Rodriguez-Arias M, Minarro J, Borras C, Garcia-Verdugo JM et al (2016) Clearing Amyloid-beta through PPARgamma/ApoE activation by genistein is a treatment of experimental Alzheimer's disease. J Alzheimer's Dis 51:701–711. https://doi.org/10.3233/JAD-151020
44. Deroo BJ, Korach KS (2006) Estrogen receptors and human disease. J Clin Investig 116:561–570. https://doi.org/10.1172/jci27987
45. Burns KA, Korach KS (2012) Estrogen receptors and human disease: an update. Arch Toxicol 86:1491–1504. https://doi.org/10.1007/s00204-012-0868-5
46. Vrtacnik P, Ostanek B, Mencej-Bedrac S, Marc J (2014) The many faces of estrogen signaling. Biochem Med 24:329–342. https://doi.org/10.11613/bm.2014.035
47. Khalil RA (2013) Estrogen, vascular estrogen receptor and hormone therapy in postmenopausal vascular disease. Biochem Pharmacol 86:1627–1642. https://doi.org/10.1016/j.bcp.2013.09.024
48. Harkonen PL, Vaananen HK (2006) Monocyte-macrophage system as a target for estrogen and selective estrogen receptor modulators. Ann N Y Acad Sci 1089:218–227. https://doi.org/10.1196/annals.1386.045
49. Kovats S (2015) Estrogen receptors regulate innate immune cells and signaling pathways. Cell Immunol 294:63–69. https://doi.org/10.1016/j.cellimm.2015.01.018
50. Klinge CM (2001) Estrogen receptor interaction with estrogen response elements. Nucleic Acids Res 29:2905–2919. https://doi.org/10.1093/nar/29.14.2905
51. Lindberg MK, Moverare S, Skrtic S, Gao H, Dahlman-Wright K, Gustafsson JA, Ohlsson C (2003) Estrogen receptor (ER)-beta reduces ERalpha-regulated gene transcription, supporting a "ying yang" relationship between ERalpha and ERbeta in mice. Mol Endocr (Baltimore, Md.) 17:203–208. https://doi.org/10.1210/me.2002-0206

52. Tsutsumi S, Zhang X, Takata K, Takahashi K, Karas RH, Kurachi H, Mendelsohn ME (2008) Differential regulation of the inducible nitric oxide synthase gene by estrogen receptors 1 and 2. J Endocr 199:267–273. https://doi.org/10.1677/joe-07-0292
53. Arias-Loza PA, Hu K, Dienesch C, Mehlich AM, Konig S, Jazbutyte V, Neyses L, Hegele-Hartung C, Heinrich Fritzemeier K, Pelzer T (2007) Both estrogen receptor subtypes, alpha and beta, attenuate cardiovascular remodeling in aldosterone salt-treated rats. Hypertension 50:432–438. https://doi.org/10.1161/hypertensionaha.106.084798
54. Lahm T, Crisostomo PR, Markel TA, Wang M, Wang Y, Tan J, Meldrum DR (2008) Selective estrogen receptor-alpha and estrogen receptor-beta agonists rapidly decrease pulmonary artery vasoconstriction by a nitric oxide-dependent mechanism. Am J Physiol. Regul Integr Comp Physiol 295:R1486–1493. https://doi.org/10.1152/ajpregu.90667.2008
55. Levin ER (2009) Plasma membrane estrogen receptors. Trends Endocrinol Metab: TEM 20:477–482. https://doi.org/10.1016/j.tem.2009.06.009
56. Kim KH, Young BD, Bender JR (2014) Endothelial estrogen receptor isoforms and cardiovascular disease. Mol Cell Endocrinol 389:65–70. https://doi.org/10.1016/j.mce.2014.02.001
57. Kim KH, Bender JR (2009) Membrane-initiated actions of estrogen on the endothelium. Mol Cell Endocrinol 308:3–8. https://doi.org/10.1016/j.mce.2009.03.025
58. Lin AH, Li RW, Ho EY, Leung GP, Leung SW, Vanhoutte PM, Man RY (2013) Differential ligand binding affinities of human estrogen receptor-alpha isoforms. PLoS ONE 8:e63199. https://doi.org/10.1371/journal.pone.0063199
59. Novella S, Perez-Cremades D, Mompeon A, Hermenegildo C (2019) Mechanisms underlying the influence of oestrogen on cardiovascular physiology in women. J Physiol 597:4873–4886. https://doi.org/10.1113/jp278063
60. Jia M, Dahlman-Wright K, Gustafsson JA (2015) Estrogen receptor alpha and beta in health and disease. Best Pract Res. Clin Endocrinol Metab 29:557–568. https://doi.org/10.1016/j.beem.2015.04.008
61. Kondo T, Hirose M, Kageyama K (2009) Roles of oxidative stress and redox regulation in atherosclerosis. J Atheroscler Thrombosis 16:532–538. https://doi.org/10.5551/jat.1255
62. Barton M (2013) Cholesterol and atherosclerosis: modulation by oestrogen. Curr Opin Lipidol 24:214–220. https://doi.org/10.1097/MOL.0b013e3283613a94
63. Usselman CW, Stachenfeld NS, Bender JR (2016) The molecular actions of oestrogen in the regulation of vascular health. Exp Physiol 101:356–361. https://doi.org/10.1113/ep085148
64. Murphy E, Steenbergen C (2014) Estrogen regulation of protein expression and signaling pathways in the heart. Biol Sex Differ 5:6. https://doi.org/10.1186/2042-6410-5-6
65. Pare G, Krust A, Karas RH, Dupont S, Aronovitz M, Chambon P, Mendelsohn ME (2002) Estrogen receptor-alpha mediates the protective effects of estrogen against vascular injury. Circ Res 90:1087–1092. https://doi.org/10.1161/01.res.0000021114.92282.fa
66. Arnal JF, Lenfant F, Metivier R, Flouriot G, Henrion D, Adlanmerini M, Fontaine C, Gourdy P, Chambon P, Katzenellenbogen B et al (2017) Membrane and nuclear estrogen receptor alpha actions: from tissue specificity to medical implications. Physiol Rev 97:1045–1087. https://doi.org/10.1152/physrev.00024.2016
67. Revankar CM, Cimino DF, Sklar LA, Arterburn JB, Prossnitz ER (2005) A transmembrane intracellular estrogen receptor mediates rapid cell signaling. Science (New York, N.Y.) 307:1625–1630. https://doi.org/10.1126/science.1106943
68. Prossnitz ER, Barton M (2011) The G-protein-coupled estrogen receptor GPER in health and disease. Nat Rev. Endocrinol 7:715–726. https://doi.org/10.1038/nrendo.2011.122
69. Miller VM, Mulvagh SL (2007) Sex steroids and endothelial function: translating basic science to clinical practice. Trends Pharmacol Sci 28:263–270. https://doi.org/10.1016/j.tips.2007.04.004
70. Sobrino A, Oviedo PJ, Novella S, Laguna-Fernandez A, Bueno C, Garcia-Perez MA, Tarin JJ, Cano A, Hermenegildo C (2010) Estradiol selectively stimulates endothelial prostacyclin production through estrogen receptor-{alpha}. J Mol Endocrinol 44:237–246. https://doi.org/10.1677/jme-09-0112

71. Mompeon A, Lazaro-Franco M, Bueno-Beti C, Perez-Cremades D, Vidal-Gomez X, Monsalve E, Gironacci MM, Hermenegildo C, Novella S (2016) Estradiol, acting through ERalpha, induces endothelial non-classic renin-angiotensin system increasing angiotensin 1–7 production. Mol Cell Endocrinol 422:1–8. https://doi.org/10.1016/j.mce.2015.11.004
72. Dubey RK, Jackson EK, Keller PJ, Imthurn B, Rosselli M (2001) Estradiol metabolites inhibit endothelin synthesis by an estrogen receptor-independent mechanism. Hypertension 37:640–644. https://doi.org/10.1161/01.hyp.37.2.640
73. Nickenig G, Baumer AT, Grohe C, Kahlert S, Strehlow K, Rosenkranz S, Stablein A, Beckers F, Smits JF, Daemen MJ et al (1998) Estrogen modulates AT1 receptor gene expression in vitro and in vivo. Circulation 97:2197–2201. https://doi.org/10.1161/01.cir.97.22.2197
74. Sumi D, Ignarro LJ (2003) Estrogen-related receptor alpha 1 up-regulates endothelial nitric oxide synthase expression. Proc Natl Acad Sci USA 100:14451–14456. https://doi.org/10.1073/pnas.2235590100
75. Rubanyi GM, Freay AD, Kauser K, Sukovich D, Burton G, Lubahn DB, Couse JF, Curtis SW, Korach KS (1997) Vascular estrogen receptors and endothelium-derived nitric oxide production in the mouse aorta. Gender difference and effect of estrogen receptor gene disruption. J Clin Invest 99, 2429–2437. https://doi.org/10.1172/jci119426
76. Rubio-Gayosso I, Sierra-Ramirez A, Garcia-Vazquez A, Martinez-Martinez A, Munoz-Garcia O, Morato T, Ceballos-Reyes G (2000) 17Beta-estradiol increases intracellular calcium concentration through a short-term and nongenomic mechanism in rat vascular endothelium in culture. J Cardiovasc Pharmacol 36:196–202. https://doi.org/10.1097/00005344-200008000-00009
77. Dantas AP, Tostes RC, Fortes ZB, Costa SG, Nigro D, Carvalho MH (2002) In vivo evidence for antioxidant potential of estrogen in microvessels of female spontaneously hypertensive rats. Hypertension 39:405–411. https://doi.org/10.1161/hy0202.102993
78. Ospina JA, Krause DN, Duckles SP (2002) 17beta-estradiol increases rat cerebrovascular prostacyclin synthesis by elevating cyclooxygenase-1 and prostacyclin synthase. Stroke 33:600–605. https://doi.org/10.1161/hs0202.102732
79. Fredette NC, Meyer MR, Prossnitz ER (2018) Role of GPER in estrogen-dependent nitric oxide formation and vasodilation. J Steroid Biochem Molr Biol 176:65–72. https://doi.org/10.1016/j.jsbmb.2017.05.006
80. Forstermann U, Sessa WC (2012) Nitric oxide synthases: regulation and function. Eur Heart J 33(829–837):837a–837d. https://doi.org/10.1093/eurheartj/ehr304
81. Sobrino A, Mata M, Laguna-Fernandez A, Novella S, Oviedo PJ, Garcia-Perez MA, Tarin JJ, Cano A, Hermenegildo C (2009) Estradiol stimulates vasodilatory and metabolic pathways in cultured human endothelial cells. PLoS ONE 4:e8242. https://doi.org/10.1371/journal.pone.0008242
82. Meyer MR, Fredette NC, Barton M, Prossnitz ER (2015) G protein-coupled estrogen receptor inhibits vascular prostanoid production and activity. J Endocrinol 227:61–69. https://doi.org/10.1530/joe-15-0257
83. Farhat MY, Lavigne MC, Ramwell PW (1996) The vascular protective effects of estrogen. FASEB J: Official Publication of the Federation of American Societies for Experimental Biology 10:615–624
84. Alvarez A, Hermenegildo C, Issekutz AC, Esplugues JV, Sanz MJ (2002) Estrogens inhibit angiotensin II-induced leukocyte-endothelial cell interactions in vivo via rapid endothelial nitric oxide synthase and cyclooxygenase activation. Circ Res 91:1142–1150. https://doi.org/10.1161/01.res.0000046018.23605.3e
85. Jiang F, Yang J, Zhang Y, Dong M, Wang S, Zhang Q, Liu FF, Zhang K, Zhang C (2014) Angiotensin-converting enzyme 2 and angiotensin 1–7: novel therapeutic targets. Nat Rev Cardiol 11:413–426. https://doi.org/10.1038/nrcardio.2014.59
86. Sullivan JC. Sex and the renin-angiotensin system: inequality between the sexes in response to RAS stimulation and inhibition. Am J Physiol Regul Integr Compar Physiol 294:R1220–1226. https://doi.org/10.1152/ajpregu.00864.2007

87. Brosnihan KB, Hodgin JB, Smithies O, Maeda N, Gallagher P (2008) Tissue-specific regulation of ACE/ACE2 and AT1/AT2 receptor gene expression by oestrogen in apolipoprotein E/oestrogen receptor-alpha knock-out mice. Exp Physiol 93:658–664. https://doi.org/10.1113/expphysiol.2007.041806
88. Sobrino A, Vallejo S, Novella S, Lazaro-Franco M, Mompeon A, Bueno-Beti C, Walther T, Sanchez-Ferrer C, Peiro C, Hermenegildo C (2017) Mas receptor is involved in the estrogen-receptor induced nitric oxide-dependent vasorelaxation. Biochem Pharmacol 129:67–72. https://doi.org/10.1016/j.bcp.2017.01.012
89. Komukai K, Mochizuki S, Yoshimura M (2010) Gender and the renin-angiotensin-aldosterone system. Fundam Clin Pharmacol 24:687–698. https://doi.org/10.1111/j.1472-8206.2010.00854.x
90. Hilliard LM, Sampson AK, Brown RD, Denton KM (2013) The "his and hers" of the renin-angiotensin system. Curr Hypertens Rep 15:71–79. https://doi.org/10.1007/s11906-012-0319-y
91. Stanhewicz AE, Wenner MM, Stachenfeld NS (2018) Sex differences in endothelial function important to vascular health and overall cardiovascular disease risk across the lifespan. Am J Physiol Heart Circ Physiol 315:H1569–h1588. https://doi.org/10.1152/ajpheart.00396.2018
92. Fichtlscherer S, Zeiher AM, Dimmeler S (2011) Circulating microRNAs: biomarkers or mediators of cardiovascular diseases? Arterioscler Thromb Vasc Biol 31:2383–2390. https://doi.org/10.1161/atvbaha.111.226696
93. Perez-Cremades D, Mompeon A, Vidal-Gomez X, Hermenegildo C, Novella S (2018) Role of miRNA in the regulatory mechanisms of estrogens in cardiovascular ageing. Oxid Med Cell Longev 2018:6082387. https://doi.org/10.1155/2018/6082387
94. Perez-Cremades D, Mompeon A, Vidal-Gomez X, Hermenegildo C, Novella S (2018) miRNA as a new regulatory mechanism of estrogen vascular action. Int J Mol Sci 19. https://doi.org/10.3390/ijms19020473
95. Vidal-Gomez X, Perez-Cremades D, Mompeon A, Dantas AP, Novella S, Hermenegildo C (2018) MicroRNA as crucial regulators of gene expression in estradiol-treated human endothelial cells. Cell Physiol Biochem: Int J Exp Cell Physiol, Biochem Pharmacol 45:1878–1892. https://doi.org/10.1159/000487910
96. Zhao J, Imbrie GA, Baur WE, Iyer LK, Aronovitz MJ, Kershaw TB, Haselmann GM, Lu Q, Karas RH (2013) Estrogen receptor-mediated regulation of microRNA inhibits proliferation of vascular smooth muscle cells. Arterioscler Thromb Vasc Biol 33:257–265. https://doi.org/10.1161/atvbaha.112.300200
97. Wang L, Tang ZP, Zhao W, Cong BH, Lu JQ, Tang XL, Li XH, Zhu XY, Ni X (2015) MiR-22/Sp-1 Links estrogens with the up-regulation of cystathionine gamma-Lyase in Myocardium, which contributes to estrogenic cardioprotection against oxidative stress. Endocrinology 156:2124–2137. https://doi.org/10.1210/en.2014-1362
98. Thum T, Gross C, Fiedler J, Fischer T, Kissler S, Bussen M, Galuppo P, Just S, Rottbauer W, Frantz S et al (2008) MicroRNA-21 contributes to myocardial disease by stimulating MAP kinase signalling in fibroblasts. Nature 456:980–984. https://doi.org/10.1038/nature07511
99. Queiros AM, Eschen C, Fliegner D, Kararigas G, Dworatzek E, Westphal C, Sanchez Ruderisch H, Regitz-Zagrosek V (2013) Sex- and estrogen-dependent regulation of a miRNA network in the healthy and hypertrophied heart. Int J Cardiol 169:331–338. https://doi.org/10.1016/j.ijcard.2013.09.002
100. Rider V, Li X, Peterson G, Dawson J, Kimler BF, Abdou NI (2006) Differential expression of estrogen receptors in women with systemic lupus erythematosus. J Rheumatol 33:1093–1101
101. Xing D, Feng W, Miller AP, Weathington NM, Chen YF, Novak L, Blalock JE, Oparil S (2007) Estrogen modulates TNF-alpha-induced inflammatory responses in rat aortic smooth muscle cells through estrogen receptor-beta activation. Am J Physiol. Heart Circ Physiol 292:H2607–2612. https://doi.org/10.1152/ajpheart.01107.2006
102. Novella S, Heras M, Hermenegildo C, Dantas AP (2012) Effects of estrogen on vascular inflammation: a matter of timing. Arterioscler Thromb Vasc Biol 32:2035–2042. https://doi.org/10.1161/atvbaha.112.250308

103. Novensa L, Novella S, Medina P, Segarra G, Castillo N, Heras M, Hermenegildo C, Dantas AP (2011) Aging negatively affects estrogens-mediated effects on nitric oxide bioavailability by shifting ERalpha/ERbeta balance in female mice. PLoS ONE 6:e25335. https://doi.org/10.1371/journal.pone.0025335
104. Hurtado R, Celani M, Geber S (2016) Effect of short-term estrogen therapy on endothelial function: a double-blinded, randomized, controlled trial. Climacteric: J Int Menopause Soc 19:448–451. https://doi.org/10.1080/13697137.2016.1201809
105. Sherwood A, Bower JK, McFetridge-Durdle J, Blumenthal JA, Newby LK, Hinderliter AL (2007) Age moderates the short-term effects of transdermal 17beta-estradiol on endothelium-dependent vascular function. in postmenopausal women. Arterioscler Thromb Vasc Biol 27:1782–1787. https://doi.org/10.1161/atvbaha.107.145383
106. Sood R, Faubion SS, Kuhle CL, Thielen JM, Shuster LT (2014) Prescribing menopausal hormone therapy: an evidence-based approach. Int J Women's Health 6:47–57. https://doi.org/10.2147/ijwh.s38342
107. Clarkson TB, Melendez GC, Appt SE (2013) Timing hypothesis for postmenopausal hormone therapy: its origin, current status, and future. Menopause (New York, N.Y.) 20:342–353. https://doi.org/10.1097/GME.0b013e3182843aad
108. Melton LJ (1990) 3rd. Hip fracture incidence and survival among members of a California medical care program. Clin Orthopaed Related Res 310–311

The manufacturer's authorised representative in the EU is Springer Nature Customer Service Centre GmbH, Europaplatz 3, 69115 Heidelberg, Germany. If you have any concerns regarding our products, please contact ProductSafety@springernature.com

Printed and bound by CPI Group (UK) Ltd, Croydon, CR0 4YY
30/09/2026
02225501-0004